An Introduction to Power System

Vol-I

AN INTRODUCTION TO POWER SYSTEM VOL-I

By

MANISH GOEL
B.Tech (Electrical Engg.)
PGDBM
Assistant Manager
NDPL, Delhi

GAURAV GUPTA
B.Tech (Electrical Engg.)
Lecturer
M.M. Engineering College
Mullana, Ambala

UNIVERSITY SCIENCE PRESS
(An Imprint of Laxmi Publications Pvt. Ltd.)

BANGALORE • CHENNAI • COCHIN • GUWAHATI • HYDERABAD
JALANDHAR • KOLKATA • LUCKNOW • MUMBAI • RANCHI
NEW DELHI • BOSTON, USA

Published by

UNIVERSITY SCIENCE PRESS

(An Imprint of Laxmi Publications Pvt. Ltd.)
113, Golden House, Daryaganj,
New Delhi-110002

Phone : 011-43 53 25 00
Fax : 011-43 53 25 28

www.laxmipublications.com
info@laxmipublications.com

First Edition: 2013

OFFICES

✆ **Bangalore**	080-26 75 69 30	✆ **Chennai**	044-24 34 47 26
✆ **Cochin**	0484-237 70 04, 405 13 03	✆ **Guwahati**	0361-254 36 69, 251 38 81
✆ **Hyderabad**	040-24 65 23 33	✆ **Jalandhar**	0181-222 12 72
✆ **Kolkata**	033-22 27 43 84	✆ **Lucknow**	0522-220 99 16
✆ **Mumbai**	022-24 91 54 15, 24 92 78 69	✆ **Ranchi**	0651-220 44 64

UPS-9587-195-INTRO POWER SYS VOL-I-GOE C—

Typesetted at Shubham Composer, New Delhi. *Printed at*

Dedicated to my Loving Wife
Mrs. Preeti Goel
— Manish Goel

Dedicated to my Loving Wife
Mrs. Neha Gupta
— Gaurav Gupta

CONTENTS

PREFACE

It gives us immense pleasure and satisfaction in bringing about the first edition of the book "An Introduction to Power System- Vol-I". The book is the outcome of the practical and teaching experience of the authors on the subject. Our intention is to cover all topics in the book with a number of solved numerical problems and solutions of the questions from previous question papers of different universities. We have tried to cover each and every topic in a easy-to-understand manner so that each student can grasp the matter without any difficulty.

It has been the constant endeavour of the author to understand the difficulties of the students in the classroom and accordingly prepare the notes after consulting various journals and books on electrical power systems. The present book is an outcome of these notes.

This book is intended primarily for the students preparing for diploma, A.M.I.E., degree and other engineering examinations. It also meets the needs of those readers who want to gain a sound understanding of the principles of power system.

A suitable number of problems have been solved to help understand the relevant theory. At the end of each chapter, unsolved problems and a large number of objective type questions have been added to help the reader to test himself.

The author wishes to thank Sh. Sunil Singh (G.M.), NDPL for his valuable suggestion, helpful criticism and the encouragement to bring about this book. Last, but not the least, we would like to thank Mr. Ashish Goel [Executive, BSNL], Mrs. Ritika Goel and Mr. Sourabh Gupta [Executive, TCS] for their constant assistance and continuous support without which the joint venture would have failed.

We would welcome to suggestions and comments given by our readers for further improvement of the book.

—Authors

1

INTRODUCTION ABOUT POWER SECTOR

1.1 THE ELECTRICITY SECTOR IN INDIA — AN OVERVIEW

A typical electric power supply system comprises :

(a) Generating units that produce electricity ;

(b) High voltage transmission lines that transport electricity over long distances ;

(c) Low voltage distribution lines that deliver electricity to consumers ;

(d) Sub-stations which are the part of the electricity transmission and distribution system where the voltages are transformed to lower levels for distributing power to end-users and

(e) Energy control centres to coordinate the operation of the components.

1.2 MAIN COMPONENTS OF AN ELECTRIC SUPPLY SYSTEM

- Power generating station
- Transmission lines and
- Distribution lines.

1.2.1 Power Generation

The total electricity generation by power utilities was only 1362 MW in 1947. It has increased 100 times in about 60 years. [Fig. 1.1]

Though there is a significant increase in power generation capacity in last 60 years, still our country is plagued by severe power shortages. As per official statistics, the total energy shortages during the year 2005–06 stood at 8.3% and the peak shortage was 12.3% during the same period.

Inadequate generation capacity is one of the reasons for power shortage. Over 1,00,000 MW generation capacity is required to be added during the Xth and XIth plan period *i.e.*, by 2012 to bridge the gap between supply and demand of power. Let us look at the power generation sector very briefly to understand the options available to us. Electric power generation in India relies on four basic energy sources :

(*i*) Fossil fuels such as coal, oil and natural gas
(*ii*) Hydroelectricity
(*iii*) Nuclear power
(*iv*) Renewable energy sources such as bio-fuels, solar, biomass, wind and tidal.

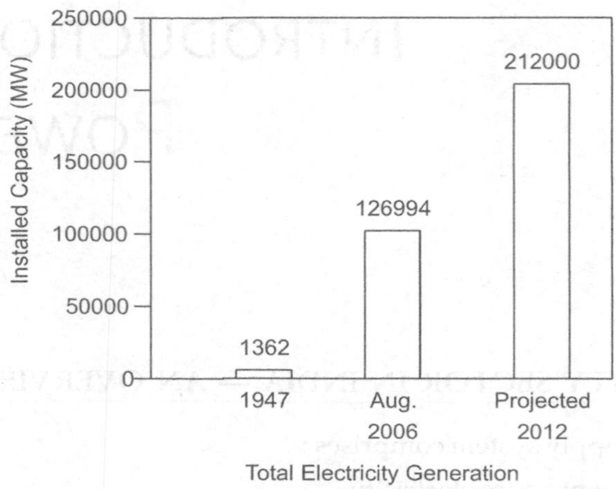

Fig. 1.1

As per data available of September 2006, about 66% of the power generated in India is from coal based thermal power plants. The hydroelectric power plants account for about 26% of total electricity generation, nuclear power plants for about 3% and the remaining 5% comes from renewable energy resources.

Source	*Power (MW)*
Thermal	83772
Hydro	33131
Renewable	6191
Nuclear	3900

Due to the large coal reserves in the country, coal has been identified as the mainstay fuel for power generation till 2012 to envisage installed capacity of 2,12,000 MW. The government has also committed to generate 10% of its power from renewable energy by year 2012. The overall potential for renewable energy generation is around 1,72,000 MW. Of this, less than 7000 MW has been realized so far. In fact, the Electricity Act, 2003 mandates all State Electricity Regulatory Commissions (SERCs) for fixing a minimum percentage of procurement from renewable energy. The entry of renewable energy can help us in meeting present peaking power and general shortages. Another step in this direction is the scheme of allowing captive power generation which refers to generation from a unit set up by an individual, co-operative society or industry for their own consumption. Industries in the manufacturing sector *e.g.*, cement, aluminium, fertilizers, steel, paper, sugar etc., have their own captive power plants to supplement the electricity purchased from utilities or for producing energy.

1.2.2 Transmission and Distribution

In India, the power plants typically produce 50 cycle / second (Hertz), alternating-current (A.C) with voltages between 11 kV and 33 kV. Electric power is brought from the power plant to the consumer through an extensive transmission and distribution (T & D) System comprising distribution networks, state grids and regional grids. The various stages of the process are given below :

(*a*) At the power plant site, the 3-phase voltage is generated and stepped up to a higher voltage for transmission on cables string on cross-country towers.

(*b*) High Voltage (HV) and Extra High Voltage (EHV) transmission is the next stage to transport A.C. power from the power plant over long distances at voltages like 220 KV, 400 KV and 760 KV. For longer distances and higher powers, higher voltages are economical. In special cases HVDC (High Voltage Direct Current transmission) is preferred.

(*c*) Sub-transmission network at 132 KV, 110 KV, 66 KV or 33 KV constitutes the next link towards the end user.

(*d*) Distribution at 11 KV / 6.6 KV / 3.3 KV constitutes the last link to the consumer who is connected directly or through step-down transformers. These transformers bring the voltage levels, down to 400 V for 3-phase, 4-wire distribution. The single phase residential lighting load is connected between any phase and neutral (230 V) and 3-phase load is connected across 3-phase lines directly.

The transmission and distribution network includes sub-stations, lines and distribution transformers. Sub-stations containing step-down transformers reduce the voltage for distribution to industrial, commercial and residential customers. These sub-stations and distribution lines are referred to as the distribution system.

Transmission lines are usually capable of transmitting large quantities of electrical energy over long distances. They operate at higher voltages.

Distribution lines carry limited quantities of power over shorter distances.

The State Electricity Boards (SEBs) or State Governments (through the State Electricity Departments) primarily own and operate the distribution networks and state grids. The Power Grid Corporation of India Limited (PGCIL) owns and operates most of the inter-regional transmission lines in order to facilitate transfer of surplus power to a deficit region.

The interconnected transmission system within each region is also called the regional Grid. Each region or zone has its load control centre.

1.3 OBJECTIVES OF DISTRIBUTION SYSTEMS

The distribution system management aims at following:

(*i*) Planning, modernisation and automation.

(*ii*) To provide service connections to various urban, rural and industrial consumers in the allocated area.

(*iii*) Maximum security of supply and minimum duration of interruptions.

(*iv*) Safety of consumers, utility personnel.

(*v*) Collection of energy payments.

To provide electricity of accepted quality in terms of :

- Balance 3-phase A.C. Supply
- Good power-factor
- Pure sinusoidal voltage waveform
- Minimum duration of interruptions
- Steady state voltage with permissible limits of variation
- Voltage dips within permissible limits.

1.4 NOMINAL VOLTAGE LEVELS IN TRANSMISSION AND DISTRIBUTION SYSTEMS

3-phase, 50 Hz A.C. systems are universally used for generation, transmission and distribution of electric power. The voltages are stepped-up at generating station for purpose of long distance high power transmission. The voltages are stepped down in receiving sub-stations for sub-transmission. The voltages are further stepped down in distribution sub-stations and finlly near load points for final utilization. Thus, the entire power system network has several A.C. Voltage levels obtained by means of power transformers located in sub-stations and distribution systems voltage levels are standardized. Choice of voltage levels is made from standard values recommended by IEC (Indian Electricity Code) or Indian Bureau of Standards.

Table 1.1 gives reference list of Standard Voltages.

Table 1.1

Nominal Voltage of 3-phase, AC system at 50 Hz, phase to phase KV, rms	440 KV	EHV	Transmission
	220 KV	HV	
	132 KV	HV	
	66 KV	HV	
	220 KV	HV	Sub-transmission
	132 KV	HV	
	66 KV	HV	
	33 KV	HV	
	33 KV	MV	Distribution primary Circuits and Direct Feeders to Large Consumers
	22 KV	MV	
	15 KV	MV	
	11 KV	MV	
	6.6 KV	MV	
	3.3 KV	MV	
3-phase AC, 50 Hz phase to phase and 1-phase AC, 50 Hz phase to neutral	400 V or 415 V	LV (ph. to ph.)	Distribution Secondaries and Service to Consumers
	230 V or 240 V	(ph. to neutral)	

where, EHV = Extra High Voltage

HV = High Voltage

MV = Medium Voltage

LV = Low voltage

The various voltage levels are obtained by means of 3-phase AC Power Transformers [Fig. 1.2]. The voltages are controlled by means of On Load Tap Changers (OLTC) fitted with the sub-station power transformers and by switching on/off shunt capacitors.

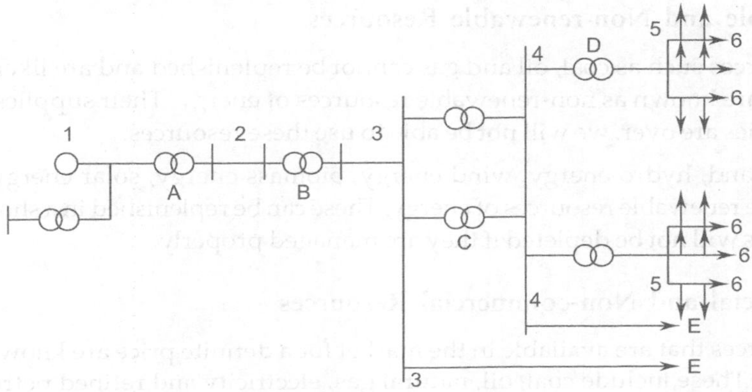

Fig. 1.2

Distribution transformers are generally without OLTCs. Voltages are controlled from Sub-station Power-Transformers.

In Fig. 1.2,

(A) Step-up transformer in generating station

(B) Step-down transformer in receiving sub-station.

(C) Step-down transformer in Distribution sub-station

(D) Small distribution transformer in distribution system.

(E) H.T. Feeders to Industry.

Voltage Levels :

1. Generator Voltage (6.6 KV to 22 KV, 3-phase AC, 50 Hz)
2. Main transmission voltage (400 KV or 220 KV)
3. Sub-transmission voltage (220 KV / 132 KV / 66 KV / 33 KV)
4. Primary distribution voltage (33 KV / 22 KV / 15 KV / 11 KV / 6.6 KV / 3.3 KV)
5. and 6. Secondary distribution voltage (400 V, 3-phase and 230 V, 1-phase)

1.5 EARTH'S ENERGY RESOURCE BASE

The Sun is the principal source of energy on our planet. Other common sources of energy are fossil fuels like coal, oil, natural gas, biomass (such as wood, agricultural waste etc.), water, wind and nuclear energy. Some uncommon energy sources include geo-thermal energy and tidal energy. These energy resources are usually classified as :

- Primary and Secondary Resources.
- Renewable and Non-renewable Resources.
- Commercial and Non-commercial Resources.

1.5.1 Primary and Secondary Resources

Primary energy resoures are those that are found in nature. These are mostly converted into secondary resources in industrial utilities. For example, coal, oil and gas (primary energy resources) are converted into steam and eletricity (secondary resources of energy).

1.5.2 Renewable and Non-renewable Resources

The energy resources such as coal, oil and gas cannot be replenished and are likely to be depleted with time. These are known as non-renewable resources of energy. Their supplies are limited and when these supplies are over, we will not be able to use these resources.

On the other hand, hydro energy, wind energy, biomass energy, solar energy, tidal and geothermal energy are renewable resources of energy. These can be replenished in a short period of time. So, these resources will not be depleted if they are managed properly.

1.5.3 Commercial and Non-commercial Resources

The energy resources that are available in the market for a definite price are known as commercial energy resources. These include coal, oil, natural gas, electricity and refined petroleum products. There are certain energy resources that are not available in the commercial market for a price. These are called non-commercial energy resources. For example, cattle dung, agricultural wastes, solar energy for drying food products, animal power for transport, threshing, lifting water etc., are not bought or sold at a price. These are also called traditional fuels.

1.6 HOW FAST ARE THE ENERGY RESOURCES BEING USED UP ?

To answer this question, we need to understand term reserve/production $\left(\frac{R}{P}\right)$ ratio. $\left(\frac{R}{P}\right)$ ratio for each type of energy resource is obtained by dividing its reserves remaining at the end of the year by its production in that year.

The result is the length of time the remaining reserves would last if production were to continue at that level. Table 1.2 gives $\left(\frac{R}{P}\right)$ ratios for coal, oil and natural gas for the world.

Table. 1.2

Resource Base	*Reserves*	*Production*	$\frac{R}{P}$ *ratio*
Coal	909 Gt	4.8 Gt/Year	200 Years
Oil	148 Gt	3.5 Gt/Year	41.2 Years
Natural Gas	171 Gt	2.6 t cm/Year	59.8 Years

From Table 1.2, it is clear that we have about 200 years of coal left. The information is summed up below for the entire world :

Oil > 40 Years

Natural Gas > 60 Years

Coal > 200 Years

Although India has huge coal reserves, it cannot be used due to high ash content and low calorific value.

1.7 CONCEPT OF ENERGY

Energy is defined as the capacity to do work. We use energy to every thing we do in our lives – right from using our muscles to lighting up our homes, travelling to work, running machines building houses and so on.

In the power generation sector, energy is needed to run turbines that move the electric generators. For example, in thermal power plants, coal is burnt to heat water and produce steam, which in turn moves the turbine. In hydropower plants, the mechanical energy of water is used to move turbines. The turbine is connected to a generator that produces electricity. The electricity is transmitted to our homes and is used to run appliances that provide us light and heat.

1.8 FORMS OF ENERGY

Energy exists in different forms such as :

(*a*) **Kinetic Energy :** The kinetic energy of an object is the energy possessed by it due to its motion. All material bodies in motion, for example, waves, wind, water in a stream and the steam turning the turbine in a power plant possess kinetic energy.

(*b*) **Potential Energy :** The potential energy of an object is the energy possessed by an object due to its position or the arrangement of its parts. For example, water stored in a dam possess potential energy and charges stored in a capacitor possess potential energy.

(*c*) **Mechanical Energy :** It is the energy possessed by an object due to its motion (kinetic) and due to its position or its configuration (potential). Water stored in a dam has gravitational potential energy which is essentially mechanical energy water falling from the top of the dam as well as water flowing in a stream possesses kinetic energy.

(*d*) **Thermal (Heat) Energy :** It is the energy possessed by an object due to the random motion or vibration of the atoms or molecules it is made up of. Steam in a thermal power plant has thermal energy.

(*e*) **Chemical Energy :** It is the energy stored in the bonds of atoms or molecules of an object. Coal, petroleum, biomass, natural gas, batteries have chemical energy stored in them.

(*f*) **Gravitational Energy :** It is the energy possessed by an object due to the force of gravitation *i.e.*, the force of attraction between all objects. All objects in the universe possess gravitational energy.

(*g*) **Electrical Energy :** The energy stored in an eletric field, the potential energy of a charged particle in an electric field or the energy available due to the flow of charged particles is known as electrical energy. For example, electricity, energy stored in the capacitors and solar photovoltaic cells.

(*h*) **Nuclear Energy :** It is the energy stored in the atomic nucleus. It results from the force between the particles in the nucleus. For example, energy in the Sun and Stars and energy in the atomic nuclei.

1.9 UNITS OF ENERGY

Most electrical appliances have a certain power rating. For example, a 60 W light bulb or a 1 KW room heater. It is therefore very common to measure electrical energy consumption in terms of power and time.

Power is the rate at which energy is used.

$$\text{Power} = \frac{\text{Energy}}{\text{Time}}$$

or $$\text{Energy} = \text{Power} \times \text{Time}$$

The unit of energy at the moment is kilowatt-hour (KWH). It is the energy consumed in one hour by an electrical equipment of 1 KW (1000 W) rating. This is one unit of electrical energy measured by the meter.

Table 1.3 gives conversions of some energy units.

Table. 1.3

1 KW	1000 W
1 MW	1000 KW
1 KWh	3.6×10^6 Joules
1 KWh	3412 BTU (British Thermal Units)
1 BTU	1053 Joules
1 Horse Power	746 W

1.10 TRANSMISSION AND DISTRIBUTION LOSSES (T & D LOSSES)

T and D loss is the difference between units injected into the system and the units billed to the ultimate consumer and is generally expressed as percentage of units injected.

$$\text{T and D loss (\%)} = \frac{(\text{Energy input} - \text{Energy billed}) \times 100}{\text{Energy input}}$$

T and D loss is generally calculated for a period of one year which is usually the financial year. It accounts for the loss of energy in this period from the point of generation to the point of billing.

T and D losses occur on account of technical and commercial reasons. The losses occurring on account of technical factors are known as technical losses. The losses occurring on account of commercial factors are known as commercial losses.

1.10.1 Technical Losses

Technical loss is inherent in electrical systems ; as all electrical devices have some resistance and the flow of currents causes a power loss (I^2R loss). Integration of this power loss over time *i.e.*, $\int I^2R \cdot dt$ is the energy loss. Every element in a power system offers resistance to power flow and thus consumes some energy while performing the duty expected of it. The cumulative energy consumed by all these elements is classified as Technical Loss. Technical losses are due to energy dissipated in the conductors and equipment used for Transmission, Transformation, Sub-transmission and Distribution of Power. These occur at many places in a distribution system in lines, mid-span joints and terminations, transformers, service cables and connections etc.

Some reasons for technical losses are :

(*a*) Ill maintained equipment and sub-stations, ageing transformers.

(*b*) Overloading of system elements like transformers, feeders, conductors, etc.

(*c*) Non-usage of smaller size energy efficient distribution transformers.

(*d*) Insufficient reactive compensation, *e.g.*, non-inclusion of appropriate capacitor banks at appropriate places.

(*e*) No re-configuration of feeder lines and distribution transformers so as to reduce the length of LT lines.

(*f*) Inadequate investments for infrastructure improvement.

1.10.2 Commercial Losses

Commercial losses are caused by pilferage, theft, defective meters, un-metered supply of energy and errors in meter reading.

Some reasons for commercial losses are :

(*a*) Faulty meter reading

(*b*) Inefficient billing

(*c*) Theft and pilferage

(*d*) Inadequate revenue collection

(*e*) Non-reading of meters

(*f*) Low metering efficiency

(*g*) Under billing

(*h*) Faulty bill distribution.

1.11 AGGREGATE TECHNICAL AND COMMERCIAL LOSS (AT & C LOSS)

It is unfortunate that in addition to T and D losses, there is also a loss in revenue due to non-realisation of billed demand which leaves a portion of the units billed as unrealised units. The aggregate of T and D loss and loss due to non-realisation of billed demand is termed as AT & C loss.

AT & C loss to the utility is the sum total of technical losses, commecial losses and loss due to non-realisation of total billed units. It is the difference between units injected into the system and the units for which the payment is collected.

$$\text{AT \& C loss (\%)} = \frac{(\text{Energy input} - \text{Energy realised}) \times 100}{\text{Energy input}}$$

$$\text{Energy realised} = \text{Energy billed} \times \text{Collection efficiency}$$

where, $$\text{Collection efficiency (\%)} = \frac{\text{Amount realised} \times 100}{\text{Amount billed}}$$

1.12 REACTIVE POWER CONTROL IN DISTRIBUTION SYSTEMS

Reactive power (KVAR) control represents an efficient method of reducing the cost of utility operation. The savings brought about by KVAR/voltage control are not confined to the monetary value of the energy saved, the released system capacity can serve to delay a costly expansion and reduce the ageing of components. KVAR control provides appropriate placement of compensation devices to ensure a satisfactory voltage profile while minimizing the power losses and the cost of compensation.

Other ancillary benefits gained by correcting the power factor are :

- lower energy losses
- better voltage regulation and
- released system capacity.

All electric equipment requires "Vars", a term used by electric power engineers to describe the reactive or magnetizing power required by the inductive characteristics of electrical equipment. These inductive characteristics are more pronounced in motors and transformers, and therefore, can be quite significant in industrial facilities.

The flow of vars or reactive power through a watt-hour meter will not affect the meter reading but the flow of vars through the power system will result in energy losses on both the utility and the industrial facility.

Some utilities charge for these vars in the form of a penalty or KVA demand charge to justify the cost for lost energy and the additional conductor and transformer capacity required to carry the vars. In addition to energy losses, var flow can also cause excessive voltage drop which may have to be corrected by either the application of shunt capacitors or other more expensive equipment such as load-tap changing transformers, synchronous motors and synchronous condensers.

The power factor triangle shown in Fig. 1.3 is the simplest way to understand the effects of reactive power. The longest leg of the triangle (on the upper or lower triangle), labelled total power represents the vector sum of the reactive power and real power components. Mathematically,

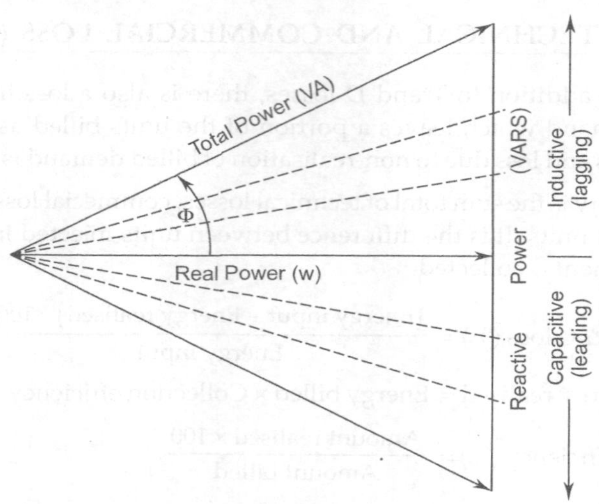

Fig. 1.3

$$\text{Total Power} = \sqrt{(\text{Real Power})^2 + (\text{Reactive Power})^2}$$

The angle ϕ in the power triangle is called the power factor angle and is mathematically equal to:

$$\cos \phi = \frac{\text{Real Power (KW)}}{\text{Total Power (KVA)}}$$

The ratio of the real power to the total power is called the power factor.

The advantages of power factor improvement are :

(*a*) Reactive component of the network and total current in the system from the source end are reduced.

(*b*) I^2R power losses are reduced in system because of reduction in current.

(*c*) Voltage level at the load end is increased.

(*d*) A high power factor can help in utilizing the full capacity of electrical system.

SELF TEST

Fill in the blanks by inserting appropriate words/figures :

1. The most important form of energy is the __________.
2. The basic unit of energy is __________.
3. The primary source of energy is the __________.
4. 1 KWh = __________ K cal.
5. The __________ power plants do not require any fuel.
6. Useful life of hydroelectric power plants is about __________ years.
7. Standard frequency for power supply in India is __________ Hz.

Answers to Self Test

1. electrical energy
2. Joule
3. Sun
4. 860
5. hydroelectric
6. 50
7. 50.

OBJECTIVE QUESTIONS

1. In SI system the frequency is expressed in :

   (*a*) Hz (*b*) Hz/S

   (*c*) Cycles (*d*) None of these.

2. In an ac circuit the ratio of active power to apparent power is called the ________ factor :

   (*a*) load (*b*) Q

   (*c*) form (*d*) power.

3. The standard supply frequency in India is :

   (*a*) 100 Hz (*b*) 60 Hz

   (*c*) 50 Hz (*d*) 25 Hz.

4. Which of the following relations between apparent power (KVa), true power (KW) and reactive power KVar is true ?

   (*a*) KVa = KW + KVar (*b*) KVa = KW – KVar

   (*c*) $KVa = \sqrt{(KW)^2 - (KVar)^2}$ (*d*) $KVa = \sqrt{(KW)^2 + (KVar)^2}$.

5. Which is a conventional source of energy ?
   (*a*) Solar (*b*) Wind
   (*c*) Geothermal (*d*) Radio-active substances.
6. Which is the non-conventional source of energy ?
   (*a*) Water (*b*) Radio-active substances
   (*c*) Fossile Fuels (*d*) Geothermal, ocean tides and waves.
7. As per Indian Electricity rules the maximum difference between declared voltage and actual voltage at consumer premises should be :
   (*a*) ± 10% (*b*) ± 8%
   (*c*) ± 6% (*d*) ± 5%.
8. Which of the following power plants is free from environmental problems ?
   (*a*) Nuclear (*b*) Steam
   (*c*) Hydroelectric (*d*) Diesel engine.
9. Which of the following power plants is the most reliable ?
   (*a*) Diesel (*b*) Tidal
   (*c*) Steam (*d*) Hydroelectric.
10. The disadvantage (*s*) of renewable source of energy is / are :
    (*a*) lack of dependability (*b*) availability in low energy densities
    (*c*) intermittency (*d*) all of the above.
11. The commercial sources of energy are :
    (*a*) Solar, wind, biomass
    (*b*) Fossile fuels, water and radio-active substances,
    (*c*) Wood, animal wastes and agricultural wastes,
    (*d*) None of the above.
12. Which of the following power plants is the least reliable ?
    (*a*) Tidal (*b*) Geothermal
    (*c*) Solar (*d*) Wind.
13. The higest transmission voltage in India is :
    (*a*) 132 KV (*b*) 220 KV
    (*c*) 400 KV (*d*) 750 KV.
14. The domestic consumers are supplied power at :
    (*a*) 230 V (*b*) 400 V
    (*c*) 440 V (*d*) 600 V.
15. Large industrial consumers are supplied power at :
    (*a*) 11 KV (*b*) 66 KV
    (*c*) 132 KV (*d*) 400 KV.
16. Transmission lines connect :
    (*a*) Stepdown transformer station to service tansformer banks,
    (*b*) Distribution transformer to consumer premises,
    (*c*) Generating station to a switching station / stepdown transformer station,
    (*d*) Service points to consumer premises.
17. The electric power can be transmitted by :
    (*a*) Underground system (*b*) Overhead system
    (*c*) Either (*a*) or (*b*) (*d*) None of the above.

18. The main advantage of ac transmission over dc transmission system is :

(*a*) less insulation problems

(*b*) less instability problem

(*c*) low corona loss and reduced interference with communication circuits

(*d*) low losses in transmission over long distances.

19. Which of the following distribution systems is not normally used ?

(*a*) single phase, 3-wire (*b*) single phase, 2-wire

(*c*) 3 phase, 3-wire (*d*) 3 phase, 4-wire.

Answers to Objective Questions

1. (*a*)	**2.** (*d*)	**3.** (*c*)	**4.** (*d*)
5. (*d*)	**6.** (*d*)	**7.** (*c*)	**8.** (*c*)
9. (*d*)	**10.** (*d*)	**11.** (*b*)	**12.** (*d*)
13. (*c*)	**14.** (*a*)	**15.** (*a*)	**16.** (*c*)
17. (*c*)	**18.** (*d*)	**19.** (*a*).	

EXERCISE

1. Why do we endeavour to use water power for the generation of electrical energy?

2. Establish the following relations:

(*i*) 1 kWh = 36×10^5 Joules

(*ii*) 1 kWh = 860 Keal

3. Enumerate the various sources of energy for generation of electrical energy.

4. What are the reasons for power crises in India?

5. What are the various sources of energy in nature?

2

LOAD MANAGEMENT

2.1 INTRODUCTION

The electrical power is drawn by various consumers from the distribution network as per their prevailing demand. The consumers demand certain MW and this demand acts as the load on supply system. Hence, the terms load and demand are used in similar context. Most of the complexities of modern power plant operation arise from the inherent variability of the load demanded by the users. Unfortunately, electrical power cannot be stored and therefore the power station must produce power as and when demanded to meet the requirements of the consumers. On one hand, the power engineer would like that the alternators in the power station should run at their rated capacity for maximum efficiency and on the other hand, the demands of the consumers have wide variations. This makes the design of a power station highly complex. In this chapter, we will study the variable loads on power system with reshaping of load curves by various management techniques.

2.2 VARIABLE LOAD ON POWER STATION

The load on a power station varies from time to time due to uncertain demands of the consumers and is known as variable load on the station. The consumers require power in accordance with the demands of their activities. Thus, the load demand of one consumer at any time may be different from that of the other consumer. The result is that load on the power system varies from time to time.

The important effects of variable load on power stations are :

(*a*) **Increase in Production Cost :** The variable load on the plant increases the cost of the production of electrical energy. An alternator operates maximum efficiency near its rated capacity. If a single alternator is used, it will have poor efficiency during periods of light loads on the plant. Thus, a number of alternators of different capacities are installed so that most of the alternators can be operated at nearly full load capacity. Hence, it increases the initial cost per KW of the plant capacity as well as floor area required.

(*b*) **Need of Additional Equipment :** Consider a steam power station in which air, coal and water are used as raw materials. In order to produce variable power, the supply of these materials will be required to be varied correspondingly. If the power demand on the plant increases, it must be followed by the increased flow of coal, air and water to the boiler in order to meet the increased demand. Therefore, additional equipment has to be installed to accomplish this job.

2.3 LOAD CURVES

The curve showing the variation of load on the power station with reference to time is known as a load curve. Load curves are very useful for planning, operation and for Supply Side Management (SSM) and Demand Side Management (DSM). SSM aims at generation, transmission and distribution of energy according to the load curves to fulfil the expectations of consumers. DSM aims at reshaping the load curves by various available methods (as discussed later) to suit the power supply capability of the supply system and economic operation.

In order to draw a load curve, Chronological time is plotted on X-axis and load in MW on a particular plant or a group of plants is plotted on Y-axis. A load curve is drawn for :-

1. 24 hours of a day (Daily load curve)
2. 7 days of a week (Weekly load curve)
3. 12 months of an year (Yearly load curve).

Figure 2.1 shows a daily load curve. The daily load variation has a certain repetitive cyclic curve depending upon life style, business hours, industrial hours etc. The load line touching the highest peak gives peak load. The total installed capacity of a plant (or a group of plants). Should be more than the peak load with adequate surplus margin for steady state stability limits of generators, planned outages, single or twin contingencies due to forced outages etc.

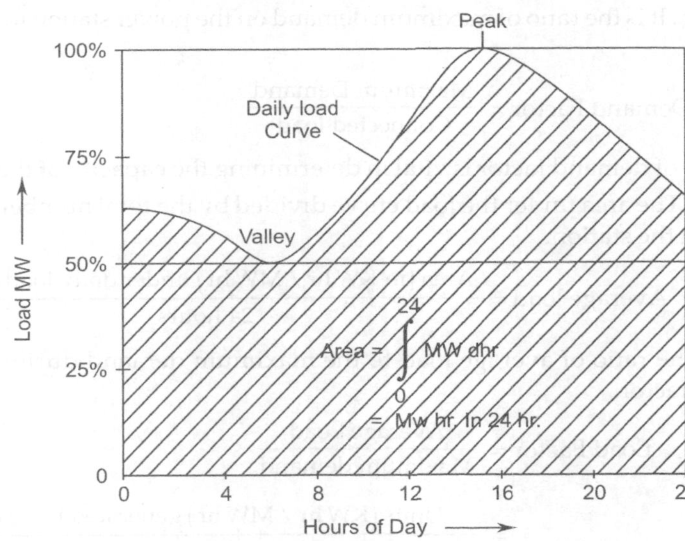

Fig. 2.1

Area under the daily load curve and corresponding daily load duration curve is given by :

$$\text{Area} = \int_0^{24} \text{MW} \cdot d\,\text{hr} = \text{MW hr}$$

The area under these curves represents electrical energy supplied during 24 hours.

From the well known law of conservation of energy.

Total MWhr Energy Generated = Total MWhr Energy Consumed + Total MWhr Losses:

- Instead of increasing installed capacity for peak load, the region may import power through interconnected line as per scheduled exchange. Peak loads in two regions are generally displaced in time. Peak load in Paris occurs at mid-night whereas peak load in London occurs at 9 P.M. Power is exchanged via interconnection by HVDC submarine cable between England and France.
- Another alternative is Energy Storage during off-peak valley periods and release of the stored energy during peak loads.

2.4 IMPORTANT TERMS

1. **Connected Load :** It is the sum of continuous ratings of all the equipments connected to supply system. For example, if a consumer has connections of ten 100 W lamps and two power point of 500 W, then connected load of the consumer is $(10 \times 100 + 2 \times 500) = 2000$ W.
2. **Maximum Demand :** It is the greatest demand of load on the power station during a given period. The load on the power station varies from time to time. The maximum of all the demands that have occured during a given period is the maximum demand. The value of maximum demand helps in determining the installed capacity of the station.
3. **Demand Factor :** It is the ratio of maximum demand on the power station to its connected load, *i.e.*,

$$\text{Demand Factor} = \frac{\text{Maximum Demand}}{\text{Connected load}}$$

   The knowledge of demand factor is vital in determining the capacity of the plant equipment.
4. **Average Load :** The area under the load curve divided by the total number of hours gives the average load on the station.

$$\text{Average load} = \frac{\text{Area (in KW hr / MW hr) under daily load curve}}{\text{24 hours}}$$

5. **Load Factor :** The ratio of average load to the maximum demand during a given period is known as load factor.

   i.e.,
$$\text{Load Factor} = \frac{\text{Average Load}}{\text{Maximum demand}}$$
$$= \frac{\text{Units (KW hr / MW hr) generated in a given period}}{\text{Maximum demand} \times \text{Hours of operation in given period}}$$

   Load Factor is always less than 1 because average load is smaller than the maximum demand. The load factor plays key role in determining the overall cost per unit generated. For economic operation of power plants and the energy supply system, load factor should be high.

6. **Diversity Factor :** The ratio of the sum of individual maximum demands to the maximum demand on power station is known as diversity factor *i.e.,*

$$\text{Diversity Factor} = \frac{\text{Sum of individual maximum demands}}{\text{Maximum demand on power station}}$$

Since the maximum demand on the power station is always less than the sum of individual maximum demands of the consumers, diversity factor will always be greater than 1. For economic operation of power plants and the energy supply system, diversity factor should be high as greater diversity factor means lesser maximum demand. This in turn means that lesser plant capacity is required. Thus, the capital investment on the plant is reduced.

7. **Plant Capacity Factor :** It is the ratio of actual energy produced to the maximum possible energy that could have been produced during a given period.

i.e.,
$$\text{Plant Capacity Factor} = \frac{\text{Actual energy produced}}{\text{Maximum energy that could be produced}}$$

The capacity factor is an indication of the reserve capacity of te plant.

$$\text{Reserve Capacity} = \text{Plant Capacity} - \text{Maximum demand}$$

8. **Plant Use Factor :** It is ratio of KW hr generated to the product of plant capacity and the number of hours for which the plant was in operation.

i.e.,
$$\text{Plant use factor} = \frac{\text{Station output in KW hr}}{\text{Plant capacity} \times \text{Hours of use}}$$

9. **Firm Power :** Power which should always be readily available even during emergency state.
10. **Cold Reserves :** Reserve generating capacity available but not in operation.
11. **Hot Reserves :** Reserve capacity available with thermal process in operational readyness.
12. **Spinning Reserves :** Generating capacity connected to bus and ready for taking load.

2.5 UNITS GENERATED PER ANNUM

$$\text{Load factor} = \frac{\text{Average load}}{\text{Maximum demand}}$$

or

$$\text{Average Load} = \text{Maximum demand} \times \text{Load factor}$$

$$\text{Units generated per annum} = \text{Average load (in KW)} \times \text{Hours in a year}$$

$$= \text{Maximum demand (in KW)} \times \text{L.F.} \times 8760$$

Example 2.1. *The maximum demand on a power station is 150 MW. If the annual load factor is 40%. Calculate the total energy generated in a year.*

Solution.
$$\frac{\text{Energy generated}}{\text{Year}} = \text{Maximum demand} \times \text{L.F.} \times \text{Hours in a year}$$

$$= 150 \times 10^3 \times 0.4 \times 8760$$

$$= 5256 \times 10^5 \text{ KWh.}$$

Example 2.2. *A power station has a maximum demand of 15000 KW. The annual load factor is 60% and capacity factor is 40%. Determine plant capacity as well as reserve capacity of the plant.*

Solution. Energy generated/annum

$$= \text{Maximum demand} \times \text{L.F.} \times \text{Hours in a year}$$
$$= 15000 \times 0.6 \times 8760$$
$$= 78.84 \times 10^6 \text{ KWh.}$$

$$\text{Plant capacity factor} = \frac{\text{Actual energy produced}}{\text{Maximum energy that could be produced}}$$

If considered period is one year, then

$$\text{Annual capacity factor} = \frac{\text{Annual KWh output}}{\text{Plant capacity} \times 8760 \text{ (Hours in a year)}}$$

$$\therefore \quad \text{Plant capacity} = \frac{78.84 \times 10^6}{0.4 \times 8760}$$
$$= 22500 \text{ KW.}$$

$$\text{Reserve Capacity} = \text{Plant capacity} - \text{Maximum Demand}$$
$$= 22500 - 15000 = 7500 \text{ KW.}$$

Example 2.3. *A 100 MW power station delivers 100 MW for 2 hours, 50 MW for 6 hours and is shut down for the rest of each day. It is also shut down for maintenance for 65 days each year. Calculate the annual load factor.*

Solution. Energy supplied for each working day

$$= (100 \times 2) + (50 \times 6) = 500 \text{ MWh.}$$

$$\text{Station operates for} = 365 - 65 = 300 \text{ days in a year}$$

$$\therefore \quad \text{Energy supplied/year} = 500 \times 300$$
$$= 150{,}000 \text{ MWh.}$$

$$\text{Annual load factor} = \frac{\text{MWh supplied per annum}}{\text{Maximum demand (MW)} \times \text{Working hours}} \times 100$$

$$= \frac{150000}{100 \times (300 \times 24)} \times 100 = 20.8\%.$$

Example 2.4. *It has been desired to install a diesel power station to supply power in a sub-urban area having the following particulars :*

(a) *1000 houses with average connected load of 1.5 KW in each house. The demand factor and diversity factor being 0.5 and 2.5.*

(b) *15 factories having overall maximum demand of 100 KW.*

(c) *10 tubewells of 10 KW each and operating together in the morning.*

The diversity factor among above three types of consumers is 1.5. What should be the minimum capacity of power station.

Solution.

(a) Sum of maximum demands of houses

$$= 1.5 \times 0.5 \times 1000 = 750 \text{ KW}$$

$$\text{Maximum demand of domestic load} = \frac{750}{2.5} = 300 \text{ KW}$$

[As per formula of Diversity factor]

(*b*) Maximum demand for factories = 100 KW

(*c*) Maximum demand for tubewells = 10×10 = 100 KW

[Since the tube-wells operate together, the diversity factor is 1]

Sum of maximum demands of three types of load.

$$= 300 + 100 + 100 = 500 \text{ KW}$$

Diversity factor among three types of loads = 1.5

$$\therefore \quad \text{Maximum demand on station} = \frac{500}{1.5} = 333 \text{ KW}$$

Minimum capacity of station required = 333 KW.

Example 2.5. *A diesel station supplies the following loads to various consumers :-*

Industrial consumer = 1500 KW

Commercial establishment = 1000 KW

Domestic power = 200 KW

Domestic Light = 800 KW

If the maximum demand on the station is 3000 KW and the number of KWh generated per year is 45×10^5. Determine (1) Diversity factor ; (2) Annual load factor.

Solution.

1. Diversity Factor $= \dfrac{\text{Sum of individual maximum demands}}{\text{Maximum demand on power station}}$

$$= \frac{1500 + 1000 + 200 + 800}{3000} = 1.16$$

2. Average demand $= \dfrac{\text{KWh generated / annum}}{\text{Hours in a year}}$

$$= \frac{45 \times 10^5}{8760} = 513.7 \text{ KW.}$$

$$\text{Load factor} = \frac{\text{Average load}}{\text{Maximum demand}}$$

$$= \frac{513.7}{3000} = 17.12\%.$$

Example 2.6. *The daily demands of three consumers are given below :*

Time	*Consumer 1*	*Consumer 2*	*Consumer 3*
12 midnight to 8 A.M.	*No load*	*200 W*	*No load*
8 A.M. to 2 P.M.	*600 W*	*No load*	*200 W*
2 P.M. to 4 P.M.	*200 W*	*1000 W*	*1200 W*
4 P.M. to 10 P.M.	*800 W*	*No load*	*No load*
10 P.M. to midnight	*No load*	*200 W*	*200 W*

Plot the load curve and find : (1) Maximum demand of individual consumer ; (2) load factor of individual consumer ; and (3). Diversity factor.

Solution. The load curve is drawn below : as shown in Fig 2.2

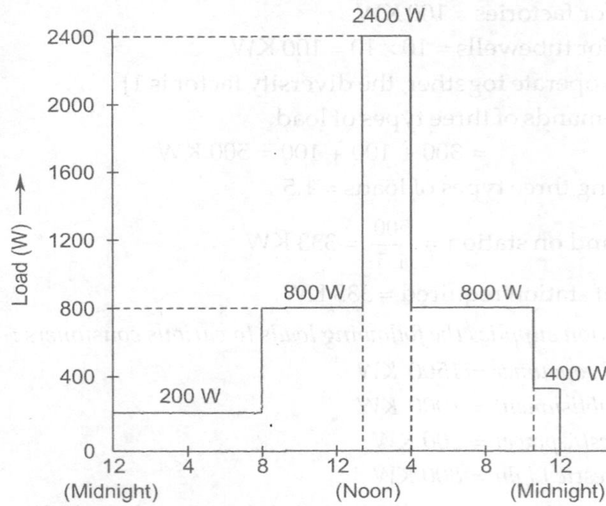

Fig. 2.2

1. Maximum demand of consumer 1 = 800 W
   Maximum demand of consumer 2 = 1000 W
   Maximum demand of consumer 3 = 1200 W
2. Load factor of consumer 1

$$= \frac{\text{Energy consumed/day}}{\text{Maximum demand} \times \text{Hours in a day}} \times 100$$

$$= \frac{600 \times 6 + 200 \times 2 + 800 \times 6}{800 \times 24} \times 100$$

$$= 45.8\%.$$

Load factor of consumer 2 $= \dfrac{200 \times 8 + 1000 \times 2 + 200 \times 2}{1000 \times 24} \times 100 = 16.7\%$

Load factor of consumer 3 $= \dfrac{200 \times 6 + 1200 \times 2 + 200 \times 2}{1200 \times 24} \times 100 = 13.8\%.$

3. Sum of individual maximum demand = 800 + 1000 + 1200 = 3000 W
   Maximum demand on power station = 2400 W

$\therefore$ Diversity factor $= \dfrac{3000}{2400} = 1.25$

2.6 LOAD DURATION CURVE

When the elements of a load curve are arranged in the order of descending magnitudes, the curve thus obtained is called a load duration curve. The load duration curve is obtained from the same data as the load curve but the ordinates are arranged in the order of descending magnitudes. Hence, the area under the load duration curve and the area under the load curve are equal. Figure 2.3 shows the daily load curve.

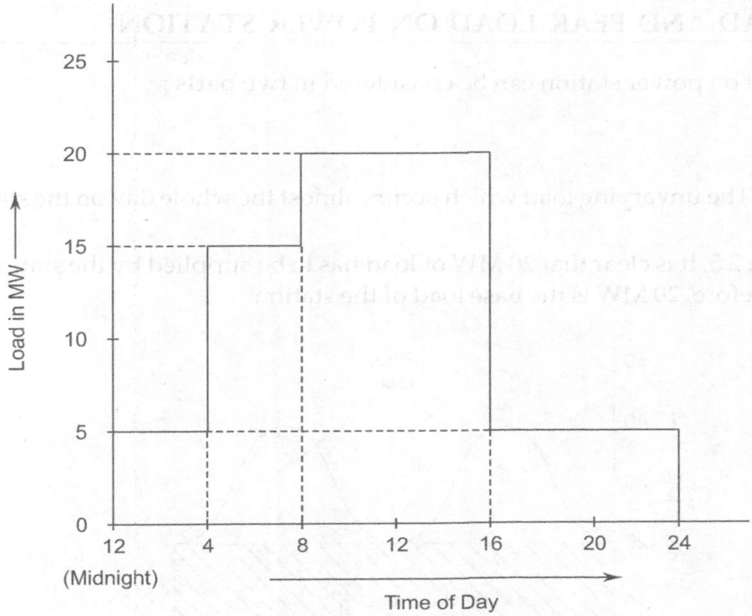

Fig. 2.3

The daily load duration curve can be obtained from daily load curve as shown in Fig. 2.4. The load elements in order of descending magnitude are : 20 MW for 8 hours, 15 MW for 4 hours and 5 MW for 12 hours.

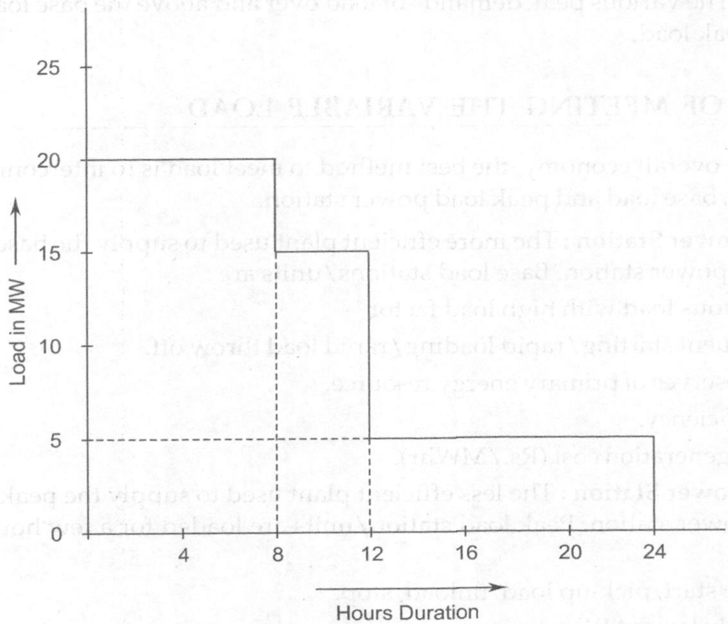

Fig. 2.4

2.7 BASE LOAD AND PEAK LOAD ON POWER STATION

Normally the load on power station can be considered in two parts :

(*a*) **Base Load :**

(*b*) **Peak Load :**

(*a*) **Base Load :** The unvarying load which occurs almost the whole day on the station is known as base load.

Refer to Fig. 2.5, it is clear that 20 MW of load has to be supplied by the station throughout 24 hours. Therefore, 20 MW is the base load of the station.

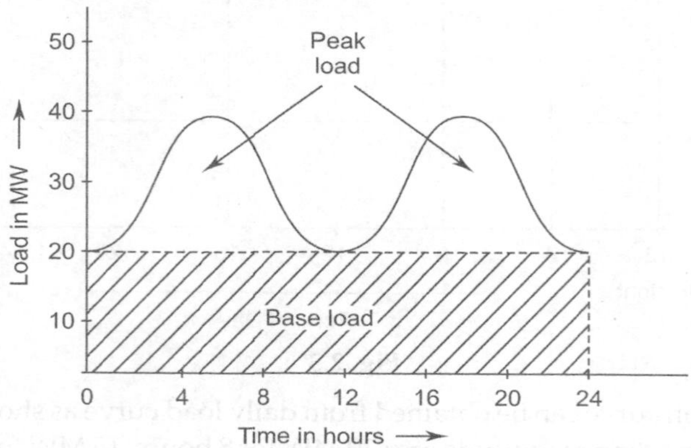

Fig. 2.5

(*b*) **Peak Load :** The various peak demands of load over and above the base load of the station is known as peak load.

2.8 METHOD OF MEETING THE VARIABLE LOAD

In order to achieve overall economy, the best method to meet load is to interconnect two different power stations, *i.e.,* base load and peak load power station.

(*a*) **Base Load Power Station :** The more efficient plant used to supply the base load is known as base load power station. Base load stations/units are :

- Continuous load with high load factor
- No frequent starting/rapid loading/rapid load throw off.
- Large reserves of primary energy resource.
- High efficiency.
- Lowest generation cost (Rs./MWhr).

(*b*) **Peak Load Power Station :** The less efficient plant used to supply the peak load is known as peak load power station. Peak load station/units are loaded for a few hours in a day. They should be :

- Quick to start, pick-up load, unload, stop.
- Relatively lesser MW rating
- Cost of generation Rs./MWhr may be higher but is justified due to lesser MWhr produced by peaking station/unit.

For example, when water is available in sufficient quantity as in summer and rainy season, the hydroelectric plant is used to carry the base load and the steam plant supplies the peak load. However, when water is not available in sufficient quantity as in winter, the steam plant carries the base load whereas the hydroelectric plant carries the peak load.

SELF TEST

Fill in the blanks by inserting appropriate words/figures :

(*i*) In the annual load curve, __________ is taken along Y-axis and __________ along X-axis.

(*ii*) The value of diversity factor is __________ than 1.

(*iii*) The lesser the diversity factor, the __________ is the cost of generation of power.

(*iv*) Annual load factor is determined from __________ load curve.

(*v*) Area under the daily load curve divided by 24 gives __________

(*vi*) A diesel power plant is generally used as a __________

(*vii*) The area under the daily load curve gives __________

(*viii*) More efficient plants are used as __________

(*ix*) The connected load is generally __________ than the maximum demand.

(*x*) The knowledge of diversity factor helps in determining __________

(*xi*) The value of demand factor is __________ than 1.

(*xii*) Base load occurs on the power station for __________ hours in a day.

(*xiii*) The higher the load factor of a power station, the __________ is the cost per unit generated.

(*xiv*) Installed capacity of a power station is __________ than the maximum demand.

Answers to Self Test

(*i*) Load, hours (*ii*) more (*iii*) greater

(*iv*) annual (*v*) average load (*vi*) peak load station

(*vii*) units generated in a day (*viii*) base load stations (*ix*) more

(*x*) plant capacity (*xi*) less (*xii*) 24

(*xiii*) lesser (*xiv*) more.

OBJECTIVE QUESTIONS

1. The value of diversity factor is

(*a*) less than 1 (*b*) more than 1

(*c*) equal to 1 (*d*) other than 1.

2. In an interconnected grid system, diversity factor of the whole system

(*a*) increases (*b*) decreases

(*c*) remains constant (*d*) none of the above.

3. The knowledge of diversity factor helps in determining

(*a*) units generated (*b*) average load

(*c*) plant capacity (*d*) maximum demand.

4. The value of demand factor is

(*a*) equal to 1 (*b*) less than 1

(*c*) more than 1 (*d*) other than 1.

5. A generating station has a connected load of 43 MW and a maximum demand of 20 MW. The demand factor is

(*a*) 0.754 (*b*) 0.869

(*c*) 0.523 (*d*) 0.465.

6. Annual load factor is determined from

(*a*) daily load curve (*b*) annual load curve

(*c*) monthly load curve (*d*) quarterly load curve.

7. The maximum demand on a power station is 100 MW. If annual load factor is 40%, then the total energy generated in a year is

(*a*) 5256×10^5 KWh (*b*) 6529×10^5 KWh

(*c*) 3504×10^5 KWh (*d*) 4504×10^5 KWh.

8. The load curve of a system is shown in the figure. The load factor of the system is

(*a*) 1.66% (*b*) 6.013%

(*c*) 16.6% (*d*) 60.13%.

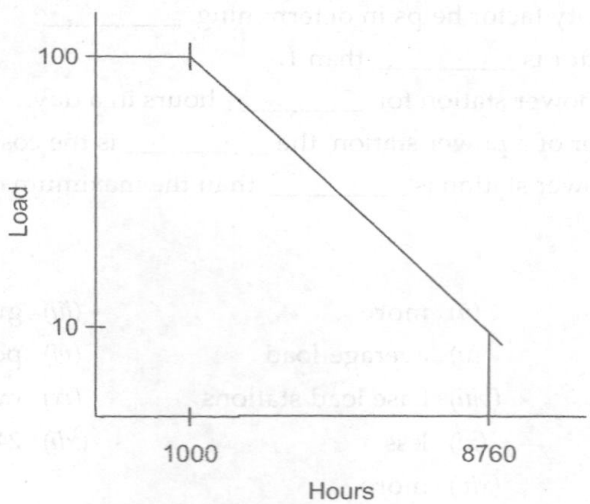

9. In order to have lower cost of electrical energy generation :

(*a*) The load factor should be low but diversity factor should be high ;

(*b*) The load factor and diversity factors should be high ;

(*c*) The load factor should be high but diversity factor low ;

(*d*) The load factor and diversity factor should be low.

10. The maximum demand of a consumer is 2 KW and his daily energy consumption is 20 units. His load factor is

(*a*) 10% (*b*) 50% (*c*) 41.6% (*d*) 60%.

11. The connected load of a consumer is 2 KW and his maximum demand is 1.5 KW. The load factor of the consumer is

(*a*) 0.375 (*b*) 1.33

(*c*) 0.75 (*d*) None of the above.

Answers to Objective Questions

1. (*b*) **2.** (*a*) **3.** (*c*) **4.** (*b*)
5. (*d*) **6.** (*b*) **7.** (*c*) **8.** (*d*)
9. (*b*) **10.** (*c*) **11.** (*d*)

[**Hint :** 8. Average load $= (1000 \times 100) + (7760 \times 10) \times \left(\frac{1}{2} \times 90 \times 7760\right)$

$= 526800$

$$\text{Load factor} = \frac{526800}{100 \times 8760} \times 100\% = 60.13\%$$].

EXERCISE

1. What do you understand by the load curve? What informations are conveyed by a load curve?
2. Why is the load on a power station variable? What are the effects of variable load on the operation of the power station?
3. Define and explain the importance of the following terms in generation:
   (*i*) connected load (*ii*) maximum demand
   (*iii*) demand factor (*iv*) average load.
4. Explain the terms load factor and diversity factor. How do these factors influence the cost of generation?
5. A generating station has a maximum demand of 35,500 kW and has a connected load of 65,000 kW. The number of units generated annually are 25.6×10^7. Calculate (*a*) the load factor (*b*) the demand factor. **[Ans:** $a = 0.822$ $b = 0.546$]
6. A generating station supplies the following loads: 15,000 kW, 12,00 kW, 6000 kW and 450 kW. The station has a maximum demand of 22,000 kW. The annual load factor of the station is 48 percent. Calculate (*a*) the number of units supplied annually (*b*) the diversity factor (*c*) the demand factor. **[Ans:** (*a*) 9.24×10^7 units (*b*) 1.905 (*c*) 0.525 or 52.5%]
7. What do you understand by (*i*) base load and (*ii*) peak load of a power station?

3

Inductance and Resistance of Transmission Lines

3.1 INTRODUCTION

An electric transmission line has four parameters namely resistance, inductance, capacitance and conductance. In this chapter, we will discuss first two parameters. These four parameters are uniformally distributed along the whole line. Each line element has its own value and it is not possible to concentrate or lump them at discrete points on the line. Thus, these parameters are known as distributed parameters. Their values are given per unit length of line and are denoted as R, L, C and G. The line resistance and inductance form the series impedance of the line. The capacitance and conductance form the shunt admittance of the line.

3.2 LINE-INDUCTANCE

When current flows in an electric circuit, a magnetic flux is set up. Variation of the current in conductors causes a change in the number of lines of magnetic flux linking the circuit, and an e.m.f. is induced in it. The magnitude of the self-induced e.m.f. is directly proportional to the rate of change of flux-linkages and its direction is such as to oppose the cause. Mathematically, the induced e.m.f. is given by

$$e = \frac{d}{dt}(\mathrm{N}\phi) = \mathrm{N}\frac{d\phi}{dt}\ \mathrm{V} \quad \text{...(1)}$$

where (ϕN) is the number of flux linkages of the circuit in weber-turns and is equal to the product of flux and the number of turns of the circuit linked.

When the current in a circuit is changing, its associated magnetic field must be changing. If constant permeability is assumed for the medium in which the magnetic field is set-up, the number of flux linkages is directly proportional to the current and therefore the induced emf is proportional to the rate of change of current. Thus, we can write,

$$e = L \frac{di}{dt} \text{ V} \qquad ...(2)$$

where L = Constant of proportionality and known as inductance of circuit, Henry (H)

Equating both equations (1) and (2)

$$N \frac{d\phi}{dt} = L \frac{di}{dt}$$

or

$$L = \frac{N d\phi}{di} \text{ H}$$

If permeability is assumed to be a constant,

$$\frac{d\phi}{di} = \frac{\phi}{i}$$

∴

$$L = \frac{N\phi}{i} \text{ H}$$

It shows that the self-inductance of an electric circuit is numerically equal to the flux linkages of the circuit per unit of the current.

3.3 INDUCTANCE OF A CONDUCTOR

Consider a solid, round infinitely long conductor of radius r situated in air and carrying a current of I amps. The flux linking with the conductor consists of two parts, internal flux and external flux. The internal flux is present inside the conductor due to the current in it and it does not link with the whole current.

The external flux is produced around the conductor due to its own current and the currents of each of the other conductors in the vicinity. Thus, the total inductance L per metre length of the conductor is given by :

$$L = (L_{int} + L_{ext}) \text{ H/m}$$

Internal inductance (L_{int}) : Let us consider the long cylindrical conductor whose cross-section is shown in Fig. 3.1.

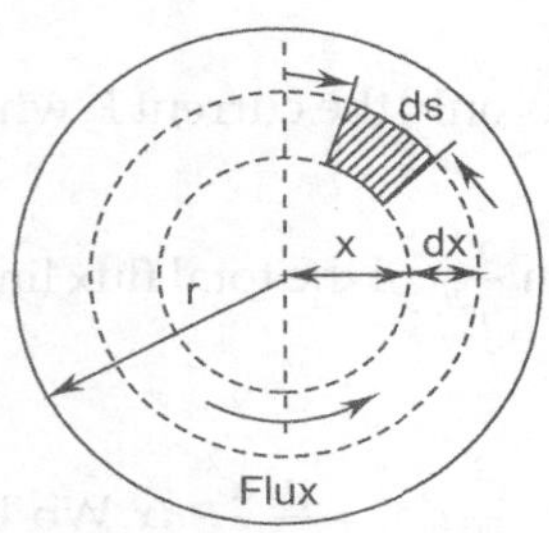

Fig. 3.1

The magnetomotive force (mmf) in ampere turns around any closed path is equal to the current in amperes enclosed by the path. Thus,

$$\text{mmf} = \oint H \cdot dS = I \text{ (Ampere's law)} \qquad ...(3)$$

where, H = Magnetic field intensity (At/m)

I = enclosed current

and S = distance along path.

Let the field intensity at a distance x meters from the center of the conductor be designated H_x. Since the field is symmetrical, H_x is constant at all points equidistant from the center of the conductor. Thus, equation (3) becomes

$$\oint H_x \cdot dS = I_x$$

or $$2\pi x\, H_x = I_x$$

or $$H_x = \frac{I_x}{2\pi x} \qquad ...(4)$$

Assuming the uniform current density

$$I_x = \frac{\pi x^2}{\pi r^2} \cdot I \qquad ...(5)$$

where I is the total current in the conductor.

Put value of I_x from equation (5) in the equation (4)

$$\therefore \quad H_x = \frac{x}{2\pi r^2} \cdot I \text{ At/m}$$

The flux-density x meter from the center of conductor is

$$B_x = \mu H_x = \frac{\mu x I}{2\pi r^2} \text{ Wb/m}^2.$$

Where μ is the permeability of the conductor.

The flux through a cylindrical shell distant x from the centre with radial thickness dx and one metre in length is given by :

$$d\phi_x = (\text{flux-density at } x) \cdot (\text{elemental area normal to flux path})$$

$$= B_x \cdot A_x = B_x\,(dx \cdot 1) = \frac{\mu I x}{2\pi r^2}\,dx \text{ Wb.}$$

This flux in the cylindrical shell links only the current I_x which is $\frac{x^2}{r^2}$ of the total current I so that each weber of flux produces the fraction $\frac{x^2}{r^2}$ of the total flux linkages. Thus, the flux linkages $d\psi$ due to flux $d\phi_x$ is given by :

$$d\psi = \frac{x^2}{r^2}\,d\phi_x = \frac{\mu I x^3}{2\pi r^4}\,dx \text{ Wb T/m.}$$

The total internal flux linkages are obtained by integrating from the centre of the conductor to its surface.

i.e., $$\psi_{int} = \int_0^r \frac{\mu I x^3}{2\pi r^4}\,dx$$

$$= \frac{\mu I}{2\pi r^4}\left[\frac{x^4}{4}\right]_0^r = \frac{\mu I}{8\pi} \text{ Wb T/m.}$$

But $$\mu = \mu_0 \mu_r$$

For a relative permeability of 1, $\mu = 4\pi \times 10^{-7}$ H/m.

$$\therefore \quad \psi_{int} = \frac{4\pi \times 10^{-7}}{8\pi} \text{ I Wb T/m.}$$

$$= \frac{1}{2} \times 10^{-7} \text{ I Wb T/m}$$

or $$L_{int} = \frac{\psi_{int}}{I} = \frac{1}{2} \times 10^{-7} \text{ H/m.}$$

It indicates that the internal inductance is independent of the conductor dimensions if the current distribution is uniform.

External Inductance (L_{ext}) : The cross-section of a conductor and its magnetic field are shown in Fig. 3.2.

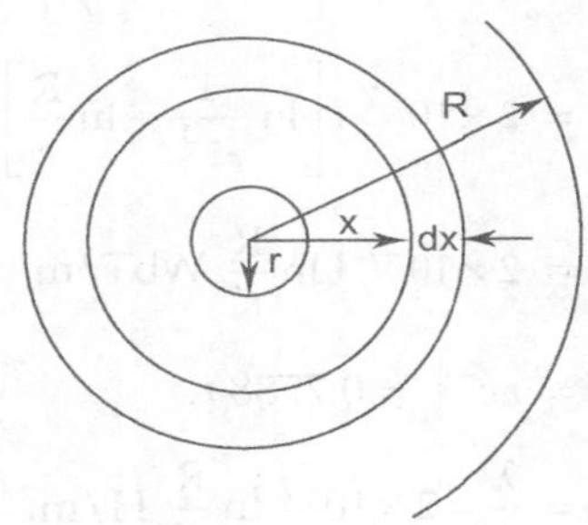

Fig. 3.2

The flux density within a cylindrical shell at a distance x metres from the centre and with radial thickness dx outside the conductor is given by :

$$B_x = \frac{\mu_0 I}{2\pi x} \text{ Wb/m}^2.$$

Now, flux $d\phi$ through a cylindrical shell of thickness dx and axial length 1 metre is :

$$d\phi_x = B_x \cdot dx = \frac{\mu_0 I}{2\pi x} dx \text{ Wb/m.}$$

The flux linkages $d\psi$ per meter are numerically equal to the flux $d\phi$, since flux external to the conductor links all the current in the conductor once and only once. *i.e.*,

$$d\psi = d\phi_x = \frac{\mu_0 I}{2\pi x} dx$$

The total flux linkages can be obtained by integrating $d\psi$ from $x = r$ to $x = R$.

$$\psi_{ext} = \int_r^R \frac{\mu_0 I}{2\pi x} dx$$

$$= \frac{\mu_0 I}{2\pi} \ln \frac{R}{r} \text{ Wb T/m.}$$

$$= 2\times10^{-7} I \ln \frac{R}{r} \text{ Wb-T/m}$$

or $$L_{ext} = \frac{\psi_{ext}}{I} = 2\times10^{-7} \ln \frac{R}{r} \text{ H/m.}$$

The total linkages per metre length of conductor produced by the entire flux with in a cylinder of radius R is the sum of the internal and external linkages.

$$\psi = \psi_{ext} + \psi_{int}$$

$$= 2\times10^{-7} I \ln \frac{R}{r} + \frac{1}{2}\times10^{-7} I$$

$$= 2\times10^{-7} I \left[\frac{1}{4} + \ln \frac{R}{r}\right]$$

$$= 2\times10^{-7} I \left[\ln e^{\frac{1}{4}} + \ln \frac{R}{r}\right]$$

$$= 2\times10^{-7} I \left[\ln \frac{1}{e^{-\frac{1}{4}}} + \ln \frac{R}{r}\right] = 2\times10^{-7} I \ln \frac{R}{re^{-\frac{1}{4}}}$$

$$= 2\times10^{-7} I \ln \frac{R}{r'} \text{ WbT/m}$$

Where, $$r' = re^{-\frac{1}{4}} = 0.7788\, r.$$

The total inductance, $$L = \frac{\lambda}{I} = 2\times10^{-7} \ln \frac{R}{r'} \text{ H/m.}$$

3.4 INDUCTANCE OF A SINGLE-PHASE TWO-WIRE LINE

As in Fig. 3.3 consider a single-phase line consisting of two conductors a and b of equal radius r. They are situated at a distance of D metres. Let these conductors carry the same current in opposite directions so that one serves return path for the other.

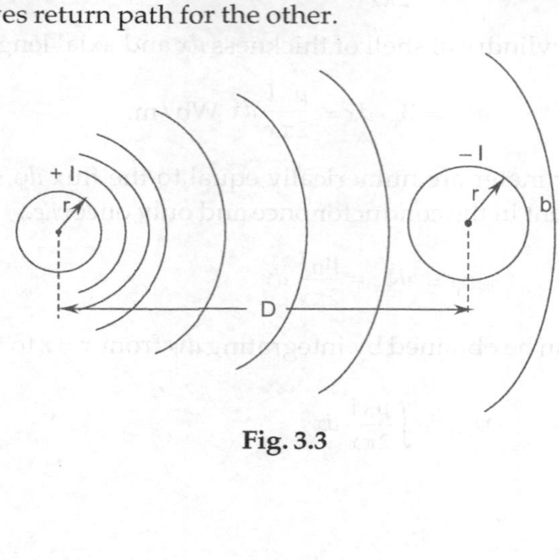

Fig. 3.3

The flux linkages of conductor a may be written as :

$$\psi_a = 2\times10^{-7}\left[I_a \ln\frac{1}{D_{aa}} + I_b \ln\frac{1}{D_{ab}}\right] \quad ...(6)$$

Since, $I_a = +I$

or $I_b = -I$

and $D_{aa} = re^{-\frac{1}{4}} = r'$

and $D_{ab} = D$

Put all these values in equation (6)

$$\psi_a = 2\times10^{-7}\left[I\ln\frac{1}{r'} - I\ln\frac{1}{D}\right] = 2\times10^{-7}\, I\ln\frac{D}{r'}.$$

Similarly, flux linkages of conductor b

$$\psi_b = 2\times10^{-7}\, I\ln\frac{D}{r'}$$

Inductance per conductor $L = L_a = L_b = 2\times10^{-7}\ln\frac{D}{r'}$ H/m.

Loop Inductance

$$= L_a + L_b$$

$$= 2\times10^{-7}\ln\frac{D}{r'} + 2\times10^{-7}\ln\frac{D}{r'}$$

$$= 4\times10^{-7}\ln\frac{D}{r'} \text{ H/m.}$$

It means that the inductance of individual conductor is one-half the total inductance of a two-wire line.

Example 3.1. *A single-phase line has two parallel conductors 2 metres apart. The diameter of each conductor is 1.2 cm. Calculate (a) the inductance of each conductor per km. (b) the loop inductance of the line per km. (c) inductive reactance of the line.*

Solution. Given : $D = 2\text{ m} = 200\text{ cm}$

$$r = \frac{1.2}{2} = 0.6\text{ cm.}$$

$$\therefore \quad r' = 0.778\, r = 0.778\times0.6 = 0.4668\text{ cm.}$$

(*a*) Inductance of each conductor – $L = 2\times10^{-7}\ln\frac{D}{r'}$

$$= 2\times10^{-7}\ln\frac{200}{0.4668}\text{ H/m}$$

$$= 12.12\times10^{-7}\text{ H/m} = 12.12\times10^{-4}\text{ mH/m}$$

Inductance of each conductor per km $= 12.12\times10^{-4}\times10^{3}$ mH/km

$$= 1.212\text{ mH/km.}$$

(*b*) Loop inductance of line per km $= 2\times1.212 = 2.424$ mH/km.

(*c*) Inductive reactive per km $= 2\pi fL$

$$= 2\times3.142\times50\times2.424\times10^{-3}$$

$$= 0.761\ \Omega/\text{km.}$$

Example 3.2. *A two-conductor single-phase line operates at 50 Hz. The diameter of each conductor is 20 mm and spacing between the conductors is 3 m. Calculate the loop inductance per km of the line if the material of the conductor is (a) copper with relative permeability of 1 ; (b) steel with relative permeability of 50.*

Solution. Given :

$$D = 3\,\text{m}$$

$$r = \frac{20}{2}\ \text{mm} = 10\ \text{mm} = 0.01\ \text{m}.$$

$$r' = 0.778\,r = 7.78 \times 10^{-3}\ \text{m}.$$

(*a*) Loop inductance per km with copper-conductor

$$= 2\,(L_{int} + L_{ext})$$

$$= 2\left(\frac{1}{2}\times 10^{-7}\,\mu_{in} + 2\times 10^{-7}\ln\frac{D}{r}\right)$$

$$= 10^{-7}\left(1 + 4\ln\frac{3}{0.01}\right)$$

$$= 10^{-7}\,(1 + 22.81) = 23.8 \times 10^{-7}\ \text{H/m}.$$

$$= 23.8 \times 10^{-7} \times 10^{6}\ \text{mH/km} = 2.38\ \text{mH}.$$

(*b*) Loop inductance per km with steel-conductor.

$$= 10^{-7}\left(50 + 4\ln\frac{3}{0.01}\right)$$

$$= 10^{-7}\,(50 + 22.81)\ \text{H/m}$$

$$= 7.281\ \text{mH/km}.$$

3.5 FLUX LINKAGES OF ONE CONDUCTOR IN A GROUP

As shown in Fig. 3.4, consider a group of n-parallel round conductors carrying phasor currents $I_1, I_2, \ldots\ldots, I_n$ whose sum equals zero. The distances of these conductors from a remote point P are indicated as $D_{1P}, D_{2P}, \ldots\ldots, D_{nP}$.

Let us determine ψ_{1P_1}, the flux-linkages of conductor 1 due to I_1 including internal flux linkages but excluding all the flux beyond the point P.

$$\psi_{1P_1} = 10^{-7}\left(\frac{1}{2}I_1 + 2I_1\ln\frac{D_{1P}}{r_1}\right)\ \text{H/m}$$

$$= 2\times 10^{-7}\,I_1\ln\frac{D_{1P}}{r_1'}\ \text{H/m} \qquad \text{...(7)}$$

The flux linkages ψ_{1P_2} with conductor 1 due to I_2 –

$$\psi_{1P_2} = 2\times 10^{-7}\,I_2\ln\frac{D_{2P}}{D_{12}}\ \text{H/m} \qquad \text{...(8)}$$

The flux-linkages ψ_{1P} with conductor 1 due to all conductors in the group but excluding flux beyond point P is

$$\psi_{1P} = 2\times 10^{-7}\left(I_1\ln\frac{D_{1P}}{r_1'} + I_2\ln\frac{D_{2P}}{D_{12}} + I_3\ln\frac{D_{3P}}{D_{13}} + \cdots\cdots + I_n\ln\frac{D_{nP}}{D_{1n}}\right)$$

Expand by logarithimic terms

$$\psi_{1P} = 2\times10^{-7}\left(I_1 \ln\frac{1}{r_1'} + I_2 \ln\frac{1}{D_{12}} + I_3 \ln\frac{1}{D_{13}} + \cdots + I_n \ln\frac{1}{D_{1n}} + I_1 \ln D_{1P} + I_2 \ln D_{2P} + I_3 \ln D_{3P} + \cdots + I_n \ln D_{nP}\right) \quad ...(9)$$

Since, $I_1 + I_2 + I_3 + \ldots\ldots + I_n = 0$

or $I_n = -(I_1 + I_2 + I_3 + \ldots\ldots + I_{n-1})$

Substituting the value of I_n in equation (9)

$$\psi_{1P} = 2\times10^{-7}\left(I_1 \ln\frac{1}{r_1'} + I_2 \ln\frac{1}{D_{12}} + I_3 \ln\frac{1}{D_{13}} + \cdots\cdots + I_n \ln\frac{1}{D_{1n}} + I_1 \ln D_{1P} + I_2 \ln D_{2P} + I_3 \ln D_{3P} + \cdots\cdots + \left(-I_1 - I_2 - I_3 - \cdots\cdots I_{n-1}\right)\ln D_{nP}\right)$$

$$= 2\times10^{-7}\left[I_1 \ln\frac{1}{r_1'} + I_2 \ln\frac{1}{D_{12}} + I_3 \ln\frac{1}{D_{13}} + \cdots\cdots + I_n \ln\frac{1}{D_{1n}} + I_1 \ln\frac{D_{1P}}{D_{nP}} + I_2 \ln\frac{D_{2P}}{D_{nP}} + I_3 \ln\frac{D_{3P}}{D_{nP}} + \cdots\cdots + I_{n-1} \ln\frac{D_{(n-1)P}}{D_{nP}}\right]$$

If the distances from all the conductors to point P be assumed to be very large, all the ratios will approach unity and the logarithm of such a ratio becomes zero.

Thus, the flux linkages of conductor 1

$$\psi_1 = 2\times10^{-7}\left(I_1 \ln\frac{1}{r_i'} + I_2 \ln\frac{1}{D_{12}} + I_3 \ln\frac{1}{D_{13}} + \cdots\cdots + I_n \ln\frac{1}{D_{1n}}\right)$$

This equation may be written as

$$\psi_1 = 2\times10^{-7}\left[I_1 \ln\frac{1}{D_{11}} + I_2 \ln\frac{1}{D_{12}} + I_3 \ln\frac{1}{D_{13}} + \cdots\cdots + I_n \ln\frac{1}{D_{1n}}\right] \quad ...(10)$$

where, $r_1' = 0.778\, r_1 = D_{11}$ (say)

or

$$\psi_1 = 2\times10^{-7}\left[\sum_{x=1}^{x=n} I_x \ln\frac{1}{D_{1x}}\right] \text{ H/m}$$

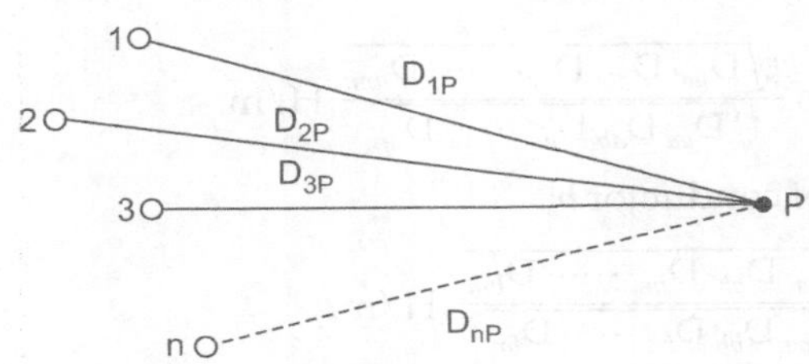

Fig. 3.4

Therefore, equation (10) expresses all the flux linkages of conductor 1 in a group of conductors provided the sum of all the currents is zero.

3.6 INDUCTANCE OF COMPOSITE-CONDUCTOR LINES

Composite conductors means conductors composed of two or more elements or strands electrically in parallel. We will now study the inductance of transmission lines composed of composite conductors.

Figure 3.5 shows a single-phase line composed of two conductors. Conductor X is composed of n-identical, parallel filaments, each of which carries the current I/n. Conductor Y which is the return circuit for the current in conductor X, is composed of m-identical, parallel filaments, each of which carries the current $-I/m$. Distances between the elements will be designated by the letter D with appropriate subscripts.

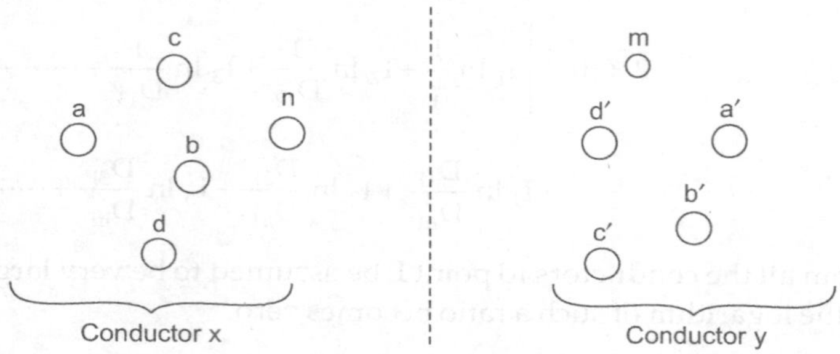

Fig. 3.5

Applying equation (10), the flux linkages about any conductor, say a of group A are :

$$\psi_a = 2\times10^{-7}\,\frac{I}{n}\left(\ln\frac{1}{D_{aa}}+\ln\frac{1}{D_{ab}}+\ln\frac{1}{D_{ac}}+\cdots\cdots+\ln\frac{1}{D_{an}}\right)$$

$$-\,2\times10^{-7}\,\frac{I}{m}\left(\ln\frac{1}{D_{aa'}}+\ln\frac{1}{D_{ab'}}+\ln\frac{1}{D_{ac'}}+\cdots\cdots+\ln\frac{1}{D_{am}}\right)$$

$$= 2\times10^{-7}\, I\ln\frac{\sqrt[m]{D_{aa'}D_{ab'}D_{ac'}\cdots\cdots D_{am}}}{\sqrt[n]{D_{aa}D_{ab}D_{ac}\cdots\cdots D_{an}}}\ \text{Wbt/m.}$$

Since conductor a carries $\frac{I}{n}$ amperes, the inductance of conductor a :

$$L_a = \frac{\psi_a}{I/n} = 2n\times10^{-7}\ln\frac{\sqrt[m]{D_{aa'}D_{ab'}D_{ac'}\cdots\cdots D_{am}}}{\sqrt[n]{D_{aa}D_{ab}D_{ac}\cdots\cdots D_{an}}}\ \text{H/m.}$$

Similarly the inductance of conductor b

$$L_b = 2n\times10^{-7}\ln\frac{\sqrt[m]{D_{ba'}D_{bb'}D_{bc'}\cdots\cdots D_{bm}}}{\sqrt[n]{D_{ba}D_{bb}D_{bc}\cdots\cdots D_{bn}}}\ \text{H/m.}$$

It is found that the conductors of the group have different inductances. The average inductance of a conductor of group X is

$$L_{av} = \frac{L_a + L_b + L_c + \cdots\cdots + L_n}{n}$$

The inductance of conductor × Consisting of n-conductors connected in parallel electrically will be $\frac{1}{n}$ times the average inductance *i.e.,*

$$L_x = \frac{L_{av}}{n} = \frac{L_a + L_b + L_c + \cdots\cdots + L_n}{n^2}$$

Substituting the values of L_a, L_b, L_c, ... L_n, the inductance of conductor X is given by

$$L_x = 2\times10^{-7}\ln\frac{\sqrt[mn]{(D_{aa'}D_{ab'}D_{ac'}\cdots D_{am})(D_{ba'}D_{bb'}D_{bc'}\cdots D_{bm})\cdots(D_{na'}D_{nb'}D_{nc'}\cdots D_{nm})}}{\sqrt[n^2]{(D_{aa}D_{ab}D_{ac}\cdots D_{an})(D_{ba}D_{bb}D_{bc}\cdots D_{bn})\cdots(D_{na}D_{nb}D_{nc}\cdots D_{nn})}} \quad ...(11)$$

The numerator of the logarithmic term of equation (11) is a product of $(m\times n)$ distances for which the $(m\times n)^{th}$ root is taken. It is therefore the geometric mean of all the possible $(m\times n)$ mutual distances from filaments of conductor X to the filaments of conductor Y. This geometric mean is called the mutual geometric mean distance (D_m) between the two conductors.

The denominator of the logarithmic term of equation (11) represents the product of $(n\times n)$ distances for which $(n^2)^{th}$ root is taken. It may be described as the product of distances from every filament in the conductor to itself and to every other filament. It is therefore called the self geometric mean distance (D_s) or geometric mean radius (GMR) of Condcutor X.

Thus, equation (11) becomes

$$L_X = 2\times10^{-7}\ln\frac{D_m}{D_s}\ \text{H/m}$$

where, D_m = Mutual GMD

and D_s = Self GMD.

Similarly, the inductance of conductor Y is determined and the total inductance of the line is

$$L = L_X + L_Y.$$

Example 3.3. *Figure 3.6 shows a stranded conductor having 7-identical strands each of radius r. Find GMR of the conductor.*

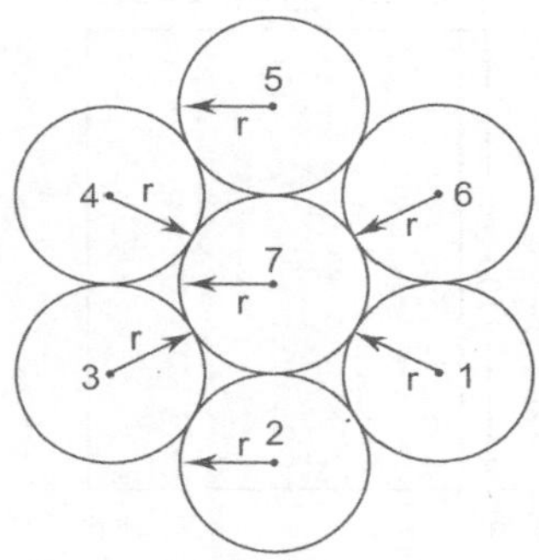

Fig. 3.6

Solution. Distances from strand 1 to other strands are :

$$D_{12} = D_{16} = D_{17} = 2r$$

$$D_{14} = 4r$$

$$D_{13} = \sqrt{D_{14}^2 - D_{34}^2}$$

$$= \sqrt{16r^2 - 4r^2} = 2r\sqrt{3} \qquad [\because \quad D_{34} = 2r]$$

$$D_{15} = D_{13} = 2r\sqrt{3}.$$

The GMR of the seven-strand conductor is the 49th root of the product of 49 distances.

$$D_s = \sqrt[49]{(r')^7 \,(D_{12}\, D_{13}\, D_{14}\, D_{15}\, D_{16}\, D_{17})^6 \,(2r)^6}$$

where r' is GMR of each strand

or

$$D_s = \sqrt[49]{(r')^7 \left(D_{12}^2\, D_{13}^2\, D_{14}\, D_{17}\right)^6 (2r)^6} \qquad [\because\ D_{12} = D_{16} \text{ and } D_{13} = D_{15}]$$

The term $(2r)^6$ accounts for the product of the distances from the central strand to every outside strand.

$$\therefore \qquad D_s = \sqrt[49]{\left[(0.7788r)^7 \left\{(2r)^2 \left(2\sqrt{3}r\right)^2 (4r)(2r)\right\}^6 (2r)^6\right]}$$

$$= 2.177\ r.$$

Example 3.4. *Two conductors of a single phase line each of 1 cm diameter are arranged in a vertical plane with one conductor mounted 1 m above the other. A second identical line is mounted at the same height as the first and spaced horizontally 0.25 m apart from it. The two upper and two lower conductors are connected in parallel. Determine the inductance per km of the resulting line.*

Solution. From Fig. 3.7,

Radius of the conductor r $= \dfrac{1}{2} = 0.5$ cm.

GMR of conductor $= 0.7788r$

$= 0.7788 \times 0.5 = 0.389$ cm.

$$\text{Self-GMD } (D_s) = \sqrt[4]{D_{aa} \times D_{aa'} \times D_{a'a'} \times D_{a'a}}$$

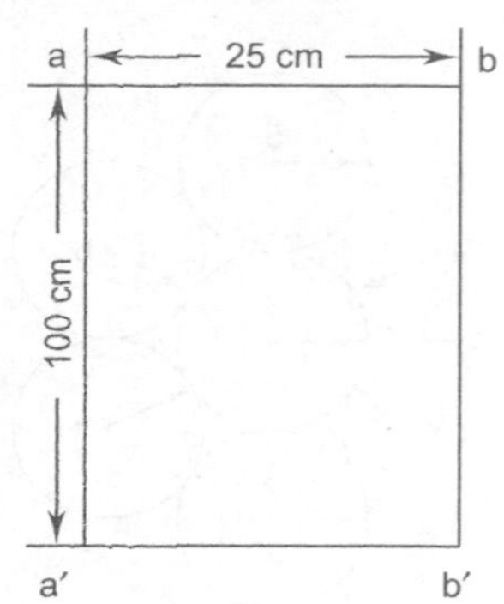

Fig. 3.7

$$= \sqrt[4]{(0.389 \times 100)^2} = 6.23 \text{ cm.}$$

Mutual GMD between *a* and *b*

$$D_m = \sqrt[4]{D_{ab} \times D_{ab'} \times D_{a'b} \times D_{a'b'}}$$

From Fig. 3.7, $D_{ab'} = D_{ba'} = \sqrt{(25)^2 + (100)^2} = 103 \text{ cm.}$

∴ $D_m = \sqrt[4]{(100 \times 103)^2} = 101.5 \text{ cm.}$

Inductance per conductor per metre $= 2 \times 10^{-7} \ln \dfrac{D_m}{D_s}$

$$= 2 \times 10^{-7} \ln \frac{101.5}{6.23} = 0.56 \times 10^{-6} \text{ H.}$$

∴ Loop inductance per km of the line.

$$= 2 \times (0.56 \times 10^{-6}) \times 1000 \text{ H} = 1.12 \text{ mH.}$$

Example 3.5. *One circuit of a single-phase transmission line is composed of three solid 0.25 cm radius wires. The return circuit is composed of two 0.5 cm radius wires. The arrangement of conductors is shown in Fig. 3.8. Find the inductance due to the current in each side of the line and the inductance of the complete line in henry's per metre.*

Solution. Mutual GMD between sides X and Y is

$$D_m = \sqrt[6]{D_{ad} \times D_{ae} \times D_{bd} \times D_{be} \times D_{cd} \times D_{ce}}$$

$$D_{ad} = D_{be} = 9 \text{ m.}$$

$$D_{ae} = D_{bd} = D_{ce} = \sqrt{(6)^2 + (9)^2} = \sqrt{117}$$

$$= 10.82 \text{ m.}$$

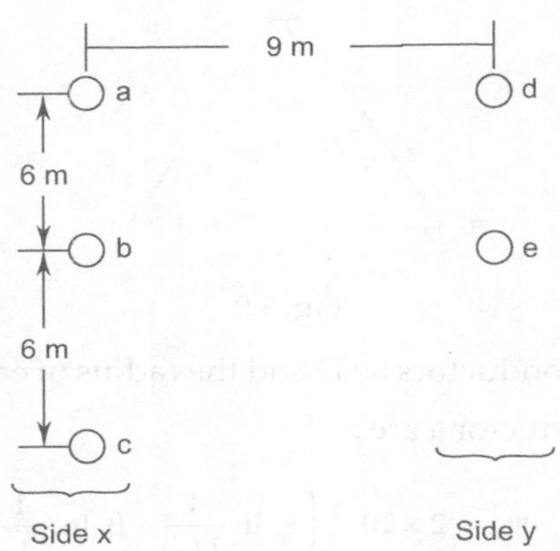

Fig. 3.8

$$D_{cd} = \sqrt{(9)^2 + (12)^2} = 15 \text{ m.}$$

$$D_m = \sqrt[6]{9 \times 10.82 \times 10.82 \times 9 \times 15 \times 10.82}$$
$$= 10.745 \text{ m.}$$

GMR for Side X:
$$D_s = \sqrt[9]{D_{aa} \times D_{ab} \times D_{ac} \times D_{ba} \times D_{bb} \times D_{bc} \times D_{ca} \times D_{cb} \times D_{cc}}$$
$$D_{aa} = D_{bb} = D_{cc} = 0.7788 \times 0.25 \times 10^{-2} = 0.1947 \times 10^{-2} \text{ m.}$$
$$\therefore \quad D_s = \sqrt[9]{\left(0.1947 \times 10^{-2}\right)^3 \times 6 \times 12 \times 6 \times 6 \times 12 \times 6}$$
$$= 0.481 \text{ m.}$$

GMR for side Y:
$$D_s = \sqrt[4]{D_{dd} \times D_{de} \times D_{ed} \times D_{ee}}$$
$$D_{dd} = D_{ee} = 0.7788 \times 0.5 \times 10^{-2} = 0.3894 \times 10^{-2} \text{ m.}$$
$$\therefore \quad D_s = \sqrt[4]{\left(0.3894 \times 10^{-2}\right)^2 \times 6^2} = 0.153 \text{ m.}$$
$$\therefore \quad L_X = 2 \times 10^{-7} \ln \frac{10.745}{0.481} = 6.212 \times 10^{-7} \text{ H/m.}$$
$$L_Y = 2 \times 10^{-7} \ln \frac{10.745}{0.153} = 8.503 \times 10^{-7} \text{ H/m.}$$
$$L = L_X + L_Y = 14.715 \times 10^{-7} \text{ H/m.}$$

3.7 INDUCTANCE OF THREE-PHASE LINES WITH EQUILATERAL SPACING OR INDUCTANCE OF SYMMETRICAL THREE-PHASE LINE

A three-phase line is said to be symmetrical when its conductors are situated at the corners of an equilateral triangle. Such an arrangement of conductors is shown in Fig. 3.9.

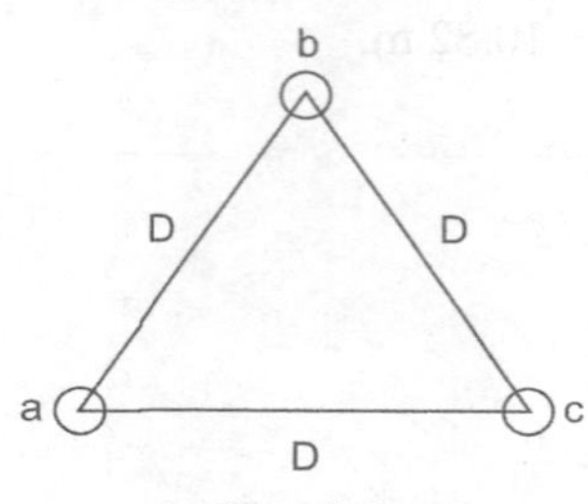

Fig. 3.9

Let the spacing between the conductors be D and the radius of each conductor r.

Then, the flux linkages of conductor a are :

$$\psi_a = 2 \times 10^{-7} \left(I_a \ln \frac{1}{D_{aa}} + I_b \ln \frac{1}{D_{ab}} + I_c \ln \frac{1}{D_{ac}} \right)$$

but $\quad D_{ab} = D_{ac} = D$

and $\quad D_{aa} = r' = 0.7788r = re^{-\frac{1}{4}}.$

$$\therefore \qquad \psi'_a = 2\times10^{-7}\left(I_a \ln\frac{1}{r'} + I_b \ln\frac{1}{D} + I_c \ln\frac{1}{D}\right) \qquad ...(12)$$

If we assume balanced three-phase phasor currents, then

$$I_a + I_b + I_c = 0$$

or

$$I_a = -(I_b + I_c)$$

Thus, the equation (12) becomes

$$\psi_a = 2\times10^{-7}\left(I_a \ln\frac{1}{r'} + (I_b + I_c)\ln\frac{1}{D}\right)$$

$$= 2\times10^{-7}\left(I_a \ln\frac{1}{r'} - I_a \ln\frac{1}{D}\right)$$

$$= 2\times10^{-7}\, I_a \ln\frac{D}{r'} \text{ Wb t/m.}$$

The inductance of conductor a is

$$L_a = \frac{\psi_a}{I_a} = 2\times10^{-7} \ln\frac{D}{r'} \text{ H/m.}$$

Because of symmetry, the inductances of conductors *b* and *c* are same as the inductance of conductor *a*.

Example 3.6. *A three-phase transmission line has its conductors at the corners of an equilateral triangle with side 4 m. The diameter of each conductor is 1.63 cm. Find the inductance per phase per km of the line.*

Solution.

$$L = 2\times10^{-7}\frac{D}{r'} \text{ H/m}$$

$$r' = 0.7788 \times 0.815 = 0.635 \text{ cm.}$$

$$\therefore \qquad L = 2\times10^{-7} \ln\frac{400}{0.635} = 1.289\times10^{-6} \text{ H/m.}$$

or

$$L = 1.289 \text{ mH/km.}$$

3.8 INDUCTANCE OF THREE-PHASE LINE WITH UNSYMMETRICAL SPACING OR INDUCTANCE OF UNSYMMETRICAL THREE-PHASE LINE

When three-phase line conductors are not equidistant from each other, the conductor spacing is said to be unsymmetrical. For an unsymmetrically spaced three-phase line, the inductances and therefore the voltage drops will be different for all the phases even under balanced current conditions. Therefore, the voltage at the receiving end will not be same for all phases. Moreover, if communication lines are also running adjacent to the power line, an unbalance of voltage is also produced in them. This results in disturbances in them. In order to reduce the inequality of inductance and inductive interference with parallel running communication lines, the line is transposed. The transposition is done by exchanging the positions of the conductors at regular intervals along the line so that each conductor occupies the original position of every other conductor over an equal distance. Figure 3.10 is an unsymmetrical line with its transposition cycle.

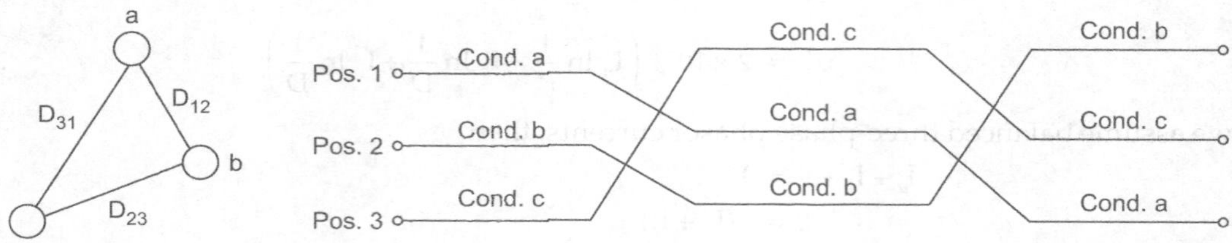

Fig. 3.10

In Fig. 3.10, the phase conductors are designated as *a*, *b* and *c*, and the positions occupied are numbered, 1, 2 and 3. The effect of transposition is that each conductor has the same average inductance.

When a non-symmetrically spaced line is properly transposed, the average value of flux-linkages of a conductor may be found be adding up its flux linkages for each of its three positions in the transposition cycle and dividing the sum by 3. Let us find the flux linkages of conductor a.

When *a* is in position 1, *b* in position 2 and *c* in position 3, the flux-linkages of a are :

$$\psi_{a_1} = 2\times10^{-7}\left[I_a \ln\frac{1}{r'} + I_b \ln\frac{1}{D_{12}} + I_c \ln\frac{1}{D_{31}}\right] \quad ...(13)$$

When *a* is in position 2, *b* is in position 3 and *c* is in position 1, then flux linkages of a are :

$$\psi_{a_2} = 2\times10^{-7}\left(I_a \ln\frac{1}{r'} + I_b \ln\frac{1}{D_{23}} + I_c \ln\frac{1}{D_{12}}\right) \quad ...(14)$$

When *a* is in position 3, *b* is in position 1, and *c* is in position 2, then flux linkages of conductor a are :

$$\psi_{a_3} = 2\times10^{-7}\left(I_a \ln\frac{1}{r'} + I_b \ln\frac{1}{D_{31}} + I_c \ln\frac{1}{D_{23}}\right) \quad ...(15)$$

The average value of flux-linkages of a is

$$\psi_a = \frac{1}{3}\left(\psi_{a_1} + \psi_{a_2} + \psi_{a_3}\right)$$

$$= \frac{2}{3}\times10^{-7}\left[\left(I_a \ln\frac{1}{r'} + I_b \ln\frac{1}{D_{12}} + I_c \ln\frac{1}{D_{31}}\right) + \left(I_a \ln\frac{1}{r'} + I_b \ln\frac{1}{D_{23}} + I_c \ln\frac{1}{D_{12}}\right) + \left(I_a \ln\frac{1}{r'} + I_b \ln\frac{1}{D_{31}} + I_c \ln\frac{1}{D_{23}}\right)\right]$$

$$= \frac{2}{3}\times10^{-7}\left[3I_a \ln\frac{1}{r'} + I_b \ln\frac{1}{D_{12}D_{23}D_{31}} + I_c \ln\frac{1}{D_{12}D_{23}D_{31}}\right] \quad ...(16)$$

Let us assume balanced conditions *i.e.*, $I_a + I_b + I_c = 0$

or $\quad I_a = -(I_b + I_c)$

Thus, equation (16) becomes

$$\psi_a = \frac{2}{3}\times10^{-7}\left[3I_a \ln\frac{1}{r'} + \ln\frac{1}{D_{12}D_{23}D_{31}}(I_b + I_c)\right]$$

$$= \frac{2}{3} \times 10^{-7} \left[3 I_a \ln \frac{1}{r'} - I_a \ln \frac{1}{D_{12} D_{23} D_{31}} \right]$$

or $$\psi_a = 3 \times \frac{2}{3} \times 10^{-7} \left[I_a \ln \frac{1}{r'} - \frac{1}{3} I_a \ln \frac{1}{D_{12} D_{23} D_{31}} \right]$$

$$= 2 \times 10^{-7} I_a \ln \frac{(D_{12} D_{23} D_{31})^{\frac{1}{3}}}{r'}$$

Average inductance of phase a is :

$$L_a = \frac{\psi_a}{I_a} = 2 \times 10^{-7} \ln \frac{(D_{12} D_{23} D_{31})^{\frac{1}{3}}}{r'} \text{ H/m}$$

Similarly, $$L_b = L_c = 2 \times 10^{-7} \ln \frac{(D_{12} D_{23} D_{31})^{\frac{1}{3}}}{r'} \text{ H/m}$$

Thus, it is found that the values of the inductance for the three phases are equalized by transposition.

The quantity $(D_{12} D_{23} D_{31})^{\frac{1}{3}}$ is the geometric mean of the three unequal spacings and is called the equivalent equilateral spacing between the conductors of the line and the corresponding value of inductance found will be called the equivalent inductance and can be written as :

$$L = 2 \times 10^{-7} \ln \frac{D_{eq}}{r'} \text{ H/m}$$

Where, $D_{eq} = \sqrt[3]{D_{12} D_{23} D_{31}}$.

Example 3.7. *The three conductors of a 3-phase line are arranged at the corners of a triangle of sides 2 m, 2.5 m and 4.5 m. Calculate the inductance per km of the line when the conductors are regularly transposed. The diameter of each conductor is 1.24 cm.*

Solution. From Fig. 3.11, equivalent equilateral spacing-

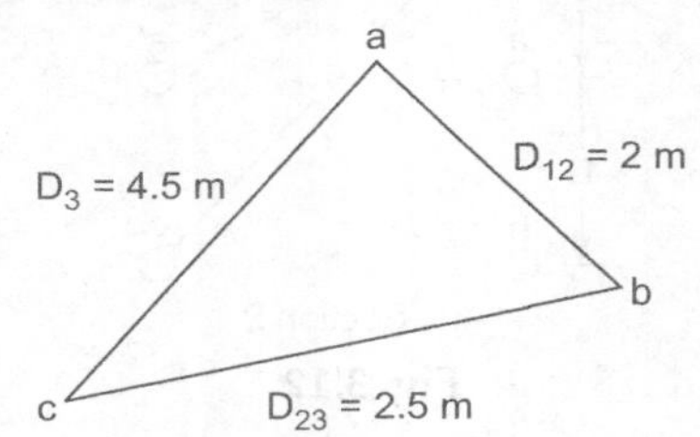

Fig. 3.11

$$D_{eq} = \sqrt[3]{D_{12} D_{23} D_{31}}$$

$$= \sqrt[3]{2 \times 2.5 \times 4.5} = 2.82 \text{ m} = 282 \text{ cm.}$$

$$r' = 0.7788r = 0.7788 \times \left(\frac{1.24}{3}\right) = 0.4828 \text{ cm}$$

$$L = 2\times10^{-7}\ \ln\frac{282}{0.4828} = 2\times10^{-7}\times6.37 = 12.74\times10^{-7}$$

Inductance/phase/km = $1.274\times10^{-6}\times10^{3} = 1.274\times10^{-3}$ H = 1.274 mH.

Example 3.8. *A three-phase transmission line has a horizontal configuration with a spacing of 6 m between adjacent conductors and 12 m between outer conductors. The radius of each conductor is 2.4 m. Find the inductance per phase per km of the line.*

Solution. $D_{eq} = \sqrt[3]{6\times6\times12} = 7.5595$ m.

$r' = 0.7788\times2.4 = 1.869$ m.

$$L = 2\times10^{-7}\ \ln\frac{7.5595}{1.869} = 2\times10^{-7}\times1.397 = 2.79\ \text{mH/km}.$$

3.9 INDUCTANCE OF DOUBLE CIRCUIT THREE-PHASE LINE

A double circuit three-phase line has two parallel conductors for each phase. The use of two three-phase circuits on the same towers leads to a greater reliability and a higher transmission capacity. The method of GMD can be used to find the inductance per phase by taking the conductors of one phase to be the strands of one composite conductor. It is desirable to have a configuration which results in minimum inductance (hence maximum transmission capability, good voltage regulation, greater power limit and high power factor). This is possible if GMD is low and GMR is high. Therefore, the individual conductors of a phase should be widely separated (to give high GMR) and the distance between phases should be kept low (to give low GMD).

Figure 3.12 shows the three sections of the transposition cycle of two parallel circuit three-phase lines with vertical spacing. The conductors a and *a′* belong to one phase, *b* and *b′* to second phase and *c* and *c′* to the third phase. In each case the conductors of two phases are placed diametrically opposite to each-other and those of the third phase are placed horizontally to each other. This configuration gives high value of GMR.

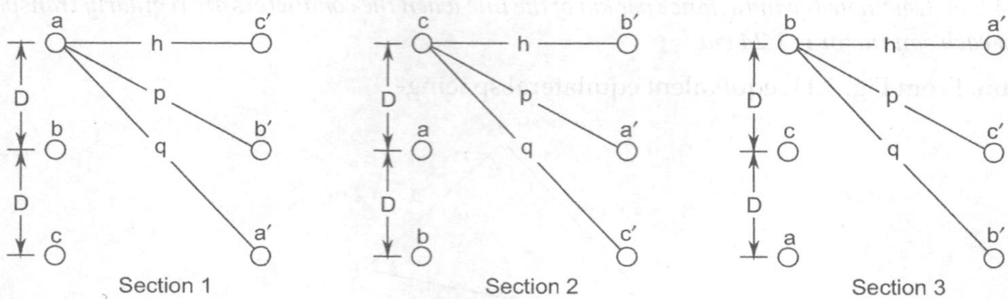

Fig. 3.12

The inductance per phase is given by

$$L = 2\times10^{-7}\ \ln\frac{D_{eq}}{D_s} \quad ...(17)$$

The equivalent equilateral spacing, $D_{eq} = \sqrt[3]{D_{ab}\,D_{bc}\,D_{ca}}$

where D_{ab} = Mutual GMD between phases *a* and *b* in section 1 of the transposition cycle = $(Dp\,Dp)^{\frac{1}{4}}$ $= \sqrt{Dp}$

$$D_{bc} = \text{Mutual GMD between phases } b \text{ and } c \text{ in Section 1}$$

$$= (Dp\, Dp)^{\frac{1}{4}} = \sqrt{Dp}$$

and $$D_{ca} = \text{Mutual GMD between phases } c \text{ and } a \text{ in Section 1}$$

$$= (2Dh\, 2Dh)^{\frac{1}{4}} = \sqrt{2Dh}$$

Hence, $$D_{eq} = \left(\sqrt{Dp}\cdot\sqrt{Dp}\cdot\sqrt{2Dh}\right)^{\frac{1}{3}}$$

$$= 2^{\frac{1}{6}}\, D^{\frac{1}{2}}\, p^{\frac{1}{3}}\, h^{\frac{1}{6}} \qquad ...(18)$$

Self GMD or equivalent GMR is given by

$$D_s = \sqrt[3]{D_{sa}\cdot D_{sb}\cdot D_{sc}}$$

GMR of conductors of phase a in section 1 is :

$$D_{sa} = (r'q\, r'q)^{\frac{1}{4}} = \sqrt{r'q}$$

GMR of conductors of phase b in section 1 is :

$$D_{sb} = (r'h\, r'h)^{\frac{1}{4}} = \sqrt{r'h}$$

GMR of conductors of phase c in section 1 is :

$$D_{sc} = (r'q\, r'q)^{\frac{1}{4}} = \sqrt{r'q}$$

Hence, $$D_s = \sqrt[3]{D_{sa}\cdot D_{sb}\cdot D_{sc}}$$

$$= \left(\sqrt{r'q}\,\sqrt{r'h}\,\sqrt{r'q}\right)^{\frac{1}{3}}$$

$$= (r')^{\frac{1}{2}}\, q^{\frac{1}{3}}\, h^{\frac{1}{6}} \qquad ...(19)$$

Putting the values of D_{eq} and D_s from equations (18) and (19) into the equation (17)

$$L = 2\times10^{-7}\ln\frac{2^{\frac{1}{6}}\, D^{\frac{1}{2}}\, p^{\frac{1}{3}}\, h^{\frac{1}{6}}}{(r')^{\frac{1}{2}}\, q^{\frac{1}{3}}\, h^{\frac{1}{6}}}$$

$$= 2\times10^{-7}\ln\left(2^{\frac{1}{6}}\left(\frac{D}{r'}\right)^{\frac{1}{2}}\left(\frac{p}{q}\right)^{\frac{1}{3}}\right) \text{ H/phase/m.}$$

Example 3.9. *Find inductance per phase per km of double circuit 3-phase line shown in Fig. 3.13. The radius of each conductor is 6 mm.*

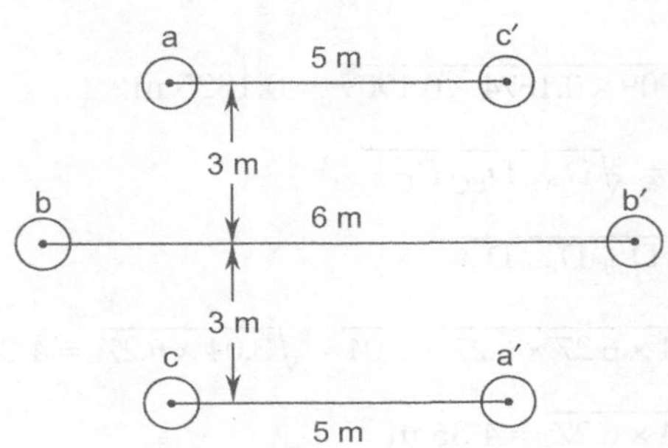

Fig. 3.13 (*a*)

Solution. GMR of conductor $r' = 0.7788 \times 6 = 4.67$ mm.

From figure, $$D_{ab} = \sqrt{(3)^2 + (0.5)^2} = 3.04 \text{ m}$$

$$D_{ab'} = \sqrt{(3)^2 + (5.5)^2} = 6.27 \text{ m}$$

$$D_{aa'} = \sqrt{(6)^2 + (5)^2} = 7.81 \text{ m}$$

$\therefore$ $$D_s = \sqrt[3]{D_{sa} \cdot D_{sb} \cdot D_{sc}}$$

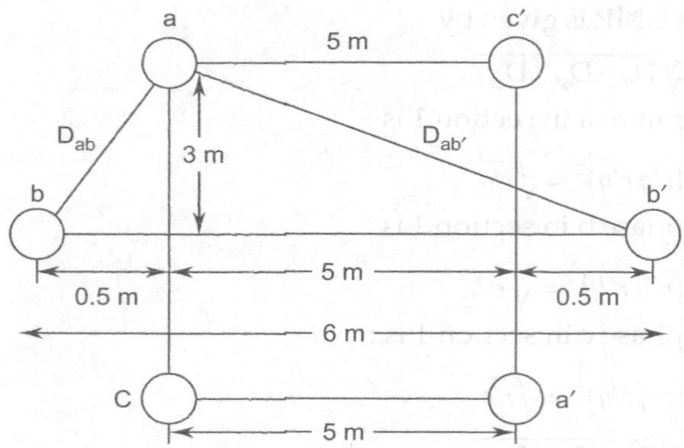

Fig. 3.13 (*b*)

$$D_{sa} = \sqrt[4]{D_{aa} D_{aa'} D_{a'a'} D_{a'a}}$$

$$= \sqrt[4]{\left(4.67 \times 10^{-3} \times 7.81\right)^2}$$

$$= \sqrt{4.67 \times 10^{-3} \times 7.81} = 0.1909 \text{ m.}$$

$$D_{sb} = \sqrt[4]{\left(4.67 \times 10^{-3} \times 6\right)^2}$$

$$= \sqrt{4.67 \times 10^{-3} \times 6} = 0.1674 \text{ m.}$$

$$D_{sc} = \sqrt[4]{\left(4.67 \times 10^{-3} \times 7.81\right)^2} = 0.1909 \text{ m.}$$

$\therefore$ $$D_s = \sqrt[3]{0.1909 \times 0.1674 \times 0.1909} = 0.1825 \text{ m.}$$

Now, we will calculate : $D_{eq} = \sqrt[3]{D_{AB} D_{BC} D_{CA}}$

$$D_{AB} = \sqrt[4]{D_{ab} D_{ab'} D_{a'b} D_{a'b'}}$$

$$= \sqrt[4]{3.04 \times 6.27 \times 6.27 \times 3.04} = \sqrt{3.04 \times 6.27} = 4.36 \text{ m}$$

$$D_{BC} = \sqrt{3.04 \times 6.27} = 4.36 \text{ m}$$

$$D_{CA} = \sqrt[4]{D_{ca} D_{ca'} D_{a'c} D_{a'c'}} = \sqrt[4]{6 \times 5 \times 6 \times 5} = 5.48 \text{ m}$$

$$\therefore \quad D_{eq} = \sqrt[3]{4.36 \times 4.36 \times 5.48} = 4.71 \text{ m}$$

Hence,
$$L = 2 \times 10^{-7} \ln \frac{4.71}{0.1825} = 6.5 \times 10^{-7} \text{ H/m.}$$

or
$$L = 0.65 \text{ mH/km.}$$

3.10 BUNDLED CONDUCTORS

A bundled conductor consists of two or more parallel sub-conductors at a spacing of several diameters. These groups of subconductors form the phase conductors. The number of subconductors per phase can be 2, 3, 4 or even 8. Bundled conductors are also called grouped or multiple conductors. 400 KV lines use only 2 or 4 subconductors per phase. The 765 KV lines in USA and other countries use upto 8 subconductors per phase.

A three-conductor bundle has its subconductors situated at the vertices of an equilateral triangle of side d and a four-conductor bundle has its subconductors situated at the corners of a square of side d. These configuration are shown in Fig. 3.14. The intra-bundle spacing d is around 40 cms.

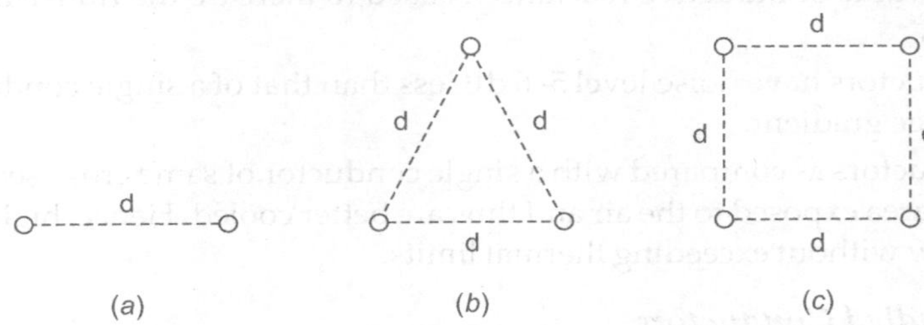

Fig. 3.14

The division of current and charge between the subconductors is not equal unless sub-conductors are transposed and GMD method is accurate for calculations. The GMR of bundled conductors can be found in same manner as that for standard conductors.

For a two-conductor (duplex) arrangement as in Fig. 3.14

$$D_s^b = \sqrt[4]{(r'd \cdot r'd)} = \sqrt{r'd}$$

For a three-conductor (triplex) arrangement as in Fig. 3.14.

$$D_s^b = \sqrt[9]{(r'd \cdot d)^3} = \sqrt[3]{r'd^2}$$

For a four-conductor (quadruplex) arrangement as in Fig. 3.14.

$$D_s^b = \sqrt[16]{\left(r'd \cdot d\sqrt{2}d\right)^4} = 1.09\sqrt[4]{r'd^3}$$

Where D_s^b denotes the GMR of bundled conductors.

The GMD of a bundled conductor line can be found by taking the root of the product of distances from each conductor of a bundle to every other conductor of the other bundles.

Advantages of Bundled Conductors

1. The use of bundled conductors is economical for transmission of larger blocks of power over longer distances at EHV and UHV, so the bundled conductors may be considered as an integral part of EHV and UHV transmission lines.
2. The bundle acts as far as the electric field is concerned, like a conductor of diameter much larger than that of the component conductors. This reduces the voltage gradient, corona and radio interference aspects. If a single conductor per phase is used and the size of conductor is selected on the basis of current carrying capacity alone, the gradient at the surface of conductor would be higher than the acceptable value (between 20–30 KV peak per cm).
3. The bundled conductors have higher capacitance and therefore lower surge-impedance $\left(Z = \sqrt{\frac{L}{C}}\right)$ as compared to single conductors of equivalent diameters. This lower surge impedance gives rise to an increase in the current carrying capacity and boost up of operating voltage. On the other hand, due to higher capacitance of bundled conductors, total generated reactive power capacity of the lines is also reduced.
4. The GMR of bundled conductors is high and therefore the inductive reactance of the line is low. This low value of inductive reactance is used to increase the transient stability limit of very long lines.
5. Bundled conductors have noise level 5–6 dB less than that of a single conductor operating at the same voltage gradient.
6. Bundled conductors as compared with a single conductor of same cross-sectional area have a larger surface area exposed to the air and thus are better cooled. Hence, higher current may be allowed to flow without exceeding thermal limits.

Limitations of Bundled Conductors

1. Bundled conductors are subject to greater ice and wind loadings than single conductors of the same total cross-sectional area. In order to avoid this situation, spacers have to be installed at intervals of 80–100 m in each span to maintain the spacing between the sub-conductors when bundled conductors are used.
2. Due to the complex mechanical structure of bundled conductors, the clearance requirement at structure increases and therefore require somewhat stronger towers and have higher initial costs.

Example 3.10. *A bundled conductor line has 4 conductors per bundle. The four sub-conductors are placed at the corners of square of side 25 cm. The ratios of each sub-conductor is 1.46 cm. Find GMR of this configuration.*

Solution. As for 4-conductor configuration,

$$D_s = 1.09 \times \sqrt[4]{r' \times d^3}$$

$$r' = 0.7788r = 0.7788 \times 1.46 = 1.137 \text{ cm.}$$

$$\therefore \quad D_s = 1.09 \times \sqrt[4]{1.137 \times (25)^3} = 12.58 \text{ cm.}$$

Example 3.11. *Calculate the inductance per phase of a three-phase double circuit line whose phase conductors have a radius of 5.3 cm with the horizontal conductor arrangement as shown in Fig. 3.15.*

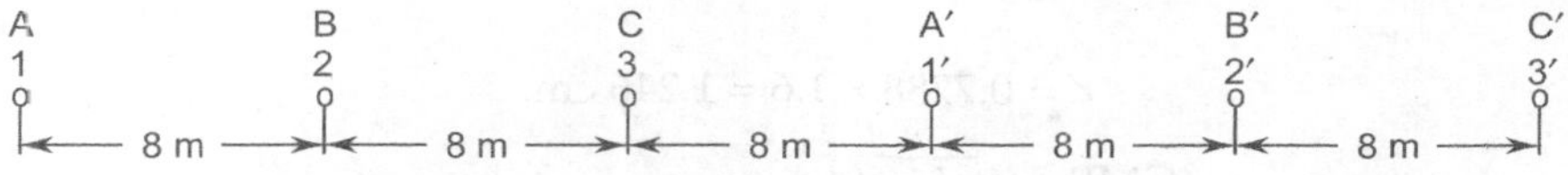

Fig. 3.15

Solution. GMR of conductor $r' = 0.7788r$

$$r' = 0.7788 \times 5.3 \times 10^{-2} = 0.0413 \text{ m.}$$

Equivalent self-GMD of one phase

$$D_s = \left(D_{s_1} \times D_{s_2} \times D_{s_3}\right)^{\frac{1}{3}}$$

where,

$$D_{s_1} = \sqrt[4]{D_{AA} \times D_{AA'} \times D_{A'A} \times D_{A'A'}}$$

$$= \sqrt[4]{0.0413 \times 24 \times 24 \times 0.0413} = 0.995 \text{ m.}$$

Similarly,

$$D_{s_2} = D_{s_3} = 0.995 \text{ m.}$$

$\therefore$

$$D_s = (0.995 \times 0.995 \times 0.995)^{\frac{1}{3}} = 0.995 \text{ m.}$$

Equivalent Mutual GMD – $D_m = (D_{AB} \times D_{BC} \times D_{CA})^{\frac{1}{3}}$

where,

$$D_{AB} = \sqrt[4]{D_{AB} \times D_{AB'} \times D_{BA} \times D_{BA'}}$$

$$= \sqrt[4]{8 \times 32 \times 8 \times 16} = 13.45 \text{ m.}$$

$$D_{BC} = \sqrt[4]{D_{BC} \times D_{BC'} \times D_{CB} \times D_{CB'}}$$

$$= \sqrt[4]{8 \times 32 \times 8 \times 16} = 13.45 \text{ m.}$$

$$D_{CA} = \sqrt[4]{D_{AC} \times D_{AC'} \times D_{CA} \times D_{CA'}}$$

$$= \sqrt[4]{16 \times 40 \times 16 \times 8} = 16.917 \text{ m.}$$

$\therefore$

$$D_m = (13.45 \times 13.45 \times 16.917)^{\frac{1}{3}} = 14.518 \text{ m.}$$

Inductance/phase/m

$$= 2 \times 10^{-7} \ln \frac{D_m}{D_s}$$

$$= 2 \times 10^{-7} \ln \frac{14.518}{0.995} = 5.36 \times 10^{-7} \text{ H/m.}$$

Example 3.12. *A 400 KV, 3-phase bundled conductor line with two sub-conductors per phase has a horizontal configuration as shown in Fig. 3.16. The radius of each sub-conductor is 1.6 cm. (a) Find the inductance per phase per km of the line. (b) Compute the inductance of the line with only one conductor per phase having the same cross-sectional area of the conductor of each phase.*

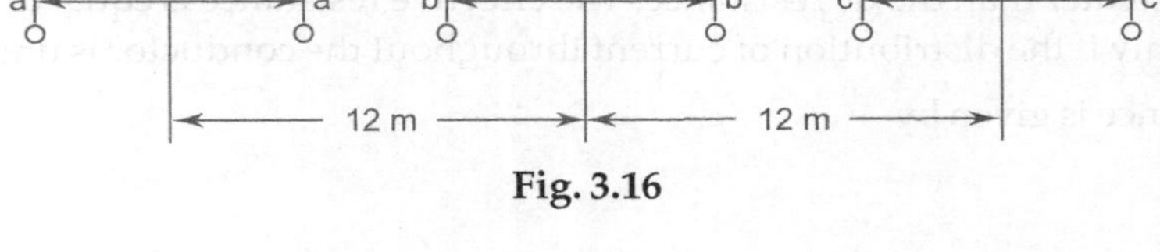

Fig. 3.16

Solution.

(*a*)
$$r' = 0.7788 \times 1.6 = 1.246 \text{ cm.}$$

$$\text{GMR} = \sqrt[4]{r'd \cdot r'd}$$

$$= \sqrt[4]{1.246 \times 45 \times 1.246 \times 45} = 7.49 \text{ cm.}$$

$$D_{eq} = \sqrt[3]{D_{ab} \times D_{bc} \times D_{ca}}$$

From Fig. 3.16
$$D_{ab} = \sqrt[4]{12 \times (12 + 0.45) \times 12 \times (12 - 0.45)}$$

$$= 11.996 \text{ m} = D_{bc}$$

$$D_{ca} = \sqrt[4]{24 \times (24 + 0.45) \times 24 \times (24 - 0.45)} = 23.998 \text{ m.}$$

$\therefore$
$$D_{eq} = \sqrt[3]{11.996 \times 11.996 \times 23.998} = 15.115 \text{ m.}$$

Alternatively, we can take centre to centre distances as :

$$D_{eq} = \sqrt[3]{D_{ab} \times D_{bc} \times D_{ca}} = \sqrt[3]{12 \times 12 \times 24} = 15.119 \text{ m.}$$

Hence,
$$L = 2 \times 10^{-7} \ln \frac{D_{eq}}{D_s}$$

$$= 2 \times 10^{-7} \ln \frac{15.115 \times 100}{7.49} = 1.06 \text{ mH/km.}$$

(*b*) Conductor radius for same cross-sectional area = $\sqrt{2} \times 1.6 = 2.2624$ cm.

$\therefore$
$$r' = 0.7788 \times 2.2624 = 1.762 \text{ cm.}$$

Hence,
$$l = 2 \times 10^{-7} \ln \frac{15.115 \times 100}{1.762} = 1.351 \text{ mH/km.}$$

Comment **:** The inductance of the bundled conductor line is less than that of the line with one conductor per phase. The lower inductance leads to a greater transmission capacity of the bundled conductor line.

3.11 RESISTANCE

The resistance of transmission-line conductors is most important cause of power loss in a transmission line. We generally use the term effective resistance of a conductor which is the ratio of power loss in a conductor in watts to the square of rms current in amperes and is given by :

$$R = \frac{\text{Power loss in conductor}}{|I|^2} \Omega.$$

This is slightly greater than the dc resistance. The effective resistance is equal to the dc resistance of the conductor only if the distribution of current throughout the conductor is uniform.

The D.C. resistance is given by

$$R = \rho \frac{l}{A} \Omega \quad ...(20)$$

where r = Resistivity of the conductor, Ω-m

l = length, m

A = cross-section area, m^2.

The d.c. resistance of a stranded conductor is greater than the value given by equation (20) because spiralling of the strands makes them longer than the conductor itself. The increase in resistance due to spiralling is around 1% for three-strand conductors and around 2% for concentrically stranded conductors.

The resistivity of the conducting material is compared with the annealed copper having a conductivity of 100%. The International Annealed Copper Standard (IACS) has a resistivity of 0.017241 ohm square millimeter per meter at 20°C is said to have a conductivity of 100 percent. The conductivity of hard-drawn copper varies with the hardness. Hard-drawn copper used in practice as a conductivity of about 97 percent at 20°C. The following formula may be used for hard-drawn copper.

$$P = \frac{T}{16}$$

Where P is the percentage increase in resistivity over annealed copper and T is the tensile-strength of hard-drawn copper in kg f/mm^2. The formula gives sufficiently accurate results within range of 30–50 kg f/mm^2 tensile strength. The average and maximum resistivity of hard-drawn aluminium may be taken as 2.845 Ω-cm and 2.873 Ω-cm.

The resistivity of conductor varies with the temperature. It is given by the formula :

$$\rho_t = \rho_{20} [1 + \alpha_{20} (t - 20)]$$

where ρ_t = resistivity at t° C.

ρ_{20} = resistivity at 20° C.

α_{20} = temperature coefficient of resistance at 20°C.

and t = the operating temperature in °C.

Similarly, $R_t = R_{20} [1 + \alpha_{20} (t - 20)]$

Where R_t = Resistance at t° C

and R_{20} = Resistance at 20° C.

α_{20} for hard-drawn copper of 97 percent conductivity is 0.00381/°C and α_{20} for hard-drawn aluminium is 0.004/°C.

3.12 SKIN EFFECT

If D.C. is passed in a conductor, the current density is uniform over the cross-section of the conductor. In a conductor carrying A.C., the distribution tends to become non-uniform and the degree of non-uniformity increases with increase in frequency. This effect is known as skin effect and is due to a magnetic flux set up by the A.C. inside the conductor.

To explain the skin-effect assume the conductor to be made up of a number of concentric cylinders. The magnetic flux linking a cylindrical element near the centre of the conductor is greater than that linking another element near the surface of conductor. This is due to the fact that the former element is surrounded by the internal as well as external flux, while the latter by the external flux only. Thus, the flux linking the elements decreases progressively from the centre to the surface. As an outcome, the inner element will possess a greater self-inductance and therefore will offer a greater inductive reactance than the outer element. This difference in the conductive reactance gives a tendency to the current to crowd towards the surface or skin of the conductor. The current density is maximum at the surface and minimum at the centre of the conductor. This effect is virtually equivalent to a reduction of cross-sectional area of the conductor and therefore the effective resistance of the conductor is increased. At 50 Hz, the increase in resistance due to skin effect is around 3–5 percent.

At low frequency such as 50 Hz, there is a small increase in the current density near the surface of conductor but at high frequency such as radio frequency, practically the whole of the current flows on the surface of the conductor.

Skin effect increases with increase in frequency, conductor diameter and permeability. The concentration of current near the conductor surface has enabled the use of ACSR conductor. In the ACSR conductor, the current flows mostly in the outer layer made of Aluminium while the steel near the centre carries practically no current and gives the high tensile strength to the conductor.

3.13 PROXIMITY EFFECT

Like skin effect, proximity effect also results in non-uniformity of current in a conductor cross-section. In proximity effect, there is increment of the apparent resistance of the conductor due to the presence of other conductors carrying current in its vicinity.

Suppose we have two conductors A and B placed near each-other as in Fig. 3.17. When one conductor (A) carries current, its flux links with other conductor (B). More flux, however, links with the nearer half of the conductor B (as shown shaded) than with the other half.

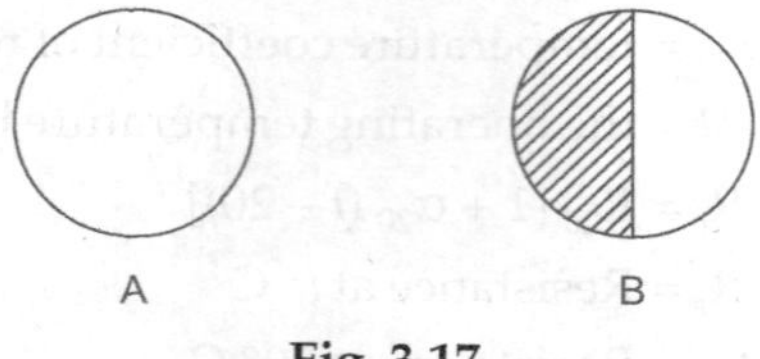

Fig. 3.17

If the conductors carry currents in the opposite directions, the magnetic fields set up will tend to cause an increase in the current density in the adjacent portions of the conductors. On the other hand, if the currents are in the same direction, the current density is increased in remote parts of the conductor. In both cases, there is an increase in losses in the conductors. So the effective resistance of the conductor is more than the ohmic resistance.

This effect is negligible for overhead lines but causes an appreciable increase in the resistance of underground cables where conductors are located close to each-other. Like skin effect, proximity effect also depends on the conductor size, frequency of the current, resistivity and permeability of the material.

SELF TEST

Fill in the blanks with appropriate words / figures :

1. The D.C. resistance of a line conductor is ________ than its A.C. resistance.
2. If the supply frequency increases, then skin effect is ________.
3. If the conductor diameter decreases, inductance of the line is ________.
4. The power loss in an overhead transmission line is mainly due to ________.
5. The skin effect is ________ for stranded conductor than the solid conductor.
6. If the spacing between the conductors is increased, the inductance of the line ________.
7. If the length of a transmission line increases, its inductance is ________.

Answers to the Self Test

1. Less
2. Increased
3. Increased
4. Line conductor resistance
5. Less
6. Increases
7. Increased.

OBJECTIVE QUESTIONS

1. Surge impedance loading a three phase 50 Hz 150 km long overhead line has :

$$L = 2.6 \text{ mH/km},$$
$$C = 0.012 \text{ µF/km}.$$

Surge impedance load is

(*a*) 465.5 Ω (*b*) 365.5 Ω
(*c*) 565.5 Ω (*d*) 462.5 Ω.

2. As the frequency is increased, the skin effect :

(*a*) decreases (*b*) increases
(*c*) remains the same (*d*) any of above.

3. The inductance of a line is minimum when :

(*a*) GMD is high (*b*) GMR is high
(*c*) Both GMD and GMR are high. (*d*) GMD is low and GMR is high.

4. As frequency increases, the magnetic flux inside a conductor :

(*a*) increases (*b*) decreases
(*c*) remains constant (*d*) any of above.

5. As the spacing between the phase conductors increases, the line resistance :

(*a*) increases (*b*) decreases
(*c*) remains the same (*d*) any of above.

6. As the spacing between the phase conductors increases, the line resistance :

(*a*) increases (*b*) decreases
(*c*) remains constant (*d*) any of above.

7. As the line inductance increases, the transmission capacity :

(*a*) increases (*b*) decreases
(*c*) remains the same (*d*) any of above.

8. The dc resistance of a conductor is R. The resistance at 50 Hz frequency is likely to be :
   (*a*) R (*b*) 1.04 R
   (*c*) 1.2 R (*d*) 1.4 R.
9. For a single circuit 3-phase line, the inductance per phase per km is of the order of :
   (*a*) 1 mH (*b*) 10 mH
   (*c*) 0.2 mH (*d*) 0.4 mH.
10. For a line with conductor spacings of 6 m, 6 m and 12 m, D_{eq} is :
   (*a*) 7.56 m (*b*) 10.23 m
   (*c*) 8.02 m (*d*) 9.05 m.
11. The increase in resistance due to non-uniform distribution of current in the conductor is known as :
   (*a*) proximity effect (*b*) skin effect
   (*c*) ferranti effect.
12. The surge impedance of 50 miles long underground cable is 50 ohms. For a 25 miles length it will be :
   (*a*) 20 ohms (*b*) 50 ohms
   (*c*) 100 ohms (*d*) None of the above.
13. The inductance of single-phase two wire power transmission line per kilometer gets doubled when the :
   (*a*) distance between the wires is doubled.
   (*b*) distance between the wires is increased four fold.
   (*c*) distance between the wires is increased as square of original distance.
   (*d*) radius of the wire is doubled.
14. The self-inductance of a long cylindrical conductor due to its internal flux linkages is k H/m. If the diameter of the conductor is doubled, then the self-inductance of the conductor due to its internal flux linkages would be :
   (*a*) 0.5 k H/m (*b*) k H/m
   (*c*) 1.414 k H/m (*d*) 4 k H/m.
15. When bundle conductors are used in place of single conductors, the effective inductance and capacitance will respectively :
   (*a*) increase and decrease (*b*) remains unaffected and increase
   (*c*) decrease and remain unaffected (*d*) decrease and increase.
16. The inductance of a three-phase transmission line is 1.2 mH/phase/km. If the spacing of conductors and the radius of the conductor are doubled, then the inductance of the line will be:
   (*a*) 4.8 mH / phase / km (*b*) 1.2 mH / phase / km
   (*c*) (In 2) × 1.2 mH / phase / km (*d*) (In 4) × 1.2 mH / phase / km.
17. A bundled conductor line compared to a single conductor line (with same conductor cross-sectional area and same mean distance between conductors) has :

	Self GMD	Mutual GMD	Per Phase Inductance
(*a*)	Lower	Nearly same	Higher
(*b*)	Higher	Lower	Nearly same
(*c*)	Higher	Nearly same	Lower
(*d*)	Lower	Higher	Higher.

18. A conductor has aluminium wire of 2.59 mm diameter. The length is 188 m and resistivity is 2.8×10^{-8} Ωm. Then resistance is :
   (*a*) 1 Ω (*b*) 2 Ω (*c*) 3 Ω (*d*) 4 Ω.

19. A conductor has copper wire of 0.5 mm diameter and 200 m length and resistivity at 20° C is 1.7×10^{-8} Ω-m. Then the resistance of wire is :

(*a*) 2.73 ohms at 20°C (*b*) 1.73 ohms at 20°C

(*c*) 1.74 ohms at 20°C (*d*) 2.74 ohms at 20°C.

20. The resistance of a copper wire at 0°C is 3.35 Ω and the temperature coefficient for given copper is $\alpha = 4.3 \times 10^{-3}$ per °C. Then the resistance at 50°C is :

(*a*) 2.07025 Ω (*b*) 3.07025 Ω

(*c*) 4.07025 Ω (*d*) 5.07025 Ω.

21. Skin effect in conductor is proportional to :

(*a*) $(\text{diameter of conductor})^{1/2}$ (*b*) diameter of conductor

(*c*) $(\text{diameter of conductor})^2$ (*d*) $(\text{diameter of conductor})^4$.

22. Skin effect exists in :

(*a*) A.C. transmission line only (*b*) D.C. transmission line only

(*c*) D.C. as well as A.C. transmission lines (*d*) cable carrying D.C. current.

23. In order to reduce the skin effect at UHF :

(*a*) anodised conductors are used

(*b*) painted conductors are used

(*c*) copper rods with silver plating are used

(*d*) copper tubes with silver plating are used.

24. Skin effect :

(*a*) reduces the effective resistance and effective internal reactance

(*b*) reduces the effective resistance but increases the effective internal reactance

(*c*) increases the effective resistance but reduces the effective internal reactance

(*d*) increases the effective resistance and effective internal reactance.

25. The conductor carries more current on the surface in comparison to its core. This phenomenon is called :

(*a*) skin effect (*b*) Ferranti effect

(*c*) Lenz's effect (*d*) Corona.

26. The skin effect of a conductor reduces with the increase in :

(*a*) supply frequency (*b*) resistivity of the conductor material

(*c*) *x*-section of conductor (*d*) permeability of conductor material.

27. Skin effect is noticeable only at :

(*a*) high frequencies (*b*) low frequencies

(*c*) audio frequencies (*d*) none of the above.

28. Which of the following parameters depends upon the supply frequency ?

(*a*) R and C only (*b*) L and C only

(*c*) R and L only (*d*) R, L, C and G.

29. Skin effect depends upon :

(*a*) Supply frequency (*b*) permeability of conductor material

(*c*) *x*-section of conductor (*d*) all of the above.

30. The effective resistance of a conductor will be the same as ohmic resistance when :

(*a*) current is true sinusoidal

(*b*) current is uniformly distributed in the *x*-section of the conductor

(*c*) voltage is low.

31. Skin effect in transmission line is due to :
(a) self inductance of conductor
(b) supply frequency
(c) high sensitivity of material in the centre
(d) both (a) and (b).

32. The inductance of line is minimum when :
(a) G.M.R. is high (b) G.M.D. is high
(c) both GMD and GMR are high (d) GMD is low but GMR is high.

33. The inductance of a single phase two wire power transmission line per km gets doubled when the :
(a) distance between the wires is doubled
(b) distance between the wires is increased as square of the original distance
(c) distance between the wires is increased four fold
(d) radius of the wire is doubled.

34. Proximity effect :
(a) increases the resistance of the conductors and reduces the self reactance
(b) is more pronounced for large conductors, high frequencies and close proximity
(c) is substantially eliminated with stranded conductors
(d) all of the above.

35. Following effects are associated with transmission lines :
1. skin effect 2. corona effect 3. proximity effect.
The effective resistance of a conductor is increased by :
(a) 1 only (b) 1 and 3 only
(c) 1, 2 and 3 (d) 2 and 3 only.

36. The self GMD method is used to evaluate :
(a) Capacitance (b) Inductance
(c) Both of the above (d) None of the above.

Answers

1. (a)	2. (b)	3. (d)	4. (b)
5. (a)	6. (c)	7. (b)	8. (b)
9. (a)	10. (a)	11. (b)	12. (b)
13. (c)	14. (b)	15. (d)	16. (b)
17. (c)	18. (a)	19. (b)	20. (c)
21. (c)	22. (a)	23. (d)	24. (c)
25. (a)	26. (b)	27. (a)	28. (d)
29. (d)	30. (b)	31. (d)	32. (d)
33. (b)	34. (d)	35. (b)	36. (b).

EXERCISE

1. What is skin effect? Why is it absent in the D.C. system?
2. What do you understand by the constants of an overhead transmission line?

3. Find an expression for the flux linkages.
   (*i*) due to a single current carrying conductor.
   (*ii*) in parallel current carrying conductors.
4. What is the proximity effect?
5. A single phase line has two parallel conductors 3 m apart. The radius of each conductor is 1.0 cm. Calculate the loop inductance of the line/km. **[Ans:** 2.3816 mH/km]
6. Calculate the loop inductance per km of a single phase transmission line comprising of two parallel conductors one metre apart and 1.25 cm in dia. Also calculate the reactance of the transmission line. Frequency is 50 Hz. **[Ans:** $L = 21.3 \times 10^{-4}$ H/km, $X_L = 0.67$ ohm]
7. Calculate the inductance of each conductor of 3-phase, 3 wire system when the conductors are arranged in a horizontal plane with spacing such that $D_{31} = 4$ mt, $D_{12} = D_{23} = 2$ mt. The conductors are transposed and have a diameter of 3.0 cm. **[Ans:** 1.11 mH]
8. A 3-phase, 3 wire system has its conductors arranged at the corners of an equilateral triangle of 2 m. side. The diameter of each conductor is 2.5 cm. Calculate the inductance of each conductor. **[Ans:** 1.065 mH/km]
9. A 3-phase transmission line 100 km long has its conductors of 0.5 cm dia spaced at the corners of an equilateral triangle of 120 cm side. Find the inductance per phase of the system. **[Ans:** 5.58×10^{-2} H]
10. A wire 4 mm in diameter is suspended at a constant height 10 m above sea level which constitutes the return conductor. Calculate the inductance of the system per km. **[Ans:** 1.89 mH]
11. A part of the transposition cycle of a 3-phase double circuit line is shown in the figure. Radius of each conductor is 0.9 cm. The conductors are solid copper. Find the inductance per phase per km. **[Ans:** 0.6136 mH/km]

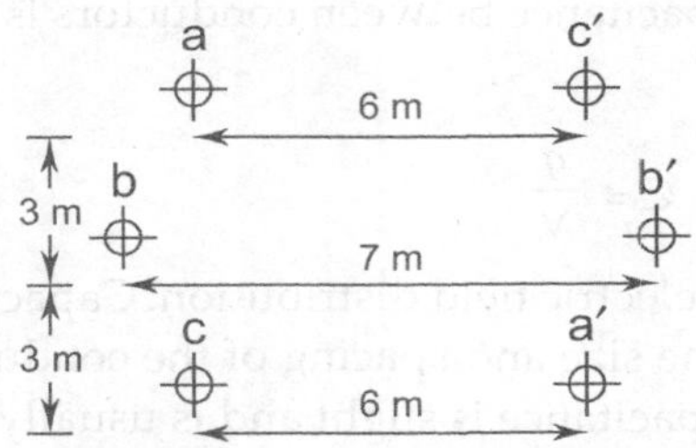

12. A 3-phase transmission line 100 km long has its conductors of 0.5 cm diameter speed at the corners of the equilateral triangle of 120 cm side. Find the inductance per phase of the system.

4

CAPACITANCE OF TRANSMISSION LINES

4.1 INTRODUCTION

Capacitance of a transmission line is the result of the potential difference between the conductors, it causes them to be charged in the same manner as the plates of a capacitor when there is a potential difference between them. The capacitance between conductors is the charge per unit of potential difference. Thus,

$$C = \frac{q}{V}.$$

Capacitance is associated with electric field distribution. Capacitance between parallel conductors is a constant depending on the size and spacing of the conductors. For power lines less than about 80 km long, the effect of capacitance is slight and is usually neglected. For longer lines, the capacitance becomes increasingly important and has to be accounted for.

4.2 CAPACITIVE REACTANCE

When alternating voltage is applied across a capacitance, current is taken by capacitance in the form of charging/discharging.

$$X_c = \frac{V_c}{I_c}\ \Omega$$

where,

V_c = Volts, rms

I_c = Charging current, rms

$$X_c = \frac{1}{2\pi f C}\ \Omega$$

X_c = Capacitive reactance of capacitor C at frequency f.

4.3 CHARGING CURRENT

An alternating voltage impressed on a transmission line causes the charge on the conductors at any point to increase and decrease with the increase and decrease of the instantaneous value of the voltage between conductors at the point. The flow of charge is current and the current caused by the alternate charging and discharging of a line due to an alternating voltage is called the charging current of the line and is equal to :

$$I_c = \frac{V}{X_c} \text{ Amp ; rms}$$

$$= \frac{V}{1/2\pi fC}$$

$$= 2\pi fC \cdot V$$

4.4 REACTIVE POWER OF SHUNT CAPACITANCE

When voltage V is applied to capacitor C, reactive power Q_c is supplied by the capacitor as given by:

$$Q_c = V \cdot I$$

$$= V \cdot \frac{V}{X_c} = \frac{V^2}{X_c}$$

$$= \omega C V^2 \quad (\text{VAr/ph})$$

where C is in farads

V is in volts

and $\omega = 2\pi f$ radians.

4.5 ELECTRIC FIELD OF A LONG STRAIGHT CONDUCTOR

If a long straight cylinderical conductor lies in a uniform medium such as air, has a uniform charge through out its length, and is isolated from other charges so that the charge is uniformly distributed around its periphery, the flux is radial. All points equidistant from such a conductor are points of equipotential and have the same electric flux density. The electric flux density on the surface is equal to the flux leaving the conductor per meter of length divided by the area of the surface in an axial length of 1 m. The electric flux density is:

$$D = \frac{q}{2\pi x} \text{ C/m}^2$$

where q is the charge on the conductor in coulombs per meter of length and x is the distance in meters from the conductor to the point where the electric flux density is computed.

The electric field intensity is equal to the electric flux density divided by the permittivity of the medium and electric field intensity is :

$$E = \frac{q}{2\pi\epsilon x} \text{ V/m}$$

where ϵ is the permittivity of the medium. In SI units, the permittivity of free space is 8.85×10^{-12} F/m.

4.6 POTENTIAL DIFFERENCE BETWEEN TWO POINTS DUE TO A CHARGE

The potential difference between two points is numerically equal to the work in Joules necessary to move a unit positive charge between the two points. Since the electric field intensity is negative of potential gradient, we can find the potential difference V_{12} between the two points by integrating the electric field intensity over the radial path between the two points.

Consider a long straight wire carrying a positive charge of q C/m as shown in Fig. 4.1. Points P_1 and P_2 are located at distances D_1 and D_2 meters from the center of the wire. The positive charge on the wire will exert a repelling force on a positive charge placed in the field. For this reason and because D_2 is greater than D_1, work must be done on a positive charge to move it from P_2 to P_1 and P_1 is at higher potential than P_2. The difference in potential is the amount of work done per coulomb of charge moved. The simplest way to compute the voltage drop between the two points is to compute the voltage between the equipotential surfaces passing through P_1 and P_2 by integrating the field intensity over a radial path between the equipotential surfaces.

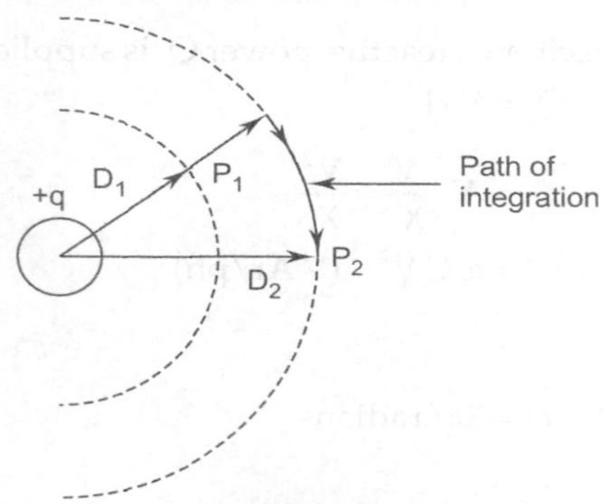

Fig. 4.1

$$V_{12} = \int_{D_1}^{D_2} E \cdot dx$$

$$= \int_{D_1}^{D_2} \frac{q}{2\pi \epsilon x} \cdot dx$$

$$= \frac{q}{2\pi \epsilon} \ln \frac{D_2}{D_1} \text{ volts} \quad \text{...(1)}$$

Thus, where q is the instantaneous charge on the wire in coulombs per meter of length.

4.7 POTENTIAL DIFFERENCE BETWEEN TWO CONDUCTORS OF A GROUP OF PARALLEL CONDUCTORS

Figure 4.2 shows a group of parallel charged conductors. We assume the distance between the conductors to be large as compared to the radii so that the charge on each conductor is uniformly distributed over its surface. By the principle of superposition, the potential difference between conductors *a* and *b* due to all the charges is the sum of potential differences due to each charge acting alone. Thus.

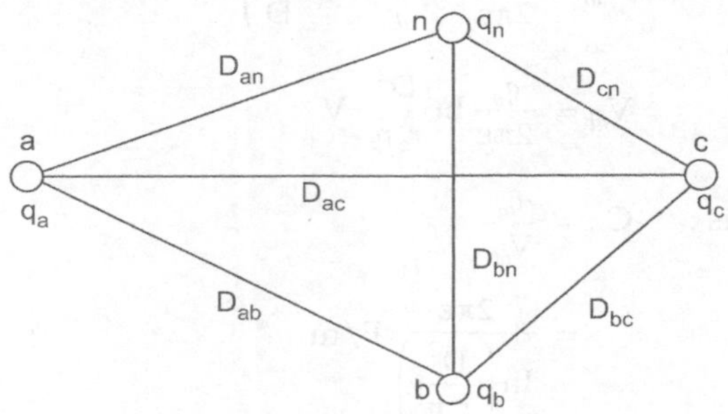

Fig. 4.2

$$V_{ab} = \frac{1}{2\pi\epsilon}\left[q_a \ln \frac{D_{ab}}{r_a} + q_b \ln \frac{r_b}{D_{ba}} + q_c \ln \frac{D_{cb}}{D_{ca}} + \cdots + q_n \ln \frac{D_{nb}}{D_{na}}\right] \text{V} \quad ...(2)$$

4.8 CAPACITANCE OF TWO-WIRE LINE

Capacitance between the two conductors of a two-wire line was defined as the charge on the conductors per unit of potential difference between them. Thus, capacitance per unit length of the line is:

$$C = \frac{q}{V} \text{F/m} \quad ...(3)$$

where q is the charge on the line in coulombs per meter and V is the potential difference between the conductors in volts. The capacitance between two conductors can be found by substituting in equation (3) the expression for V in terms of q from equation (1).

Figure 4.3 shows the two conductors of a single phase line. The voltage V_{ab} between the two conductors of the two-wire line can be found by determining the potential difference between the two conductors of the line, first by computing the voltage drop due to the charge q_a on conductor *a* and then by computing the voltage drop due to the charge q_b on conductor *b*. By the principle of superposition the voltage drop from conductor *a* to conductor b due to the charges on both conductors is the sum of the voltage drops caused by each charge alone. The potential difference V_{ab} can be written in terms of the contributions made by q_a and q_b by use of equation (2) with associated assumptions (*i.e.*, $\frac{D}{r}$ is large and ground is far away). Thus,

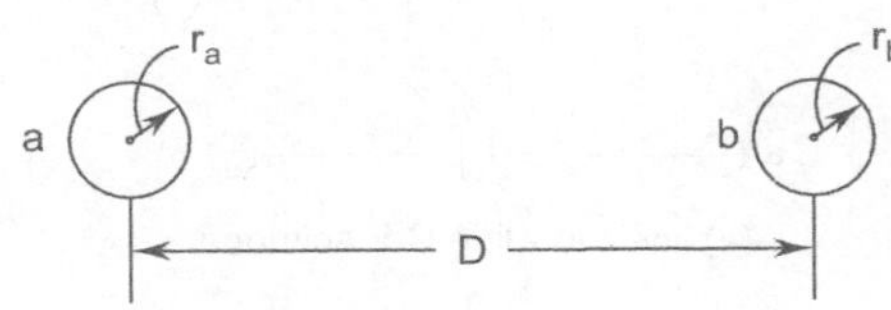

Fig. 4.3

$$V_{ab} = \underbrace{\frac{q_a}{2\pi\varepsilon} \ln \frac{D}{r_a}}_{\text{due to } q_a} + \underbrace{\frac{q_b}{2\pi\varepsilon} \ln \frac{r_b}{D}}_{\text{due to } q_b} \text{V}$$

Since, $q_a = -q_b$ for a two-wire line,

$$V_{ab} = \frac{q_a}{2\pi\varepsilon}\left(\ln\frac{D}{r_a} - \ln\frac{r_b}{D}\right) V$$

or

$$V_{ab} = \frac{q_a}{2\pi\varepsilon}\ln\frac{D^2}{r_a r_b} \ V$$

The line capacitance C_{ab} is : $C_{ab} = \frac{q_a}{V_{ab}}$

$$= \frac{2\pi\varepsilon}{\ln\left(\frac{D^2}{r_a r_b}\right)} \text{ F/m}$$

$$= \frac{\pi\varepsilon}{\ln\left(\frac{D}{\sqrt{r_a r_b}}\right)} \text{ F/m} = \frac{0.01206}{\log\left(\frac{D}{\sqrt{r_a r_b}}\right)} \ \mu\text{F/km.}$$

If

$$r_a = r_b = r$$

$$C_{ab} = \frac{\pi\varepsilon}{\ln\left(\frac{D}{r}\right)} \text{ F/m} \qquad ...(4)$$

The associated line charging current is :

$$I_c = J\omega C_{ab} V_{ab} \text{ A/km.}$$

Equation (4) gives the capacitance between the conductors. Sometimes it is desirable to know the capacitance between one of the conductors and a neutral point between them. For instance, if the line is supplied by a transformer having a grounded centre tap, the potential difference between each conductor and the ground is half the potential difference between the two conductors and the capacitance to ground or capacitance to neutral is the charge on a conductor per unit of potential difference between the conductor and ground. Thus, the capacitance to neutral for the two-wire line is twice the line-to-line capacitance (capacitance between conductors). As shown in Fig. 4.4, the line-to-line capacitance can be equivalently considered as two equal capacitances in series. The voltage across the lines divides equally between the capacitances such that the neutral point n is at the ground potential. The capacitance of each line to neutral is then given by :

$$C_n = C_{an} = C_{bn} = 2\,C_{ab} = \frac{0.02412}{\log\left(\frac{D}{r}\right)} \ \mu\text{F/km} \qquad ...(5)$$

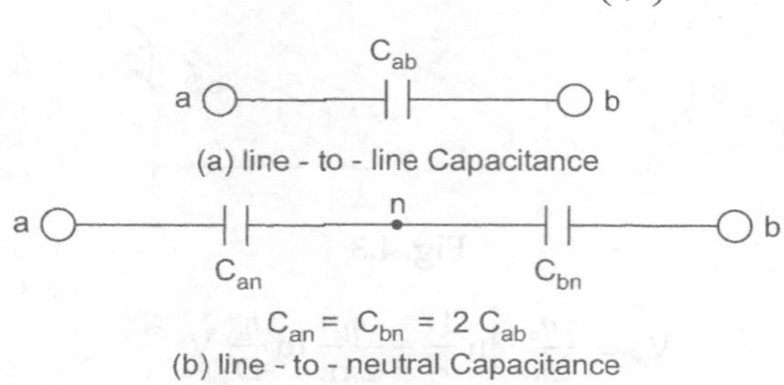

Fig. 4.4

The assumptions inherited in the above derivation are :

1. The charge on the surface of each conductor is assumed to be uniformly distributed but this is not strictly correct. If non-uniformity of charge distribution is taken into account, then,

$$C_n = \frac{0.02412}{\log\left[\frac{D}{2r} + \left(\frac{D^2}{4r^2} - 1\right)^{\frac{1}{2}}\right]} \ \mu F/km.$$

If $\frac{D}{2r} >> 1$, the above expression reduces to that equation (5) and the error caused by the assumption of uniform charge distribution is negligible.

2. In actual practice the conductor is stranded and not smooth. The electric field at the surface of a stranded conductor is not same as the field at the surface of a smooth round conductor. However, this error is again negligible and the outside radius of a stranded conductor can be used for r in equation (5).

Example 4.1. *A single phase transmission line has two parallel conductors 3 metres apart, radius of each conductor being 1 cm. Calculate the capacitance of the line per km. Given that $\varepsilon = 8.854 \times 10^{-12}$ F/m.*

Solution. Given :

$$r = 1 \text{ cm}$$
$$D = 3 \text{ m} = 300 \text{ cm}.$$

Capacitance of the line

$$= \frac{\pi\varepsilon}{\log_e \frac{D}{r}} \text{ F/m}$$

$$= \frac{\pi \times 8.854 \times 10^{-12}}{\log_e \frac{300}{1}} = 0.4875 \times 10^{-11} \text{ F/m}$$

$$= 0.4875 \times 10^{-2} \ \mu\text{F/km}.$$

Example 4.2. *A 1-phase overhead line 82 km long consists of two parallel wires each of 0.5 cm diameter, 1.5 m apart. If the line voltage be 50 KV at 50 Hz, calculate the following for line at no-load.*

(a) Total line-to-line capacitance

(b) Capacitive reactance

(c) Charging current

(d) Reactive power generated per km

Solution. Given :

$$D = 1.5 \text{ m} = 150 \text{ cm}$$
$$r = 0.25 \text{ cm}.$$

Line-to-line capacitance,

$$C_{ab} = \frac{\pi\varepsilon}{\ln \frac{D}{r}}$$

$$= \frac{3.142 \times 8.85 \times 10^{-12}}{\ln\left(\frac{150}{0.25}\right)}$$

$$= 0.004346 \ \mu\text{F/km}.$$

(*a*) Total C_{ab} for 32 km length

$$= 0.004346 \times 32 = 0.139\ \mu F.$$

(*b*) Capacitive reactance :

$$X_c = \frac{1}{2\pi f C}$$

$$= \frac{1}{2 \times 3.142 \times 50 \times 0.139 \times 10^{-6}}$$

$$= 0.0229 \times 10^{-6}\ \Omega$$

(*c*) Charging current :

$$I_c = \frac{V}{X_c}$$

$$= \frac{50 \times 10^3}{0.0229 \times 10^{-6}} = 2.182\ A$$

(*d*) Reactive Power

$$-X_c = \frac{V^2}{X_c}$$

$$= \frac{(50 \times 10^3)^2}{0.0229 \times 10^{-6}} = 109170\ VAr$$

$$= 109.17\ KVAr.$$

4.9 CAPACITANCE OF A THREE-PHASE LINE WITH EQUILATERAL SPACING

Figure 4.5 shows a three-phase line composed of three identical conductors of radius r placed in equilateral configuration with spacing between conductors be D.

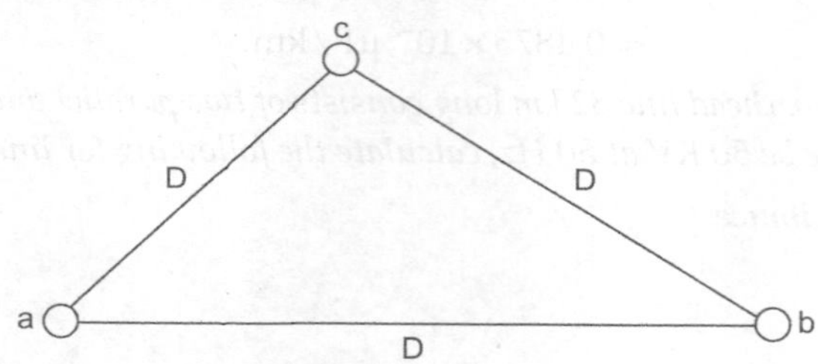

Fig. 4.5

From equation (2), we can write the expressions for V_{ab} and V_{ac} as

$$V_{ab} = \frac{1}{2\pi\varepsilon}\left[q_a \ln\frac{D}{r} + q_b \ln\frac{r}{D} + q_c \frac{D}{D}\right]$$

$$V_{ac} = \frac{1}{2\pi\varepsilon}\left[q_a \ln\frac{D}{r} + q_b \ln\frac{D}{D} + q_c \ln\frac{r}{D}\right]$$

Adding both equations:

$$V_{ab} + V_{ac} = \frac{1}{2\pi\varepsilon}\left[2q_a \ln\frac{D}{r} + (q_b + q_c) \ln\frac{r}{D}\right]$$

Since there are no other charges in the vicinity, the sum of charges on the three conductors is zero. Thus, $q_b + q_c = -q_a$ which when substituted in above equation yields.

$$V_{ab} + V_{ac} = \frac{3q_a}{2\pi\varepsilon} \ln \frac{D}{r} \qquad \text{...(6)}$$

If the three-phase voltages are V_{an}, V_{bn} and V_{cn}, then,

$$V_{an} = V \angle 0, V_{bn} = V \angle -120°, V_{cn} = V \angle +120°$$

$$V_{ab} = V_{an} - V_{bn} = V \angle 0 - V \angle -120° = \sqrt{3}\, V \angle 30°$$

$$V_{ac} = V_{an} - V_{cn} = V \angle 0 - V \angle +120° = \sqrt{3}\, V \angle -30°$$

$$\therefore \quad V_{ab} + V_{ac} = \sqrt{3}\, V \angle 30° + \sqrt{3}\, V \angle -30° = 3\,V = 3\,V_{an}$$

Thus, equation (6) becomes

$$V_{an} = \frac{q_a}{2\pi\varepsilon} \ln \frac{D}{r}.$$

Since capacitance to neutral is the ratio of the charge on a conductor to the voltage between that conductor and neutral.

$$\therefore \quad C_n = \frac{q_a}{V_{an}} = \frac{2\pi\varepsilon}{\ln\left(\frac{D}{r}\right)} \text{ F/m}$$

$$= \frac{0.02412}{\log\left(\frac{D}{r}\right)} \mu\text{F/km.}$$

The charging current per phase is

$$I_c = J\,\omega\,C_n\,V_{an}$$

Capacitive susceptance to neutral is given by :

$$b_n = \frac{1}{X_c}$$

$$= 2\pi f C_n.$$

4.10 CAPACITANCE OF THREE-PHASE LINE WITH ASYMMETRICAL SPACING

When the conductors of a three-phase line are not equilaterally spaced, the problem of calculating capacitance becomes more difficult. The capacitance of a line with asymmetrical spacing is found by assuming that the line is transposed. In the usual untransposed line, the capacitances of each phase to neutral are unequal. In a transposed line the average capacitance to neutral of any phase for the complete transposition cycle is the same as the average capacitance to neutral of any other phase, since each phase conductor occupies the same position as every other phase conductor over an equal distance along the transposition cycle.

For the line shown in Fig. 4.6, three equations are found for V_{ab} for the three different parts of the transposition cycle. With phase a in position 1, *b* in position 2 and *c* in position 3.

$$V_{ab} = \frac{1}{2\pi\varepsilon}\left[q_a \ln\frac{D_{12}}{r} + q_b \ln\frac{r}{D_{12}} + q_c \ln\frac{D_{23}}{D_{31}}\right]$$

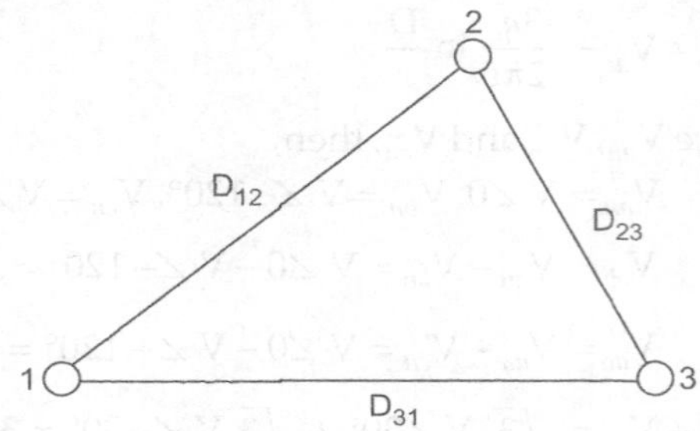

Fig. 4.6

With a in position 2, b in position 3 and c in position 1,

$$V_{ab} = \frac{1}{2\pi\varepsilon}\left[q_a \ln\frac{D_{23}}{r} + q_b \ln\frac{r}{D_{23}} + q_c \ln\frac{D_{31}}{D_{12}}\right]$$

and with a in position 3, b in position 1 and c in position 2.

$$V_{ab} = \frac{1}{2\pi\varepsilon}\left[q_a \ln\frac{D_{31}}{r} + q_b \ln\frac{r}{D_{31}} + q_c \ln\frac{D_{12}}{D_{23}}\right]$$

As the charge on each conductor is different when the position of the conductor changes with respect to other conductors. However, for usual spacings and conductors, sufficient accuracy is obtained by assuming that the charge per unit length on a conductor is the same in every part of the transposition cycle. When the above assumption is made with respect to charge, the voltage between a pair of conductors is different for each part of the transposition cycle. Then, an average value of voltage between the conductors can be found and the capacitance calculated from the average voltage. Thus, the average voltage is obtained by adding above three equations and equal to :

$$V_{ab} = \frac{1}{6\pi\varepsilon}\left[q_a \ln\frac{D_{12}D_{23}D_{31}}{r^3} + q_b \ln\frac{r^3}{D_{12}D_{23}D_{31}} + q_c \ln\frac{D_{12}D_{23}D_{31}}{D_{12}D_{23}D_{31}}\right]$$

or

$$V_{ab} = -\frac{1}{2\pi\varepsilon}\left[q_a \ln\frac{D_{eq}}{r} + q_b \ln\frac{r}{D_{eq}}\right]$$

where,

$$D_{eq} = \sqrt[3]{D_{12}\cdot D_{23}\cdot D_{31}}$$

Similarly, the average voltage drop from conductor a to conductor c is :

$$V_{ac} = \frac{1}{2\pi\varepsilon}\left[q_a \ln\frac{D_{eq}}{r} + q_c \ln\frac{r}{D_{eq}}\right]$$

As discussed in last topic,

$$V_{ac} = 3\,V_{an}$$

$$= \frac{1}{2\pi\varepsilon}\left[2q_a \ln\frac{D_{eq}}{r} + q_b \ln\frac{r}{D_{eq}} + q_c \ln\frac{r}{D_{eq}}\right]$$

Since $q_a + q_b + q_c = 0$ in a balanced three-phase circuit

$$3\,V_{an} = \frac{3}{2\pi\varepsilon} q_a \ln \frac{D_{eq}}{r}$$

or
$$V_{an} = \frac{q_a}{2\pi\varepsilon} \ln \frac{D_{eq}}{r}$$

and
$$C_n = \frac{q_a}{V_{an}} = \frac{2\pi\varepsilon}{\ln\left(\frac{D_{eq}}{r}\right)} \text{ F/m}$$

$$= \frac{0.02412}{\log\left(\frac{D_{eq}}{r}\right)} \text{ μF/km.}$$

Example 4.3. *A three-phase 50 Hz transmission line has flat horizontal spacing with 3.5 m between adajcent conductors. The conductors are No. 2/0 hard-drawn seven strand copper with outside conductor diameter equal to 1.05 cm. The voltage of the line is 110 KV. Find the capacitance to neutral and the charging current per km of line.*

Solution.
$$D_{eq} = (3.5 \times 3.5 \times 7)^{1/3} = 4.4 \text{ m} = 440 \text{ cm}$$

$$r = \frac{1.05}{2} = 0.525 \text{ cm}$$

$$C_n = \frac{0.02412}{\log\left(\frac{D_{eq}}{r}\right)}$$

$$= \frac{0.02412}{\log\left(\frac{440}{0.525}\right)} = 0.00826 \text{ μF/km}$$

$$X_n = \frac{1}{\omega C_n} = \frac{1}{2\pi \times 50 \times 0.00826}$$

$$= 0.384 \times 10^6 \ \Omega\text{/km to neutral}$$

Charging Current
$$= \frac{V_n}{X_n}$$

$$= \frac{\left(\frac{110}{\sqrt{3}}\right) \times 1000}{0.384 \times 10^6} = 0.17 \text{ A/km.}$$

Example 4.4. *A 220 KV, 50 Hz, 200 km long 3-phase line has its conductors on the corners of a triangle with sides 6 m, 6 m and 12 m. The conductor radius is 1.81 cm. Find the capacitance per phase per km.*

Solution.
$$D_{eq} = (6 \times 6 \times 12)^{1/3} = 7.56 \text{ m} = 756 \text{ cm}$$

$$r = 1.81 \text{ cm}$$

$$C_n = \frac{0.02412}{\log\left(\frac{756}{1.81}\right)} = 9.2 \times 10^{-3} \text{ μF/phase/km.}$$

4.11 CAPACITANCE OF BUNDLED CONDUCTORS

A bundled conductor line is shown in Fig. 4.7. The conductors of any one bundle are in parallel and we can assume the charge per bundle divides equally between the conductors of the bundle. Since the separation between bundles is usually more than 15 times than spacing between the conductors of the bundle. Also, since D_{12} is much greater than d, we can use D_{12} in place of the distances $D_{12} - d$ and $D_{12} + d$ with similar approximations in the equations of voltage.

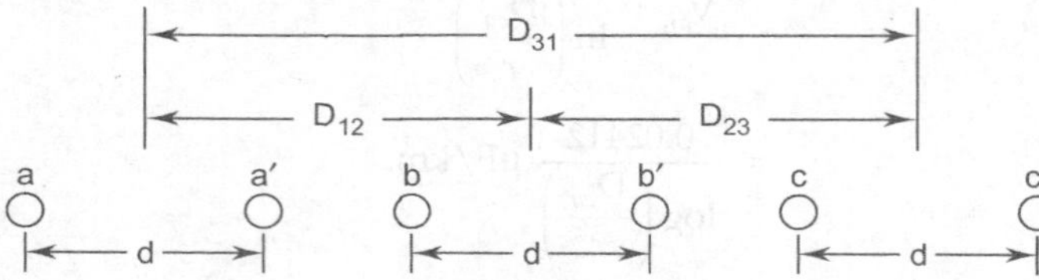

Fig. 4.7

If the charge on phase a is q_a. Conductors a and a' each have the charge $\frac{q_a}{2}$ and similar division of charge is assumed for phases b and c. Then,

$$V_{ab} = \frac{1}{2\pi\varepsilon}\left[\frac{q_a}{2}\left(\underbrace{\ln\frac{D_{12}}{r}}_{a} + \underbrace{\ln\frac{D_{12}}{d}}_{a'}\right) + \frac{q_b}{2}\left(\underbrace{\ln\frac{r}{D_{12}}}_{b} + \underbrace{\ln\frac{d}{D_{12}}}_{b'}\right) + \frac{q_c}{2}\left(\underbrace{\ln\frac{D_{23}}{D_{31}}}_{c} + \underbrace{\ln\frac{D_{23}}{D_{31}}}_{c'}\right)\right]$$

Combining terms gives

$$V_{ab} = \frac{1}{2\pi\varepsilon}\left[q_a \ln\frac{D_{12}}{\sqrt{rd}} + q_b \ln\frac{\sqrt{rd}}{D_{12}} + q_c \ln\frac{D_{23}}{D_{31}}\right]$$

If we considered the line to be transposed, then

$$C_n = \frac{2\pi\varepsilon}{\ln\left(\frac{D_{eq}}{\sqrt{rd}}\right)} \text{ F/m}$$

$$= \frac{0.02412}{\log\left(\frac{D_{eq}}{\sqrt{rd}}\right)} \ \mu\text{F/km.}$$

where, $D_{eq} = (D_{12} D_{23} D_{31})^{1/3}$

The above equation can also be written as :

$$C_n = \frac{2\pi\varepsilon}{\ln\left(\frac{D_m^b}{D_{sc}^b}\right)} \text{ F/m}$$

where, D_m^b = GMD between sub-conductors of one bundle with that of other bundles.

D_{sc}^b = Self GMR between sub-conductors of one bundle.

For a two-conductor bundle,

$$D_{sc}^{b} = \sqrt[4]{(r \times d)^2} = \sqrt{rd}$$

For a three-conductor bundle,

$$D_{sc}^{b} = \sqrt[9]{(r \times d \times d)^3} = \sqrt[3]{rd^2}$$

For a four-conductor bundle,

$$D_{sc}^{b} = \sqrt[16]{\left(r \times d \times d \times d \times \sqrt{2}\right)^4} = 1.09\sqrt[4]{rd^3}$$

Example 4.5. *A three-phase line has horizontally spaced bundle conductors as shown in Fig. 4.8. Given D = 6 m, d = 0.3 m, r = 12.5 mm. Calculate capacitance to neutral per km and capacitive reactance per ph per km. Line works at 50 Hz.*

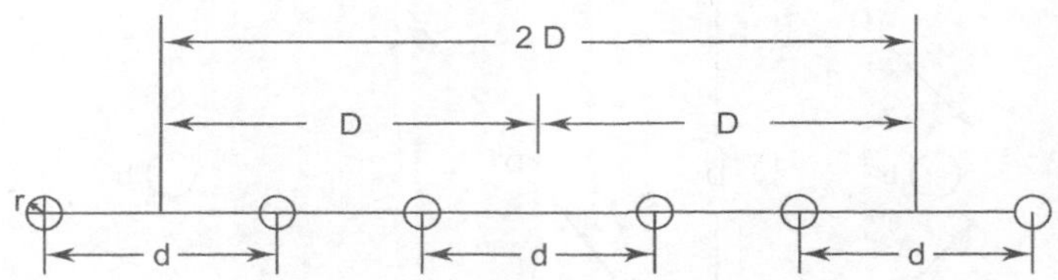

Fig. 4.8

Solution. GMR of each bundle for two sub-conductors :

$$D_{sc}^{b} = \sqrt{rd} = \sqrt{(12.5 \times 10^{-3} \times 0.3)}$$

$$= \sqrt{0.00375} = 0.06123$$

GMD between three groups of bundle conductors :

$$D_{m}^{b} = \sqrt{D \times D \times 2D}$$

$$= \sqrt[3]{2D^3} = 1.2599 \times 6 = 7.56 \text{ m}$$

$$\therefore \quad C_n = \frac{2\pi\varepsilon}{\ln\left(\frac{D_m^b}{D_{sc}^b}\right)} \text{ F/m}$$

$$= \frac{2\pi\varepsilon}{\ln\left(\frac{7.56}{0.06123}\right)}$$

$$\approx 0.0115\,\mu\text{F/km.}$$

and

$$X_c = \frac{1}{2\pi f C_n}$$

$$= \frac{1}{2\pi \times 50 \times 0.0115 \times 10^{-6}}$$

$$= 0.2769 \times 10^6\ \Omega/\text{km.}$$

4.12 DOUBLE CIRCUIT THREE-PHASE LINES

A modified GMD method has been found to apply in finding capacitance of bundled conductor lines. It can be shown that this method is equally good for transposed three-phase lines with equilateral spacing (conductors at the vertices of a hexagon) and for flat vertical spacing (conductors of the three phases of each circuit lying in the same vertical plane). It is reasonable to assume that the modified GMD method can be used to find the capacitance of a line with any configuration intermediate between these two configurations.

4.12.1 Hexagonal Spacing

Figure 4.9 shows the double circuit line having hexagonal spacing.

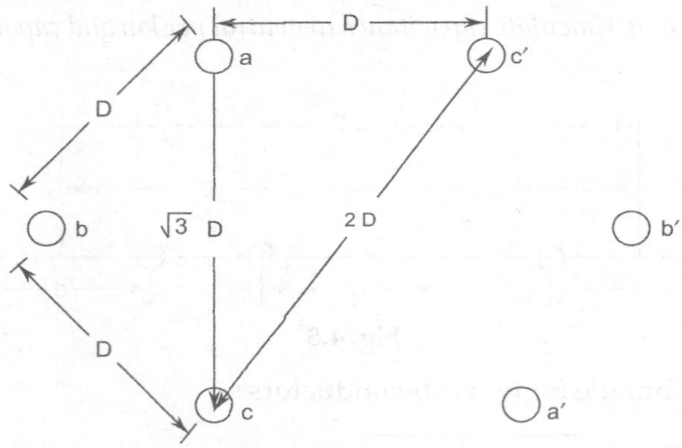

Fig. 4.9

As, $D_{eq} = \sqrt[3]{D_{ab}\, D_{bc}\, D_{ca}}$

where D_{ab} = Geometric mean distance between phases *a* and *b*

$$D_{ab} = \sqrt[4]{D \times \sqrt{3}D \times D \times \sqrt{3}D} = 3^{\frac{1}{4}}\, D$$

$$D_{bc} = \sqrt[4]{D \times \sqrt{3}D \times D \times \sqrt{3}D} = 3^{\frac{1}{4}}\, D$$

$$D_{ca} = \sqrt[4]{D \times \sqrt{3}D \times D \times \sqrt{3}D} = 3^{\frac{1}{4}}\, D$$

$$\therefore \quad D_{eq} = \left[\left(3^{\frac{1}{4}} \cdot D\right)\left(3^{\frac{1}{4}} \cdot D\right)\left(3^{\frac{1}{4}} \cdot D\right)\right]^{\frac{1}{3}} = 3^{\frac{1}{4}}\, D$$

and $D_{sc} = \sqrt{r \times 2D}$

We know that,

$$C_n = \frac{2\pi\varepsilon}{\ln\left(\dfrac{D_{eq}}{D_{sc}}\right)}$$

$$= \frac{2\pi\varepsilon}{\ln\left[\dfrac{\left(3^{\frac{1}{4}}\, D\right)}{\sqrt{2rD}}\right]}$$

$$= \frac{4\pi\varepsilon}{\ln\left(\frac{\sqrt{3}D}{2r}\right)} \text{ F/phase/m} = \frac{0.04824}{\log\left(\frac{\sqrt{3}D}{2r}\right)} \mu\text{F/phase/km.}$$

4.12.2 Flat Vertical Spacing

Refer to system of conductors in Fig. 4.10. The conductors of different phases are not symmetrically placed, therefore the derivation of capacitance expression will require the transposition of conductors as shown in Fig. 4.10.

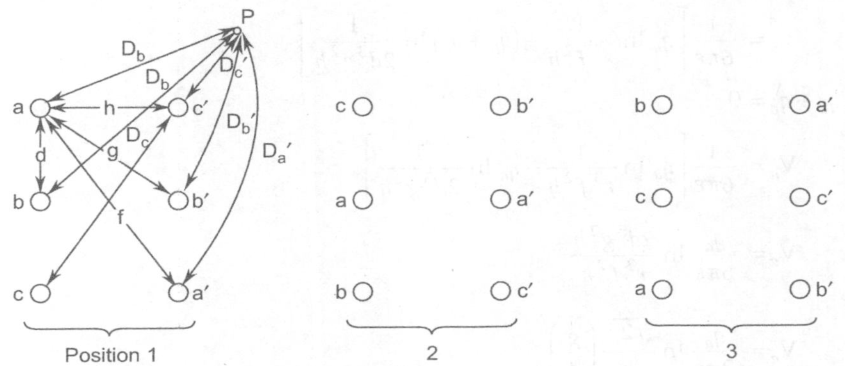

Fig. 4.10

It is required to find out average voltage of conductor a in three different positions due to the charge on conductors a, b, c, a', b' and c'. For this assume a point P very far from the system of conductors such that the distances of conductors from P are almost same. The point P corresponds to almost zero potential. Thus, $D_a \approx D_b \approx D_c \approx D_a' \approx D_b' \approx D_c'$.

Potential of conductor a in position 1

$$V_a' = \frac{1}{2\pi\varepsilon}\left[q_a \ln\frac{D_a}{r} + q_b \ln\frac{D_b}{d} + q_c \ln\frac{D_c}{2d} + q_a' \ln\frac{D_a'}{f} + q_b' \ln\frac{D_b'}{g} + q_c' \ln\frac{D_c'}{h}\right]$$

As $\quad D_a \approx D_b \approx D_c \approx D_a' \approx D_b' \approx D_c'$

and $\quad q_a + q_b + q_c = 0$

and $\quad q_a = q_a', q_b = q_b', q_c = q_c'$.

$$\therefore \quad V_a' = \frac{1}{2\pi\varepsilon}\left[q_a \ln\frac{1}{rf} + q_b \ln\frac{1}{dg} + q_c \ln\frac{1}{2dh}\right]$$

Potential of conductor a in position 2 is

$$V_a'' = \frac{1}{2\pi\varepsilon}\left[q_a \ln\frac{1}{rh} + q_b \ln\frac{1}{dg} + q_c \ln\frac{1}{dg}\right]$$

Potential of conductor a in position 3 is :

$$V_a''' = \frac{1}{2\pi\varepsilon}\left[q_a \ln\frac{1}{rf} + q_b \ln\frac{1}{2dh} + q_c \ln\frac{1}{dg}\right]$$

The average potential of conductor a in three positions is

$$V_a = \frac{V_a' + V_a'' + V_a'''}{3}$$

$$= \frac{1}{6\pi\varepsilon}\left[q_a \ln\frac{1}{r^3 f^2 h} + q_b \ln\frac{1}{2d^3 g^2 h} + q_c \ln\frac{1}{2d^3 g^2 h}\right]$$

$$= \frac{1}{6\pi\varepsilon}\left[q_a \ln\frac{1}{r^3 f^2 h} + (q_b + q_c) \ln\frac{1}{2d^3 g^2 h}\right]$$

Since, $q_a + q_b + q_c = 0$

$$\therefore \quad V_a = \frac{1}{6\pi\varepsilon}\left[q_a \ln\frac{1}{r^3 f^2 h} - q_a \ln\frac{1}{2d^3 g^2 h}\right]$$

or
$$V_a = \frac{q_a}{6\pi\varepsilon} \ln\frac{2d^3 g^2 h}{r^3 f^2 h}$$

or
$$V_a = \frac{q_a}{2\pi\varepsilon} \ln\frac{\sqrt[3]{2d}}{r}\left(\frac{g}{f}\right)^{\frac{2}{3}}$$

Thus,
$$C = \frac{q_a}{V_a}$$

$$= \frac{2\pi\varepsilon}{\ln\dfrac{\sqrt[3]{2d}}{r}\left(\dfrac{g}{f}\right)^{\frac{2}{3}}} \text{ F/metre/conductor}$$

Capacitance C per phase of the system

$$C = \frac{4\pi\varepsilon}{\ln\sqrt[3]{2}\dfrac{d}{r}\left(\dfrac{g}{f}\right)^{\frac{2}{3}}} \text{ F/metre/phase.}$$

Example 4.6. *A 3-phase double circuit line has the conductors at the vertices of a hexagon as shown in Fig. 4.9. If D = 3.5 m and r = 1.09 cm. Find capacitance per phase per km and capacitance per conductor per km.*

Solution. As
$$C_n = \frac{0.04824}{\log\dfrac{\sqrt{3}D}{2r}} \ \mu\text{F/phase/km.}$$

$$\therefore \quad C_n = \frac{0.04824}{\log\left(\dfrac{\sqrt{3}\times 350}{2\times 1.09}\right)} = 0.0197\ \mu\text{F/phase/km.}$$

Capacitance per conductor per km $= \dfrac{C_n}{2}$

$= 0.00985\ \mu\text{F/conductor/km.}$

4.13 EFFECT OF EARTH ON THE CAPACITANCE OF CONDUCTORS

Earth affects the capacitance of a transmission line because its presence alters the electric field of the line. The effect of earth on capacitance can be conveniently taken into account by the method of images.

4.13.1 Methods of Images

This method is based on realistic and practical approximations. This method was originally suggested by lord Kelvin. The method of images refers to the replacement of effect of ground on the charges and potentials of the conductors by the effect of images of conductors below the surface of the ground. Since earth is an equipotential plane which is possible only if we assume the presence of an imaginary conductor below the surface of earth at a depth equal to the height of actual conductor above the surface of earth. Also the charge on the conductor should be opposite to the charge on actual conductor. Thus, the earth can be replaced for the calculation of capacitance by a fictitious charged conductor with charge equal and opposite to the charge on the actual conductor and at a depth below the surface of the earth as the height of the actual conductor above earth. Such a fictitious conductor is the mirror image of the overhead conductor. This method of creating the same electric field as in the presence of earth is known as the method of images. It is evident that if an image conductor is located for each overhead conductor, the electric flux between the overhead conductors and their images is perpendicular to the plane which replaces the ground as shown in Fig. 4.11.

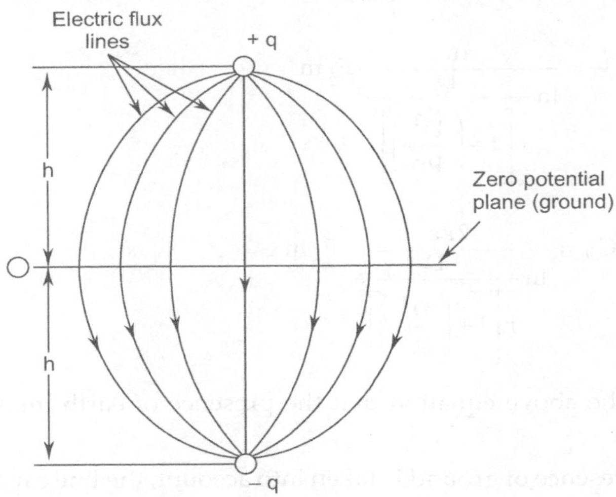

Fig. 4.11

4.14 EFFECT OF EARTH ON CAPACITANCE OF SINGLE PHASE LINE

Consider a single-phase line as shown in Fig. 4.12. The equation for the voltage drop V_{ab} as determined by the two charged conductors a and b and their images a' and b' can be written as :

$$V_{ab} = \frac{1}{2\pi\varepsilon}\left[q_a \ln\frac{D}{r} + q_b \ln\frac{r}{D} + q_a' \ln\frac{\sqrt{4h^2 + D^2}}{2h} + q_b' \ln\frac{2h}{\sqrt{4h^2 + D^2}}\right]$$

Substituting the values of different charges and simplifying,

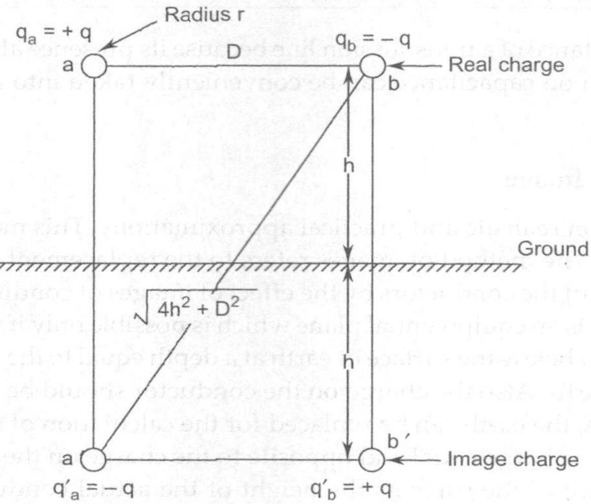

Fig. 4.12

$$V_{ab} = \frac{q}{\pi\varepsilon} \ln \frac{2hD}{r\left(4h^2 + D^2\right)^{\frac{1}{2}}}$$

$$C_{ab} = \frac{q}{V_{ab}} = \frac{\pi\varepsilon}{\ln \dfrac{D}{r\left[1 + \left(\dfrac{D^2}{4h^2}\right)\right]^{\frac{1}{2}}}} \text{ F/m}$$

and

$$C_n = 2\,C_{ab} = \frac{2\pi\varepsilon}{\ln \dfrac{D}{r\left[1 + \left(\dfrac{D^2}{4h^2}\right)\right]^{\frac{1}{2}}}} \text{ F/m}$$

It is observed from the above equation that the presence of earth modifies the radius r to $r\left(1 + \dfrac{D^2}{4h^2}\right)^{\frac{1}{2}}$. When the presence of ground is taken into account, the line capacitance increases by a small amount. In normal cases $h >> D$, thus the effect of earth on line capacitance of negligible order.

4.15 EFFECT OF EARTH ON THE CAPACITANCE OF A 3-PHASE LINE

The method of images can similarly be applied for the calculation of capacitance of a three-phase line shown in Fig. 4.13. The line is considered to be fully transposed. The conductor a, b and c carry the charges q_a, q_b and q_c and occupy positions 1, 2 and 3 respectively, in the first section of the transposition cycle. The effect of earth is simulated by image conductors with charges $-q_a$, $-q_b$ and $-q_c$ respectively.

Equations for the three parts of the transposition cycle can be written for the voltage drop from conductor a to conductor b as determined by the three charged conductors and their images. With conductor a in position 1, b in position 2 and c in position 3.

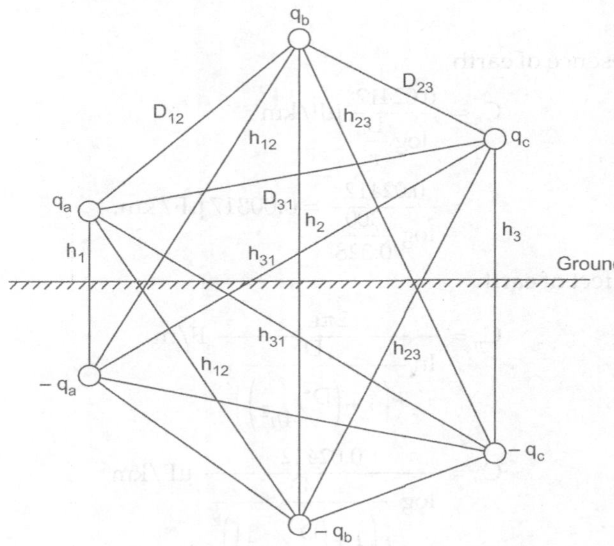

Fig. 4.13

$$V_{ab} = \frac{1}{2\pi\varepsilon}\left[q_a\left(\ln\frac{D_{12}}{r} - \ln\frac{h_{12}}{h_1}\right) + q_b\left(\ln\frac{r}{D_{12}} - \ln\frac{h_2}{h_{12}}\right) + q_c\left(\ln\frac{D_{23}}{D_{31}} - \ln\frac{h_{23}}{h_{31}}\right)\right]$$

Similar equations for V_{ab} are written for the other parts of the transposition cycle. Accepting the approximately correct assumption of constant charge per unit length of each conductor throughout the transposition cycle allows us to obtain an average value of the phasor V_{ab} and is given by :

$$V_{ab} = \frac{1}{2\pi\varepsilon}\left[q_a\left(\ln\frac{D_{eq}}{r} - \ln\frac{(h_{12}\,h_{23}\,h_{31})^{\frac{1}{3}}}{(h_1\,h_2\,h_3)^{\frac{1}{3}}}\right) + q_b\left(\ln\frac{r}{D_{eq}} - \ln\frac{(h_1\,h_2\,h_3)^{\frac{1}{3}}}{(h_{12}\,h_{23}\,h_{31})^{\frac{1}{3}}}\right)\right]$$

where, $D_{eq} = (D_{12}\,D_{23}\,D_{31})^{\frac{1}{3}}$

The equation for the average value of phasor V_{ac} is found in a similar manner. Using $V_{ab} + V_{ac} = 3\,V_{an}$ and $q_a + q_b + q_c = 0$ and proceeding as in Fig. 4.6, we obtain the following expression for the capacitance to neutral.

$$C_n = \frac{2\pi\varepsilon}{\ln\dfrac{D_{eq}}{r} - \ln\left(\dfrac{(h_{12}\,h_{23}\,h_{31})^{\frac{1}{3}}}{(h_1\,h_2\,h_3)\,h^{\frac{1}{3}}}\right)} \text{ F/m}$$

or

$$C_n = \frac{0.0242}{\log\dfrac{D_{eq}}{r} - \log\left(\dfrac{(h_{12}\,h_{23}\,h_{31})^{\frac{1}{3}}}{(h_1\,h_2\,h_3)^{\frac{1}{3}}}\right)} \ \mu\text{F/km}$$

Hence, it is evident that the effect of earth is to increase the capacitance of a line.

Example 4.7. *Calculate the capacitance to neutral / km of a single-phase line composed of No. 2 single strand conductors (radius = 0.328 cm) spaced 3 m apart and 7.5 m above the ground taking the presence and absence of ground into account.*

Solution.

(*a*) Neglecting the presence of earth

$$C_n = \frac{0.02412}{\log \frac{D}{r}} \ \mu F/km$$

$$= \frac{0.02412}{\log \frac{300}{0.328}} = 0.00817 \ \mu F/km.$$

(*b*) Considering the effect of earth

$$C_n = \frac{2\pi\varepsilon}{\ln \frac{D}{r\left(1+\left(D^2/4h^2\right)\right)^{\frac{1}{2}}}} \ F/m$$

or
$$C_n = \frac{0.02412}{\log \frac{D}{r\left(1+\left(D^2/4h^2\right)\right)^{\frac{1}{2}}}} \ \mu F/km$$

or
$$C_n = 0.082 \ \mu F/km.$$

It can be seen that the presence of earth increases the capacitance by approximately 3 parts in 800.

SELF TEST

Fill in the blanks by inserting appropriate words / figures :

1. If the length of the line is decreased, its capacitance is __________.
2. A neutral plane is one where __________ is zero.
3. Transposition of a 3-phase transmission line helps in __________.
4. If capacitance between two conductors of a 3-phase line is 4 μF, then capacitance of each conductor to neutral is __________.

Answers to Self Test

1. decreased
2. electrical intensity
3. equilising inductance and capacitance of the 3-phases
4. 8 μF.

OBJECTIVE QUESTIONS

1. The presence of earth in case of overhead lines :

   (*a*) increases the capacitance (*b*) decreases the capacitance

   (*c*) increases the inductance (*d*) decreases the inductance.

2. If the height of the transmission towers is changed, which of the following parameters is likely to change ?
   (*a*) Capacitance (*b*) Inductance
   (*c*) Resistance (*d*) None of the above.
3. With increase in its length the capacitance of a transmission line
   (*a*) increases (*b*) decreases
   (*c*) remains same.
4. The ratio of line to line capacitance and line to neutral capacitance is
   (*a*) 4 (*b*) 2 (*c*) $\frac{1}{2}$ (*d*) $\frac{1}{4}$.
5. If the effect of earth is taken into account, then the capacitance of line to ground
   (*a*) increases (*b*) decreases
   (*c*) remains same (*d*) becomes infinite.
6. Capacitance in equivalent circuit of a transmission line is due to :
   (*a*) current in the line (*b*) leakage of current
   (*c*) presence of magnetic flux (*d*) difference in potential of line.
7. In a transmission line the distributed constants are :
   (*a*) resistance and inductane only
   (*b*) resistance, inductance and capacitance only
   (*c*) resistance, inductance, capacitance and shunt conductance
   (*d*) resistance and shunt conductance only.
8. Transposition of transmission line is done to
   (*a*) reduce skin effect (*b*) reduce corona
   (*c*) reduce line loss (*d*) balance line voltage drop.
9. Bundled conductors in EHV transmission system provide :
   (*a*) increased capacitance (*b*) increased inductance
   (*c*) increased voltage gradient (*d*) reduced capacitance.
10. Two arrangements of conductors are proposed for a 3-phase transmission line : one with equilateral spacing of 4 m and the other, a flat with 4 m between the conductors as shown in the given figure,

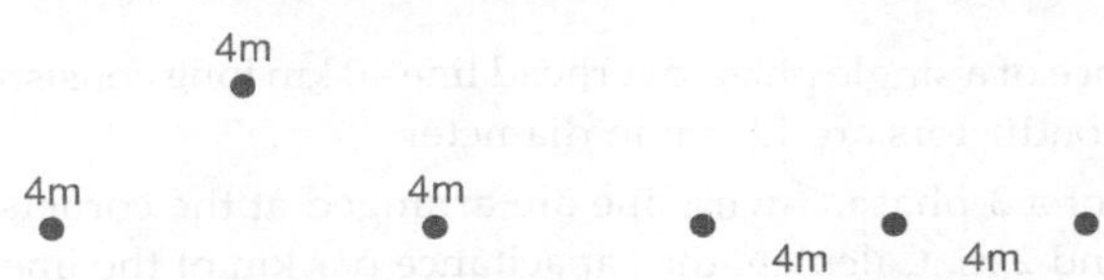

(i) (ii)

The conductor diameter in each case is 2 cm. Assuming that the line is transposed in both cases, which one of the following statements would be true ?
(C_n = Capacitance in F/m line to neutral ; L = inductance in H/m per phase)
   (*a*) $C_{n1} = C_{n2}$ and $L_1 > L_2$ (*b*) $C_{n1} = C_{n2}$ and $L_1 < L_2$
   (*c*) $C_{n1} < C_{n2}$ and $L_1 > L_2$ (*d*) $C_{n1} > C_{n2}$ and $L_1 = L_2$.
11. High voltage transmission lines are transposed because then,
   (*a*) Voltage drop in the lines can be minimized (*b*) Computation of inductance becomes easier
   (*c*) Corona losses can be minimized (*d*) Phase voltage imbalances can be minimized.

12. A 3-phase overhead transmission line has its conductors horizontally spaced with spacing between adjacent conductors equal to '*d*'. If, now the conductors of the lines are rearranged to form an equilateral triangle of sides equal to '*d*', then the
    (*a*) average capacitance will decrease but the average inductance will increase
    (*b*) surge impedance loading of the lines will increase
    (*c*) average capacitance as well as average inductance will increase
    (*d*) average capacitance will increase but the average inductance will decrease.
13. Transposition of transmission line is done to reduce the
    (*a*) line losses
    (*b*) capacitive effect
    (*c*) effect of surge voltages induced on the line
    (*d*) disturbances to nearby communication circuits.
14. The presence of ground causes the line capacitance to
    (*a*) increase by about 10% (*b*) decrease by about 10%
    (*c*) increase by about 0.2% (*d*) increase by about 2%.
15. As the spacing between the phase conductors increases, the line capacitance
    (*a*) decreases (*b*) increases
    (*c*) remains same.

Answers to Objective Questions

1. (*a*)	**2.** (*a*)	**3.** (*a*)	**4.** (*c*)	**5.** (a)
6. (*d*)	**7.** (*c*)	**8.** (*d*)	**9.** (*a*)	**10.** (*b*)
11. (*d*)	**12.** (*d*)	**13.** (*d*)	**14.** (*c*)	**15.** (*a*)

EXERCISE

1. Derive the expression for the capacitance of an unsymmetrical transposed 3-phase transmission line.
2. Derive an expression for the capacitance of a single phase overhead transmission line.
3. What do you understand by transposition in overhead lines? Explain why transposition is done in such lines.
4. Calculate the capacitance of a single phase overhead line 40 km long consisting of two parallel conductors 1.5 m apart. The conductors are 1.5 cm in diameter.
5. The three conductors of a 3-phase, 3-wire line are arranged at the corners of a triangle, the sides of which are l m, 1.3 m and 2 m. Calculate the capacitance per km of the line, when the conductors are transposed. The conductors are 1.5 cm diameter. **[Ans:** 0.0107 μF**]**
6. Will capacitance of a transmission line depend upon the ground effect?
7. A single-phase line has two parallel conductors 3 m apart. The radius of each conductor is 1.0 cm. Calculate the capacitance of each conductor/km. **[Ans:** 0.9738×10^{-2} μF/km**]**
8. What do you understand by electric potential? Derive an expression for electric potential
    (*i*) at a charged single conductor
    (*ii*) at a conductor in a group of charged conductors.
9. The 2 cm diameter conductors of a 3-phase, 3-wire transmission line are situated at the corners of a triangle of sides 3.5 cm, 5 m and 8 m. Find the capacitance/conductors/km, if the line is transposed.

10. A single phase line of 250 V, 50 Hz has conductor spacing of 1.5 meters. The diameter of each conductor is 1.5 cm. Calculate (*i*) capacitance of the line in μF for a length of 50 km and (*ii*) charging current.
[**Ans:** (*i*) 0.263 μF (*ii*) 20.66 mA]

11. Calculate the capacitance to neutral in the case of a single phase line whose conductors with radius of 0.25 cm are separated by 1.5 m and which are lying 7 m above the ground. Line length is 50 km.
[**Ans:** 0.225 μF]

12. A 3-phase, 3-wire system has its conductors arranged at the corners of an equilateral triangle of 2 m side. The diameter of each conductor is 2.5 cm. Calculate the capacitance of each conductor.
[**Ans:** 1.094×10^{-2} μF/km]

13. The conductors in a single phase transmission line are 6 m above the ground, taking the effect of the earth into account. Calculate the capacitance/km. Each conductor is of 1.5 cm diameter and the conductors are spaced 3 m apart.
[**Ans:** 0.00926 μF/km]

14. Six conductors of a double-circuit transmission line are arranged as shown below:

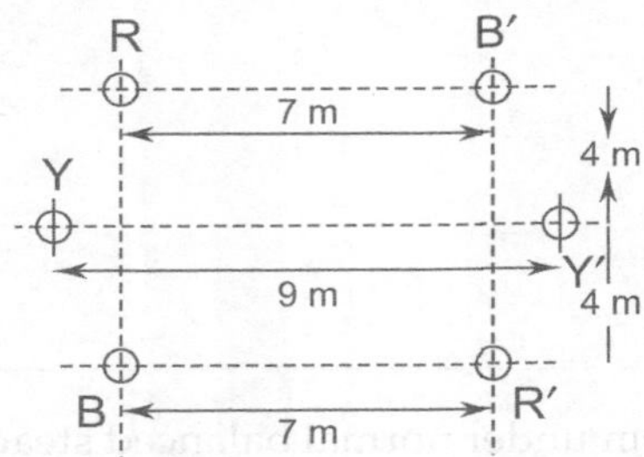

The diameter of each conductor is 2.5 cm. Find the capacitive reactance to neutral and the charging current per km per phase at 132 kV and 50 Hz, assuming that the line is regularly transposed. Neglect the effect of earth.
[**Ans:** 166.526 kΩ, 0.4575 A/km]

5

REPRESENTATION OF POWER SYSTEM COMPONENTS

5.1 INTRODUCTION

The performance of a power system under normal balanced steady state conditions is of primary importance in power system engineering. We are aware that in 3-phase circuit problems it is sufficient to compute results in one phase and subsequently predict results in the other two phases by exploiting the 3-phase symmetry. Although the lines are not spaced equilaterally and not transposed, the resulting asymmetry is slight and the phases are considered to be balanced. As such the transmission line calculations are carried out on per phase basis.

While working with balanced 3-phase circuits it is usual to assume that all the given voltages are line to line values, that all currents are line currents, that all volt-amperes, power and reactive power are 3-phase and that all power factors are phase power factors (*i.e.*, cosine of angle between phase voltage and corresponding phase current). However, all calculations are done in per phase values.

5.2 PER-UNIT (PU) SYSTEM

In a large interconnected power system with various voltage levels and various capacity equipments, it has been found quite convenient to work with per unit (p.u.) system of quantities for analysis purposes rather than in absolute values of quantities. The per-unit value of any quantity is defined as the ratio of the quantity to its base value expressed as a decimal. It can be written as :

$$\frac{\text{the actual value of the quantity (in any unit)}}{\text{the base or reference value in the same unit}}$$

Both the actual value and the base value are expressed in the same units so that the per unit value is dimensionless. Thus,

$$\text{Per unit KVA} = \frac{\text{Actual KVA}}{\text{Base KVA}}$$

$$\text{Per unit impedance} = \frac{\text{Actual impedance } (\Omega)}{\text{Base impedance } (\Omega)}$$

Out of the 4-system quantities, viz., KVA, KV, current and impedance, only two are independent. It is convenient to select the base value of KVA and KV and calculate the base values of current and impedance from the chosen base KVA and KV. In 3-phase systems, the base values usually chosen are 3-phase KVA and line to line KV. Then, Base Current = $\frac{\text{Base KVA}}{\sqrt{3} \times \text{Base KV}}$

$$\text{Base impedance} = \frac{\text{Line to neutral value of base voltage}}{\text{Base current}}$$

$$= \frac{(\text{Base KV})^2 \times 1000}{\text{Base KVA}} = \frac{(\text{Base KV})^2}{\text{Base MVA}} \text{ Ohms}$$

$$\text{p.u. impedance} = \frac{\text{Actual impedance}}{\text{Base impedance}} = \frac{\text{Z. Base MVA}}{(\text{Base KV})^2}$$

This means that the p.u. impedance is directly proportional to the base MVA and inversely proportional to square of base voltage.

Since three-phase circuits are solved as a single line with a neutral return, the bases for quantities in the impedance diagram are KVA per phase and KV from line to neutral. Data are usually given as total three-phase KVA or MVA and line-to-line KV. Because of this custom of specifying line-to-line voltage and total KVA or MVA, confusion may arise regarding the relation between the per-unit value of line voltage and the per-unit value of phase voltage. Although a line voltage may be specified as base, the voltage in the single-phase circuit required for the solution is still the voltage to neutral. The base voltage to neutral is the base voltage from line-to-line divided by $\sqrt{3}$. Since this is also the ratio between line-to-line and line-to-neutral voltages of a balanced three-phase system, the p.u. value of a line-to-neutral voltage on the line-to-neutral voltage base is equal to the p.u. value of the line-to-line voltage at the same point on the line-to-line voltage base if the system is balanced. Similarly, the three-phase KVA is three times the KVA per phase and the three phase KVA base is three times the base KVA per phase. Therefore, the p.u. value of the three-phase KVA on the three-phase KVA base is identical to the p.u. value of the KVA per phase on KVA per phase basis. The following example illustrates the above observation :

Let 3-phase Base KVA = 30,000 KVA

Line-to-line Base KV = 120 KV

$$\text{Single-phase Base KVA} = \frac{30{,}000}{3} = 10{,}000 \text{ KVA}$$

and $$\text{line-to-neutral Base KV} = \frac{120}{\sqrt{3}} = 69.2 \text{ V.}$$

For an actual line-to-line voltage of 108 KV, the line-to-neutral voltage is $\frac{108}{\sqrt{3}} = 62.3$ KV and

$$\text{p.u. voltage} = \frac{108}{120} = \frac{62.3}{69.2} = 0.90.$$

For the total 3-phase power of 18,000 KW, the power per phase is 6000 KW and

$$\text{p.u. power} = \frac{18,000}{30,000} = \frac{6,000}{10,000} = 0.6.$$

5.3 CHANGING THE BASE OF P.U. QUANTITIES

Sometimes the per unit impedance of a component of a system is expressed on a base other than the one selected as base for the part of the system in which the component is located. Since all impedances in any one part of a system must be expressed on the same impedance base when making computations, it is necessary to have a means of converting p.u. impedances from one base to another. Therefore, to change from p.u. impedance on a given base to p.u. impedance on a new base, the following equation applies :

$$\text{Per-unit } Z_{new} = \text{per-unit } Z_{old} \left(\frac{\text{Base KV}_{old}}{\text{Base KV}_{new}}\right)^2 \left(\frac{\text{Base KVA}_{new}}{\text{Base KVA}_{old}}\right).$$

This equation has nothing to do with transferring the ohmic value of impedance from one side of the transformer to another.

Example 5.1. *The reactance of a generator designed is given by as x′ = 0.25 p.u. based on the generator's name-plate rating of 18 KV, 500 MVA. The base for calculations is 20 KV, 100 MVA. Find x′ on new base.*

Solution.

$$x' = 0.25 \left(\frac{18}{20}\right)^2 \left(\frac{100}{500}\right)$$
$$= 0.0405 \text{ per unit.}$$

5.4 ADVANTAGES OF PER-UNIT SYSTEM

1. When expressed in per-unit, system parameters tend to fall in relatively narrow numerical ranges. Therefore, any erroneous data can be easily identified.
2. Calculations are simplified.
3. Per-unit data representation yields important information about relative magnitudes.
4. Power system contains a large number of transformers. The p.u. impedance of the transformer is same whether referred on to primary or secondary side of the transformer which is not the case when considering absolute value of these impedances. It is illustrated below :

Let the impedance of the transformer referred to primary side by Z_P and that on the secondary side be Z_S, then,

$$Z_P = Z_S \left(\frac{V_P}{V_S}\right)^2$$

where V_P and V_S are the primary and secondary voltages of the transformers.

The p.u. impedance of an equipment corresponding to its own rating is given by :

$$Z_{p.u.} = \frac{IZ}{V}$$

Now in case of transformer,

$$Z_{P\,p.u.} = \frac{Z_P \cdot I_P}{V_P} = Z_S \left(\frac{V_P}{V_S}\right)^2 \frac{I_P}{V_P}$$

$$= \frac{Z_S \cdot V_P\, I_P}{V_S^2}$$

$$= Z_S \frac{V_S\, I_S}{V_S^2} = \frac{Z_S\, I_S}{V_S} = Z_{S\,p.u.}$$

From this it is clear that the p.u. impedance of the transformer referred to primary side $Z_{P\,p.u.}$ is equal to the p.u. impedance of the transformer referred to the secondary side $Z_{S\,p.u.}$.

5.5 SINGLE-PHASE SOLUTION OF BALANCED THREE-PHASE NETWORKS

Consider the case where a three-phase transformer forms part of a three-phase system. If the transformer is Y/Y connected as shown in Fig. 5.1, the single-phase equivalent of the three-phase circuit can be represented by a single-phase transformer as in Fig. 5.2 with primary and secondary pertaining to phase *a* of the three-phase transformer.

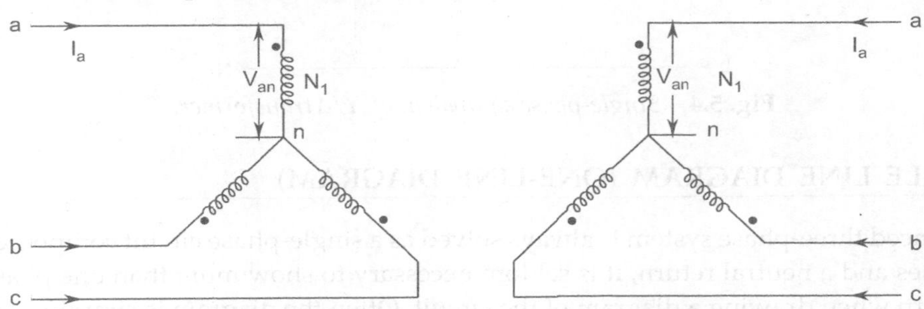

Fig. 5.1 *Three-phase Y/Y transformer.*

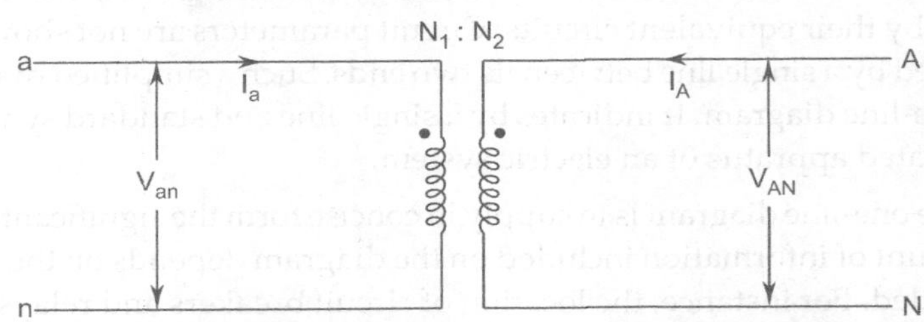

Fig. 5.2 *Single-phase equivalent of a three-phase Y/Y transformer.*

If the transformer is Y/Δ connected as in Fig. 5.3, the delta side has to be replaced by an equivalent star connection as shown dotted so as to obtain the single-phase equivalent of Fig. 5.4. On the delta side the voltage to neutral V_{AN} and line current I_A have a certain phase angle shift from the star side values V_{an} and I_a. In the single-phase equivalent, V_{AN}, I_A are respectively in phase with V_{an}, I_a. Since both phase voltage and line current shift through the same phase angle from star to delta side, the transformer per phase impedance and power flow are preserved in the single-phase equivalent.

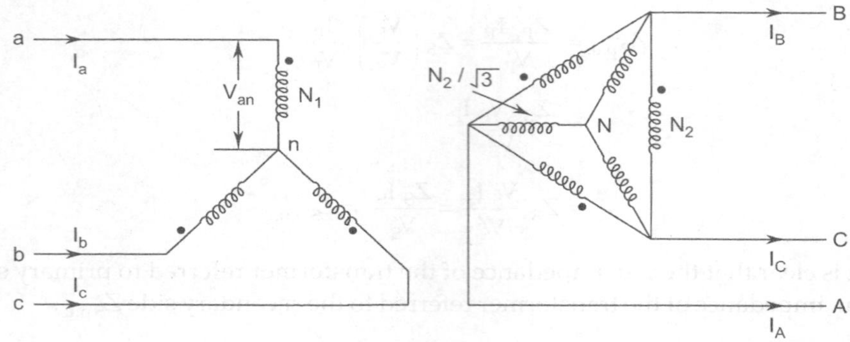

Fig. 5.3

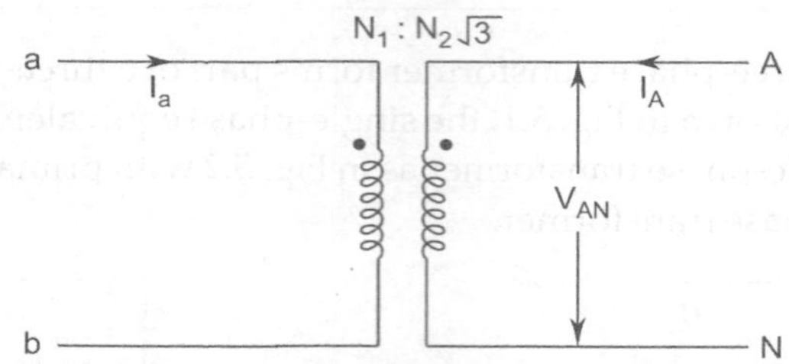

Fig. 5.4 *Single-phase equivalent of Y/Δ transformer.*

5.6 SINGLE LINE DIAGRAM (ONE-LINE DIAGRAM)

Since a balanced three-phase system is always solved as a single-phase circuit composed of one of the three lines and a neutral return, it is seldom necessary to show more than one phase and the neutral return when drawing a diagram of the circuit. Often the diagram is simplified further by omitting the completed circuit through the neutral and by indicating component parts by standard symbols rather than by their equivalent circuits. Circuit parameters are not shown and a transmission line is represented by a single line between its two ends. Such a simplified diagram of an electric system is called a one-line diagram. It indicates by a single line and standard symbols the transmission lines and associated appratus of an electric system.

The purpose of the one-line diagram is to supply in concise form the significant information about the system. The amount of information included on the diagram depends on the purpose for which the diagram is intended. For instance, the location of circuit breakers and relays is unimportant in making a load study. Breakers and relays are not shown if the primary function of the diagram is to provide information for such a study. On the other hand, determination of the stability of a system under transient conditions resulting from a fault depends on the speed with which relays and circuit breakers operate to isolate the faulted part of the system. Therefore, information about the circuit breakers may be of extreme importance. Sometimes one-line diagrams include information about the current and potential transformers which connect the relays to the system or which are installed for metering. The information found on a one-line diagram must be expected to vary according to the problem at hand and according to the practice of a particular company preparing the diagram.

The American National Standards Institute (ANSI) and the Institute of Electrical and Electronics Engineers have publised a set of standard signals for electrical diagrams. Figure 5.5 shows a few symbols which are commonly used. It is important to know the location of points where a system is connected to ground in order to calculate the amount of current flowing when an unsymmetrical fault involving the ground occurs. If a resistor a reactor is inserted between the neutral of Y and ground to limit the flow of current to ground during a fault, the appropriate symbol for resistance or inductance may be added to the standard symbol for the grounded Y. Most transformer neutrals in transmission systems are solidly grounded. Generator neutrals are usually grounded through fairly high resistances and sometimes through inductance coils.

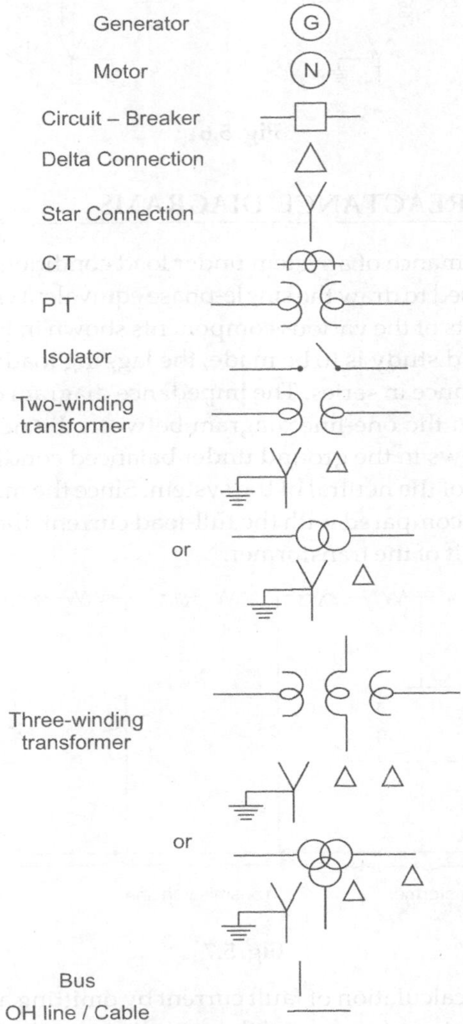

Fig. 5.5

Figure 5.6 shows the one-line diagram of a very simple power system. Two generators, one grounded through a reactor and one through a resistor are connected to a bus and through a step-up transformer to a transmission line. Another generator grounded through a reactor is connected to a bus and through a transformer to the opposite end of the transmission line. A load is connected to each bus.

A single line diagram cannot show the conditions during unbalanced operation of a power system. If it is necessary to show the currents and voltages during an unbalanced operation, it is necessary to show all the three phases and a single line diagram is not sufficient.

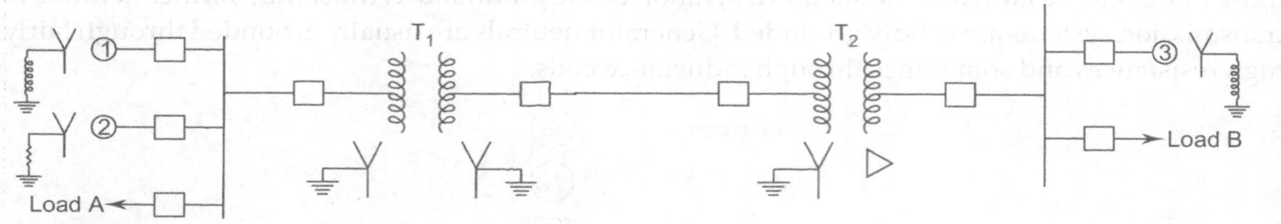

Fig. 5.6

5.7 IMPEDANCE AND REACTANCE DIAGRAMS

In order to calculate the performance of a system under load conditions or upon the occurrence of a fault, the one-line diagram is used to draw the single-phase equivalent circuit of the system. Figure 5.7 combines the equivalent circuits of the various components shown in Fig. 5.6 to form the impedance diagram of the system. If a load study is to be made, the lagging loads A and B are represented by resistance and inductive reactance in series. The impedance diagram does not include the current-limiting impedances shown in the one-line diagram between the neutrals of the generators and ground because no current flows in the ground under balanced conditions and the neutrals of the generators are at the potential of the neutral of the system. Since the magnetizing current of a transformer is usually insignificant compared with the full-load current, the shunt admittance is usually omitted in the equivalent circuit of the transformer.

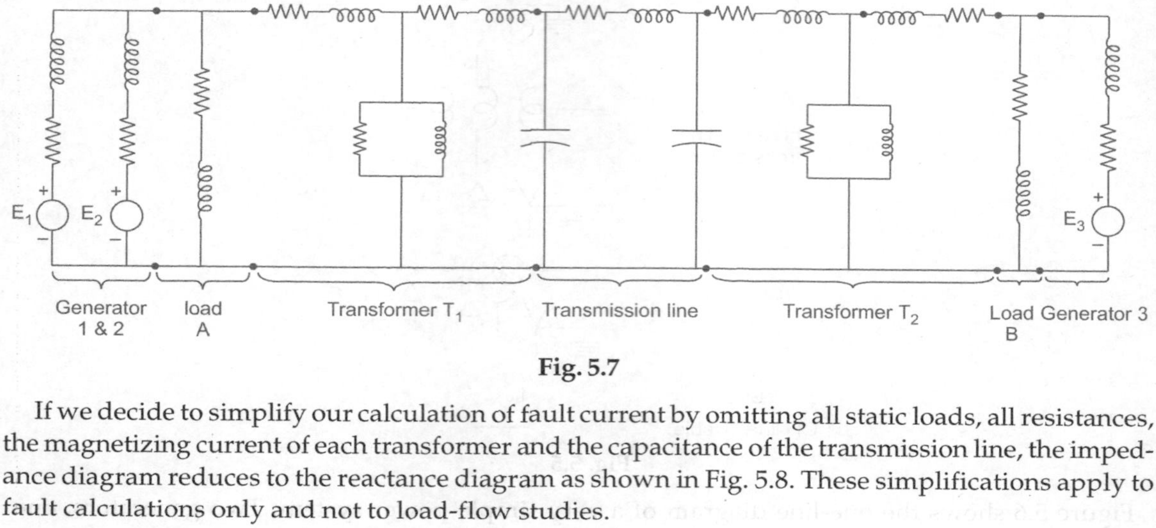

Fig. 5.7

If we decide to simplify our calculation of fault current by omitting all static loads, all resistances, the magnetizing current of each transformer and the capacitance of the transmission line, the impedance diagram reduces to the reactance diagram as shown in Fig. 5.8. These simplifications apply to fault calculations only and not to load-flow studies.

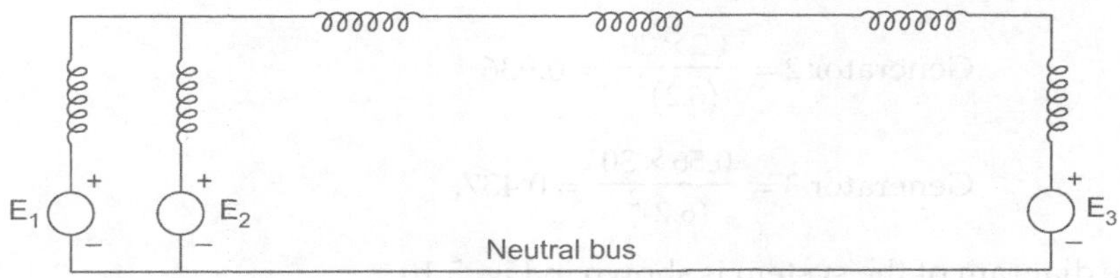

Fig. (5.8)

Example 5.2. *Obtain the per-unit impedance (reactance) diagram of the power system of Fig. 5.9.*
Generator No. 1 – 30 MVA, 10.5 KV, x" = 1.6 ohms
Generator No. 2 – 15 MVA, 6.6 KV, x" = 1.2 ohms
Generator No. 3 – 25 MVA, 6.6 KV, x" = 0.56 ohms

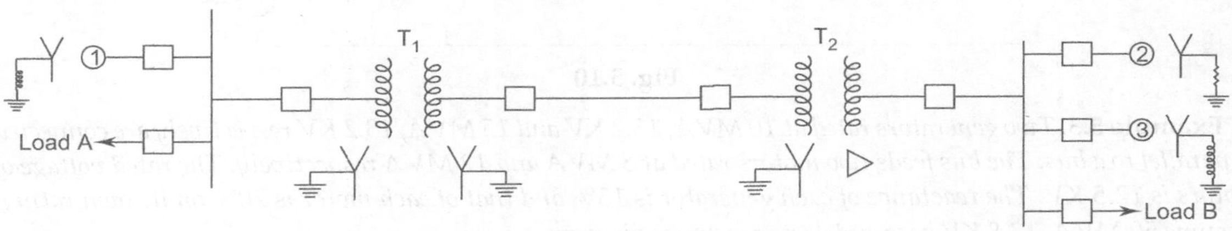

Fig. 5.9

Transformer T_1 *– 15 MVA,* $\frac{33}{11}$ *KV, x = 15.2 ohms per phase on high tension side (3-phase)*

Transformer T_2 *– 15 MVA,* $\frac{33}{6.2}$ *KV, x = 16 ohms per phase on high tension side (3-phase)*

Transmission line – 20.5 ohms/phase
Load A – 40 MW, 11 KV, 0.9 lagging p.f.
Load B – 40 MW, 6.6 KV, 0.85 lagging p.f.

Solution. Let us convert all reactances to per-unit form. Choose a common three-phase MVA base of 30 and a voltage base of 33 KV line-to-line on the transmission line. Then, the voltage base in the circuit of generator 1 is 11 KV line-to-line and that in the circuits of generators 2 and 3 is 6.2 KV.

The p.u. reactances of various components are :

$$\text{Transformer } T_1 = \frac{15.2 \times 30}{(33)^2} = 0.418$$

$$\text{Transformer } T_2 = \frac{16 \times 30}{(33)^2} = 0.44$$

$$\text{Transmission line} = \frac{20.5 \times 30}{(33)^2} = 0.564$$

$$\text{Generator 1} = \frac{1.6 \times 30}{(11)^2} = 0.396$$

$$\text{Generator 2} = \frac{1.2 \times 30}{(6.2)^2} = 0.936$$

$$\text{Generator 3} = \frac{0.56 \times 30}{(6.2)^2} = 0.437.$$

The reactance diagram of the system is shown in Fig. 5.10.

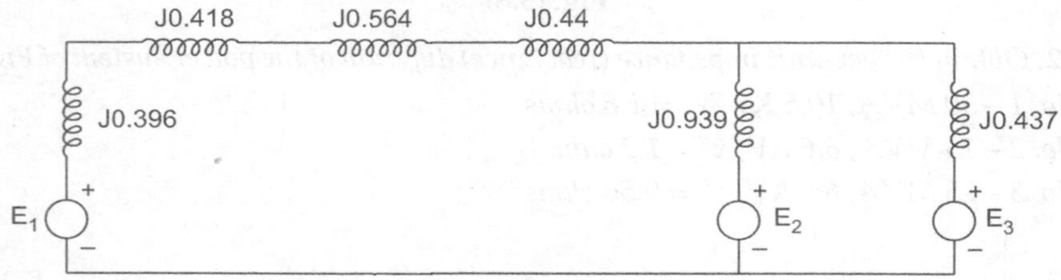

Fig. 5.10

Example 5.3. *Two generators rated at 10 MVA, 13.2 KV and 15 MVA, 13.2 KV respectively are connected in parallel to a bus. The bus feeds two motors rated at 8 MVA and 12 MVA respectively. The rated voltage of motors is 12.5 KV. The reactance of each generator is 15% and that of each motor is 20% on its own rating. Assume 50 MVA, 13.8 KV base and draw reactance diagram.*

Solution. By using equation,

$$\text{Per Unit } Z_{new} = \text{per-unit } Z_{old} \left(\frac{\text{Base KV}_{old}}{\text{Base KV}_{new}}\right)^2 \left(\frac{\text{Base KVA}_{new}}{\text{Base KVA}_{old}}\right)$$

$$\text{Reactance of } G_1 = 15 \left(\frac{13.2}{13.8}\right)^2 \left(\frac{50}{10}\right) = 68.62\%$$

$$\text{Reactance of } G_2 = 15 \left(\frac{13.2}{13.8}\right)^2 \left(\frac{50}{15}\right) = 45.75\%$$

$$\text{Reactance of } M_1 = 20 \left(\frac{12.5}{13.8}\right)^2 \left(\frac{50}{8}\right) = 102.56\%$$

$$\text{Reactance of } M_2 = 20 \left(\frac{12.5}{13.8}\right)^2 \left(\frac{50}{12}\right) = 68.37\%.$$

Figure 5.11 shows the reactance diagram.

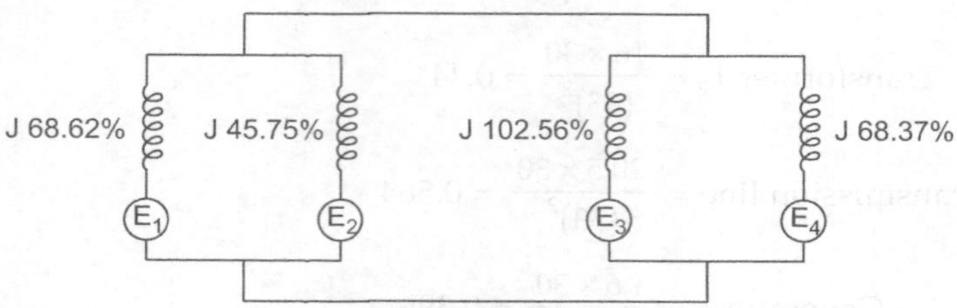

Fig. 5.11

OBJECTIVE QUESTIONS

1. The p.u. impedance value of an alternator corresponding to base values 13.2 KV and 30 MVA is 0.2 p.u. The p.u. value for the base values 13.8 KV and 50 MVA will be :
   (*a*) 0.328 p.u. (*b*) 0.318 p.u. (*c*) 0.306 p.u. (*d*) 0.33 p.u.
2. The per unit impedance of a circuit element is 0.15. If the base KV and base MVA are halved, then the new value of the per unit impedance of the circuit element will be:
   (*a*) 0.600 (*b*) 0.30 (*c*) 0.15 (*d*) 0.075.
3. The impedance value of the generator is 0.2 p.u. on a base value of 11 KV, 50 MVA. The impedance value for a base value of 22 KV, 150 MVA is:
   (*a*) 0.2 p.u. (*b*) 0.3 p.u. (*c*) 0.15 p.u. (*d*) 2.4 p.u.
4. A transmission line has 0.2 p.u. impedance on a base of 132 KV, 100 MVA, on a base of 220 KV, 50 MVA, it will have a p.u. impedance of:
   (*a*) $0.2 \times \frac{100}{50} \times \left(\frac{220}{132}\right)^2$ (*b*) $0.2 \times \frac{50}{100} \times \left(\frac{220}{132}\right)^2$
   (*c*) $0.2 \times \frac{100}{50} \times \left(\frac{132}{220}\right)^2$ (*d*) $0.2 \times \left(\frac{50}{100}\right) \times \left(\frac{132}{220}\right)^2$.
5. Base impedance of power system is given by:
   (*a*) $\frac{(\text{Base KV})^2}{\text{Base KVA}} \times 1000$ (*b*) $\frac{(\text{Base KVA})^2}{\text{Base KV}} \times 1000$
   (*c*) $\frac{(\text{Base KVA})^2}{\text{Base KV}}$ (*d*) $\frac{\text{Base KV}}{\text{Base KVA}}$.
6. The per unit impedance of a circuit element is 0.15. If the base KV and base MVA are halved, then the new value of the per unit impedance of the circuit element will be :
   (*a*) 0.60 (*b*) 0.30 (*c*) 0.15 (*d*) 0.075.
7. The per unit impedance of a line, to a 50 MVA, 132 KV base, is 0.4. The per unit impedance to a 100 MVA, 132 KV base will be :
   (*a*) 0.8 (*b*) 0.4 (*c*) 0.2 (*d*) 1.0.
8. An equipment has a per unit impedance of 0.9 p.u. to a base of 20 MVA, 33 KV. The per unit impedance to the base of 20 MVA, 11 KV will be :
   (*a*) 0.3 (*b*) 0.9 (*c*) 2.7 (*d*) 8.1.
9. The per unit impedance of a synchronous machine is 0.242. If the base voltage is increased by 1.1 times, the per unit value will be:
   (*a*) 0.200 (*b*) 0.220 (*c*) 0.242 (*d*) 0.266.

Answers to Objective Questions

1. (*c*)	**2.** (*b*)	**3.** (*c*)	**4.** (*d*)	**5.** (*a*)
6. (*b*)	**7.** (*a*)	**8.** (*d*)	**9.** (*a*).	

EXERCISE

1. Explain the terms base kVA, base kV and percentage reactance. State the relationship between them.
2. Define base impedance of a power system element and write an expression for it in terms of base MVA and kV.
3. What are the merits and demerits of per-unit systems?
4. Explain the merits and limitations of single line diagrams in power systems.
5. If the resistance in ohms is 5 Ω, find the per unit value. Given base kVA = 10 and base kV = 11. **[Ans:** 0.000413 pu**]**
6. Determine the per-unit impedance of a transmission line having an impedance of (30 + j110) Ω on 100 MVA and 132 kV base voltage. **[Ans:** (0.172 + j 0.631) pu**]**
7. Base voltage = 1100 volts. Base kVA = 10^6. What is base impedance? **[Ans:** 0.00121 Ω**]**

6

NEUTRAL GROUNDING

6.1 INTRODUCTION

Neutral grounding doesnot have any influence on the operation of a 3-phase system under balanced steady state conditions. However, the voltages and currents during ground fault conditions are greatly affected by the method used to connect neutral to earth. Protective relaying and stability analysis are also influenced by neutral grounding. In most modern HV systems, the system neutral is solidly (effectively) grounded *i.e.,* the neutral is connected directly to the ground without any intentional impedance between the neutral and the ground. Generally, the generators are grounded through a resistance to limit the stator fault currents and also for stability considerations.

6.2 ADVANTAGES OF NEUTRAL GROUNDING

1. Voltage to ground are limited to the phase voltages.
2. The high voltages due to arcing faults or transient line to ground faults are eliminated.
3. Sensitive protective relays against line to ground faults can be employed.
4. The over-voltages due to lightning are discharged to ground.

6.3 UNGROUNDED SYSTEM OR ISOLATED NEUTRAL

The main feature of an ungrounded neutral system is its ability, in some cases, to clear ground faults without interruption.

With normal balanced operation, the neutrals of an ungrounded system are held at ground voltage because of system capacitance to earth as in Fig. 6.1. The phasor diagram for balanced operation is shown in Fig. 6.2. For a fault on phase-*c*, the phasor diagram is shown in Fig. 6.3.

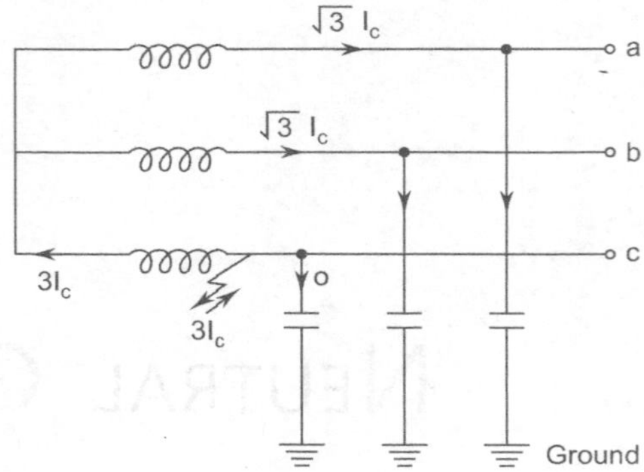

Fig. 6.1

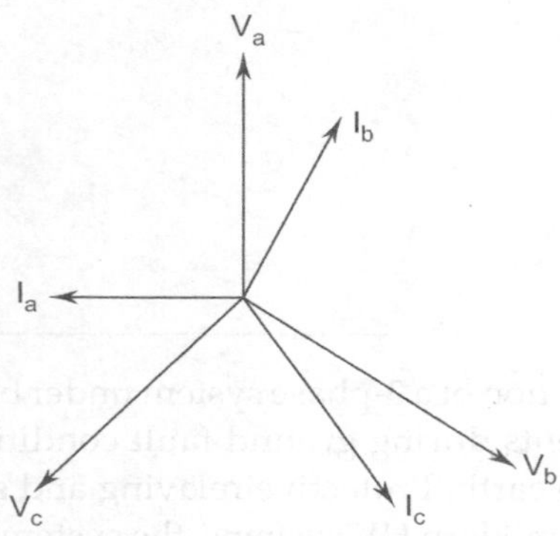

Fig. 6.2

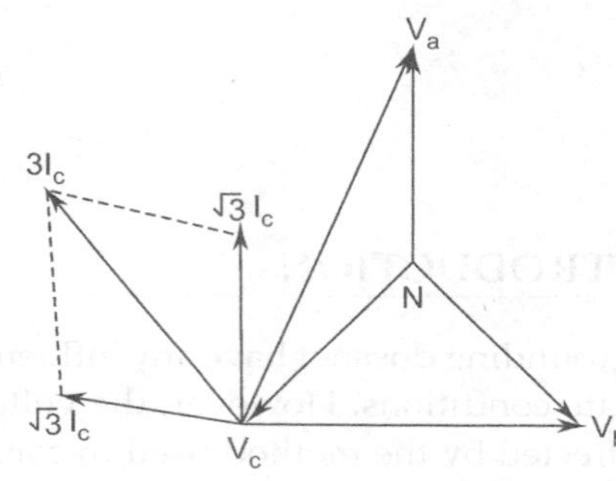

Fig. 6.3

A charging current three times that of per phase value flows. The voltage of healthy phase equals $\sqrt{3}\ V_{ph}$. The presence of inductance and capacitance in the system leads to arcing ground and the system voltage may rise to dangerously high values. These voltages can be avoided by connecting a suitable inductor between neutral and ground. If the value of inductive reactance is such that the fault current balances the charging current, then the grounding is known as resonant grounding and the grounding inductor is known as ground fault neutralizer or Peterson coil.

6.4 RESONANT GROUNDING

It was invented by W. Peterson in 1916 and is therefore also known as Peterson Coil Grounding or arc suppression coil earthing. The neutral is grounded through a tapped reactor whose value is tuned to system capacitance. By a proper selection of the value of neutral reactor, the two currents can be made almost equal so that the current in the fault is almost zero and arc cannot be maintained and is extinguished.

The general arrangement and phasor diagram of resonant grounding is shown in Fig. 6.4 and Fig. 6.5.

If V_{ph} is the line to ground voltage of the system, then voltage of the healthy phases will be $\sqrt{3}\ V_{ph}$. If C is the capacitance to ground of each phase, then charging current I_c is :

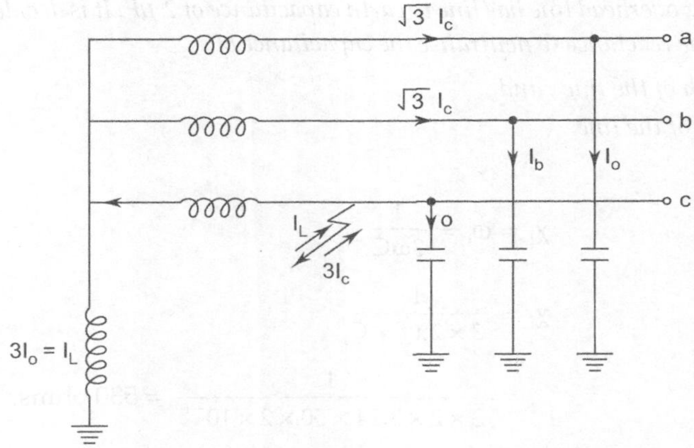

Fig. 6.4

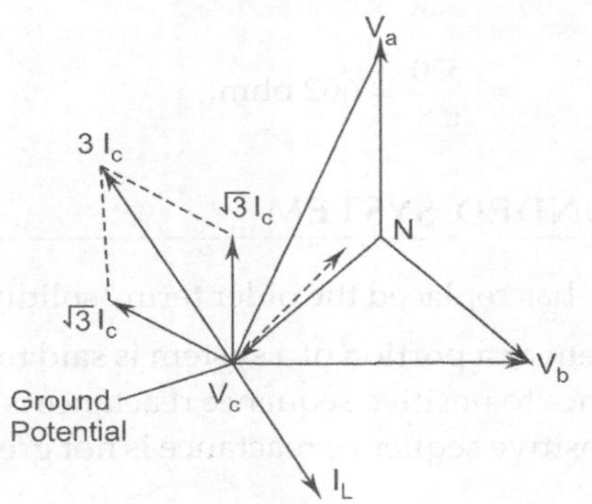

Fig. 6.5

$$I_c = 3\,V_{ph}\,\omega C$$

If L is the inductance of the Peterson coil, then

$$I_L = \frac{V_{ph}}{\omega L}$$

For compensation of arc current

$$\frac{V_{ph}}{\omega L} = 3\,V_{ph}\,\omega C$$

or $$L = \frac{1}{3\omega^2 C}$$

Also $$\omega L = \frac{1}{3\omega C}$$

$$\chi_L = \frac{1}{3\omega C}.$$

Example 6.1. *A 50 Hz overhead line has line to earth capacitance of 2 μF. It is decided to use an earth fault neutralizer. Determine the reactance to neutralise the capacitance of*

(a) 100% of the length of the line ; and

(b) 80% of the length of the line.

Solution.

(a)
$$X_L = \omega_L = \frac{1}{3\omega C}$$

∴
$$X_L = \frac{1}{3 \times 2\pi f \times C}$$

$$= \frac{1}{3 \times 2 \times 3.14 \times 50 \times 2 \times 10^{-6}} = 530 \text{ ohms.}$$

(b)
$$X_L = \frac{1}{3\omega C \times 0.8}$$

$$= \frac{530}{0.8} = 662 \text{ ohm.}$$

6.5 EFFECTIVELY GROUNDED SYSTEM

The term "effectively grounded" has replaced the older term "solidly grounded".

As per IEEE definition, a system or a portion of a system is said to be effectively grounded when the ratio of zero sequence reactance to positive sequence reactance is not greater than 3 and the ratio of zero sequence resistance to positive sequence reactance is not greater than 1 for all conditions of operation.

The main features of effectively grounded neutral system are :

1. The ground fault current is high. Its maximum value may sometimes exceed even the 3-phase short circuit current.
2. Since the ground fault current is high, danger to personnel in the vicinity of fault is high.
3. For such a system, the maximum line to ground voltage during a fault does not exceed 80% of the line voltage whereas for all other groundings, the voltage of the healthy phases rise to about 100% line-to-line voltage.
4. Arcing grounds cannot occur because the short-circuit current eliminates the influence of capacitive current.

6.6 RESISTANCE GROUNDING

Resistance grounding is normally used where the charging current is small *i.e.,* for low voltage short lines. When the neutral is grounded through a resistance, arcing grounds are eliminated. The value of resistance commonly used is quite high as compared with the system reactance in order to limit power loss in resistor during L-G fault. The value of resistance should be such that ground fault current is less than 3-phase short circuit current. The main characteristics of resistance grounded systems are :

1. Resistance grounding reduces the arcing ground hazards.
2. The ground fault relaying is simple and satisfactory.
3. The power dissipation in the grounding resistance may improve system stability.
4. Since the ground fault current is lower than that in effectively grounded system, the inductive interference with parallel communication circuits is also lesser than that in effectively grounded systems.

The resistor used for neutral earthing is usually a metallic resistance mounted on insulators in a metallic frame for voltages below 6.6 KV, liquid resistors are also used. The neutral earthing resistor is designed to carry its rated current for a short period, usually 30 seconds.

6.7 REACTANCE GROUNDING

In this system, the neutral is grounded through impedance which is highly reactive. For reactance grounded system : $\frac{\chi_0}{\chi_1} > 3$. Whereas for solid grounded system: $\frac{\chi_0}{\chi_1} < 3$. Reactance grounding lies between effective grounding and resonant grounding. The value of reactance required is to keep currents within safe limits.

Let χ_1, χ_2, χ_0 be the positive, negative and zero sequence reactances of the system and V voltage per phase. If χ_0 is negligible as compared to χ_1 $(= \chi_2)$

$$\text{Ground fault current} = \frac{3\,V}{\chi_1 + \chi_2 + \chi_0} = \frac{3\,V}{2\chi_1}$$

= 1.5 times 3-phase short circuit current.

If

$$\chi_0 = \chi_1 = \chi_2$$

$$\text{Ground fault current} = \frac{3\,V}{3\chi_1} = \frac{V}{\chi_1}$$

= 3-phase short circuit current

If a reactance χ_e is connected between neutral and earth.

$$\text{Ground fault current} = \frac{3\,V}{\chi_1 + \chi_2 + \chi_0 + 3\chi_e}.$$

As the value of χ_e is increased, the ground fault current decreaes. If χ_e is very small, the system behaves as an effectively grounded system. If χ_e is very high, the system behaves as an isolated neutral system. This method of grounding may be used for grounding the neutral of synchronous motors and capacitors and for systems having large charging currents.

The main characteristics of a system with reactance grounding are :

1. Arcing grounds are avoided.
2. The voltages across healthy phases are between 0.8 to 1 per unit of line to line voltage.
3. Ground fault relaying is simple and satisfactory.
4. Since ground fault current is reduced, inductive interference with parallel communication circuits is also reduced.

6.8 GROUNDING TRANSFORMER

Sometimes the neutral of a power transformer is not available for grounding *e.g.*, when a Δ-Δ transformer is used. In such situations, a special small size Y-Δ transformer is solely used for grounding. This transformer is known as a grounding transformer and it is a step down transformer. The star connected primaries are connected to the system and its neutral is grounded. The secondaries are in delta and generally do not supply any load but provide a closed path for triple harmonic currents to circulate in them. Under normal balanced conditions, the current in a grounding transformer is only its own exciting current.

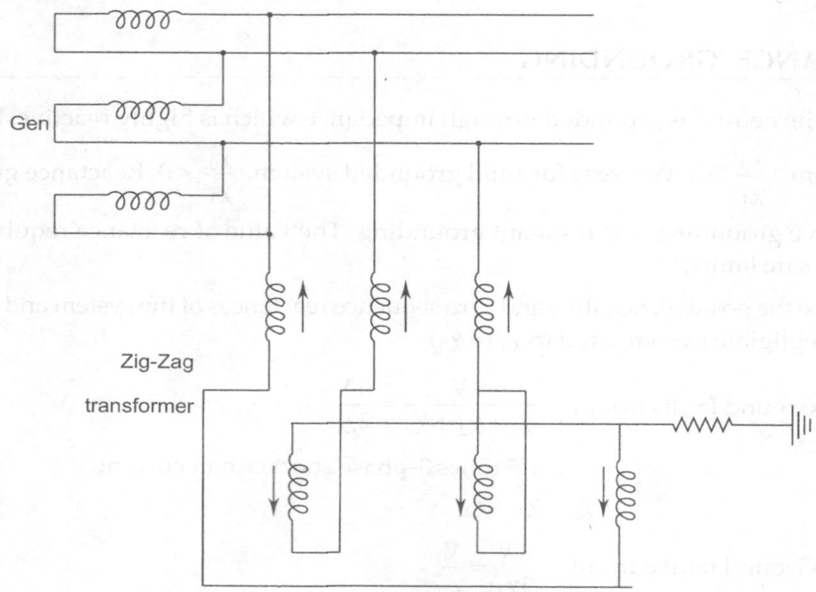

Fig. 6.6

Transformer with zig-zag connection are also used as grounding transformer as shown in Fig. 6.6. The grounding transformers are of short time rating usually 10 seconds to 1 minute. Therefore, the sizes of such transformers are small as compared to the power transformers of the same ratings.

6.9 GROUNDING PRACTICES

1. In high voltage and extra high voltage systems, the transformer neutrals are effectively grounded.
2. When several generators are operating in parallel, only one generator neutral is grounded. This is done to avoid the interference of zero sequence currents.
3. The generators are normally provided with resistance grounding and synchronous motors and capacitors are provided with reactance grounding.
4. Between generation and distribution, there are various voltage levels, it is desirable to have ground available at each voltage level.
5. For low voltages upto 600 V and for high voltages above 33 KV, solid grounding is used whereas for medium voltages between 3.3 KV and 33 KV, resistance or reactance grounding is used.

SELF TEST

Fill in the blanks by inserting appropriate words/figures :

1. Peterson coil has __________ reactance.
2. Peterson coil is used for __________ grounding.
3. Grounding by Peterson coil __________ arcing ground.
4. The arcing ground phenomenon is commonly experienced with __________ neutral system.
5. A system is said to be effectively grounded, if $\frac{\chi_0}{\chi_1}$ is __________ and $\frac{R_o}{\chi_1}$ is __________.

Answers to Self Test

1. inductive
2. resonant grounding
3. eliminates
4. isolated
5. not greater than 3, not greater than 1.

OBJECTIVE QUESTIONS

1. Earthing of electrical equipment is necessary for the protection against:
   (*a*) voltage fluctuation (*b*) high conductor temperature
   (*c*) danger of electrical shock (*d*) overloading.
2. Grounding is done generally at:
   (*a*) receiving end (*b*) supply end
   (*c*) either at receiving end or at supply end (*d*) none of these.
3. The advantage of neutral earthing is:
   (*a*) reduction of earth fault current (*b*) elimination of arcing ground
   (*c*) safety of personnel (*d*) none of the above.
4. Isolated neutral system has the disadvantage (s) of:
   (*a*) difficult earth fault relaying (*b*) voltage oscillations
   (*c*) persistent arcing ground (*d*) all of the above.
5. The neutral of the power system may be connected to earth:
   (*a*) directly (*b*) through reactance
   (*c*) through a resistor (*d*) any of these.
6. Neutral earthing is provided for:
   (*a*) The safety of equipment and personnel against lightning and voltage surges
   (*b*) the safety of personnel from electrical shock
   (*c*) controlling the earth fault current for protective relaying
   (*d*) reducing the voltage stress on lines and equipment with respect to earth under various operating and fault conditions
   (*e*) both (*c*) and (*d*).

7. Neutral earthing has the advantage (s) of:
   (*a*) low maintenance and operating costs over isolated neutral system
   (*b*) discharging of over-voltages due to lightning to earth
   (*c*) simplified design of earth fault protection
   (*d*) elimination of arcing grounds
   (*e*) all of the above.
8. Resistance earthing is employed for voltages between:
   (*a*) 66 KV and 132 KV (*b*) 33 and 66 KV
   (*c*) 11 and 33 KV (*d*) 3.3 and 11 KV.
9. Earthing of transformer neutral through reactance will improve its:
   (*a*) steady state stability (*b*) transient stability
   (*c*) both of the above (*d*) none of these.
10. Isolated neutral transmission system is not recommended as the:
   (*a*) insulation overstress may lead to its failure resulting in phase to phase faults
   (*b*) system is not adequately protected against earth fault
   (*c*) system insulation is overstressed due to over-voltages
   (*d*) all of the above.
11. A system is said to be effectively grounded only if $\frac{R_o}{X_1}$ and $\frac{X_0}{X_1}$ are respectively (symbols have the usual meanings):
   (*a*) ≤ 1 and ≤ 3 (*b*) ≥ 1 and ≥ 3
   (*c*) ≤ 2 and ≥ 2 (*d*) ≥ 2 and ≤ 2.
12. A system is said to be effectively grounded if its:
   (*a*) ratio of $\frac{X_0}{X_1} < 3.0$ (*b*) ratio of $\frac{X_0}{X_1} > 3.0$
   (*c*) ratio of $\frac{R_0}{X_1} > 2.0$ (*d*) neutral is grounded directly.

Answers to Objective Questions

1. (*c*)	2. (*b*)	3. (*b*)	4. (*d*)
5. (*d*)	6. (*e*)	7. (*e*)	8. (*c*)
9. (*b*)	10. (*d*)	11. (*a*)	12. (*a*).

EXERCISE

1. What are the different types of grounding ? Explain reactance grounding.
2. Define effective earthing. Explain why there is a trend at present towards effectively earthed systems.
3. What is Peterson coil? What protective function is performed by this device?
4. Explain the function of Peterson coil and how it helps in reducing "arcing grounds"?

5. Explain the advantages and disadvantages of the following types of earthing and indicate where each type is commonly employed:
   (*i*) Reactance earthing
   (*ii*) Resistance earthing
   (*iii*) Voltage transformer earthing.
6. A 33 KV, 50 Hz network has a capacitance to neutral of 1.0 μF per phase. Calculate the reactance of an arc suppression coil suitable for the system to avoid adverse effect of arcing ground. **[Ans:** 1061 Ω**]**
7. A 132 KV, 3-phase, 50 cycles, overhead line, 50 km long has a capacitance to earth for each line of 0.0157 μF per km. Determine the inductance. **[Ans:** 4.3 henrys**]**
8. Why should the transmission system neutrals be earthed?

7

CORONA

7.1 PHENOMENON OF CORONA

Corona phenomenon is the ionization of air surrounding the power conductor. Air is not a perfect insulator and contains a minute number of electrons and ions as a result of various effects such as ultra-violet radiations from sun, cosmic rays etc. When an electric gradient is set up in the air between two large parallel conducting planes, the electrons and ions are set in motion by the electric field and they maintain a small current by convection between the conducting planes. When the electric field intensity is less than 30 KV/cm, this current is very small and hence negligible.

This phenomenon undergoes a radical change when the electric field intensity reaches the critical value of 30 KV/cm. At this intensity, the ions will move with a high velocity and collides with the molecules of air. This produces a new electron and a positive ion which in turn are accelerated until they collide with other molecules and produce more ions. The process of ionization is thus cumulative and ultimately forms an electron avalanche. This results in ionization of the air surrounding the conductor. If the field is uniform everywhere as in the case of parallel plate electrodes, a complete electric breakdown occurs and an arc is established between the two electrodes.

If the electric field is not uniform as in the case of a vicinity of a cylinderical conductor, the intensity is maximum at the surface of conductor and then decreases in inverse proportion to the distance from the centre of the conductor. As the voltage applied to the conductor is increased, a layer adjacent to the conductor gets ionized as soon as the electric field intensity at the surface of the conductor becomes more than 30 KV/cm. The ionization will be restricted only to this layer because outside this layer, the field intensity is less than 30 KV/cm and the ions will not have sufficient energy to cause further ionization. Thus, no complete electric breakdown of the gap will take place and a region of sustained ionization will persist around the conductor. This ionization is accompanied by a luminous glow around the conductor. If the conductors are rough or dirty, the brightest parts of the luminous envelope are near the rough or dirty spots and a hissing sound can be heard. Thus, the phenomenon of corona is accompanied by a hissing sound, production of ozone, power-loss and radio interference. The higher the voltage is raised, larger luminous envelope will become and greater are the sound, power-loss and the radio-interference. If the applied voltage is increased to breakdown value, a flashover will occur between the conductors due to breakdown of air insulation.

The phenomenon of violet glow, hissing noise and production of ozone gas in an overhead transmission line is known as Corona.

When a direct voltage is applied between two conductors, there is a difference in the appearance of the positive and negative wires. The positive wire has uniform glow about it while the negative conductor has spotty glow.

7.2 CRITICAL DISRUPTIVE VOLTAGE

The minimum potential difference required between the conductors to start ionization or to generate the corona is called the critical disruptive voltage.

The critical intensity is proportional to density of air and therefore proportional to the barometric pressure and is inversely proportional to the absolute temperature. Taking the relative density at θ_0° C and 76 cm of mercury as 1.0, the relative density at a barometric pressure of p cm and a temperature of θ° C is :

$$\delta = \frac{p}{76}\left(\frac{273+\theta_0}{273+\theta}\right)$$

Taking, $\theta_0 = 20°\ \text{C}$

$$\delta = \frac{3.86p}{273+\theta}$$

The surface of the conductor also affects the critical intensity. A roughness of the conductor surface causes a distortion in electric field and gives rise to local regions of high potential gradient resulting in a lowering of the critical intensity. Thus, we take a roughness factor (m_0) into the account. After taking into account the effect of pressure, temperature and conductor surface, the rms value of critical intensity can be written as :

$$E_0 = \frac{3\times10^6}{\sqrt{2}}\,\delta\, m_0\ \text{V/m} \qquad ...(1)$$

where, m_0 = 1.0 for polished and smooth conductors.
= 0.93 to 0.98 for dirty and roughened conductors
= 0.83 to 0.87 for stranded conductors.

The electric field intensity E_r at the conductor surface of radius r metres is :

$$E_r = \frac{q}{2\pi\varepsilon_0 r}\ \text{V/m} \qquad ...(2)$$

Where q is charge on the conductor per metre length and ε_0 is permittivity of free space.

The voltage between the two conductors of a single phase line each of radius r metres and spaced at D metres is :

$$V = \frac{q}{\pi \varepsilon_0} \ln \frac{D}{r} \qquad ...(3)$$

Combining equation (2) and (3)

$$V = 2r\, E_r \ln \frac{D}{r} \qquad ...(4)$$

When E_r reaches the value E_o, the voltage between the conductors is disruptive critical voltage (V_d). Substituting the value of $E_r = E_o$ from equation (1) into equation (4)

$$V_d = \frac{6 \times 10^6}{\sqrt{2}} r \delta\, m_0 \ln \frac{D}{r} \text{ volts.}$$

For a 3-phase line, the phase voltage is

$$V = \frac{q}{2\pi \varepsilon_o} \ln \frac{D_{eq}}{r} \qquad ...(5)$$

Where D_{eq} is equivalent spacing

Combining equations (2) and (5)

$$V = r\, E_r \ln \frac{D_{eq}}{r} \qquad ...(6)$$

When E_r becomes equal to E_o, the voltage V is disruptive critical voltage V_d. Substituting the value of $E_r = E_o$ from equation (1) into equation (6)

$$V_d = \frac{3 \times 10^6}{\sqrt{2}} r\, \delta\, m_0 \ln \left(\frac{D_{eq}}{r} \right) \text{ Volts.}$$

Example 7.1. *Find the disruptive critical voltage for a 3-phase 220 KV line consisting of 22.26 mm diameter conductors spaced in a 6 m delta configuration. Following data can be assumed – Temperature 25° C, pressure 73 cm of mercury and surface factor 0.84.*

Solution.

$$r = \frac{22.26}{2} \times 10^{-3} \text{ m} = 11.13 \times 10^{-3} \text{ m}$$

$$= 1.113 \times 10^{-2} \text{ m.}$$

$$\delta = \frac{3.86p}{273 + \theta} = \frac{3.86 \times 73}{273 + 25} = 0.9456$$

$$m_0 = 0.84, D = 6 \text{ m.}$$

$$V_d = \frac{3 \times 10^6}{\sqrt{2}} r\, \delta\, m_0 \ln \frac{D}{r} \text{ volts.}$$

$$= \frac{3 \times 10^6}{\sqrt{2}} \times 1.113 \times 10^{-2} \times 0.9456 \times 0.84 \ln \frac{6.0}{1.113 \times 10^{-2}}$$

$$= 11.8 \times 10^4 \text{ V} = 118 \text{ KV.}$$

7.3 VISUAL CRITICAL VOLTAGE

It is the minimum phase-neutral voltage at which corona glow appears all along the line conductors.

It has been seen that in case of parallel conductors, the corona glow does not begin at the disruptive voltage V_d but at a higher voltage V_v called visual critical voltage. The phase-neutral effective value of visual critical voltage is given by following empirical formula :

$$V_v = \frac{6 \times 10^6}{\sqrt{2}} r \delta m_v \left(1 + \frac{0.03}{\sqrt{\delta r}}\right) \ln \frac{D}{r} \text{ volts. (For single-phase lines)}$$

$$V_v = \frac{3 \times 10^6}{\sqrt{2}} r \delta m_v \left(1 + \frac{0.03}{\sqrt{\delta r}}\right) \ln \frac{D_{eq}}{r} \text{ volts (For 3-phase lines)}$$

Where m_v is the surface factor for calculating visual critical voltage and called irregularity factor. This factor has following values :

m_v = 1.0 for smooth conductors and polished wires
= 0.93 to 0.98 for rough conductors exposed to air
= 0.72 for local corona on stranded conductors
= 0.82 for decided corona or general corona.

Example 7.2. *Calculate the visual corona voltage for above mentioned example. Assume the irregularity factor for local and general corona be 0.72 and 0.82.*

Solution. For local corona, $m_v = 0.72$

$$V_v = \frac{3 \times 10^6}{\sqrt{2}} r \delta m_v \left(1 + \frac{0.03}{\sqrt{\delta r}}\right) \ln \frac{D_{eq}}{r} \text{ volts}$$

$$= \frac{3 \times 10^6}{\sqrt{2}} \times 1.113 \times 10^{-2} \times 0.9456 \times 0.72 \times \left(1 + \frac{0.03}{\sqrt{(0.9456 \times 1.113 \times 10^{-2})}}\right) \ln \frac{6.0}{1.113 \times 10^{-2}}$$

= 130.7 KV per phase.

For general corona, $m_v = 0.82$

$$V_v = \frac{3 \times 10^6}{\sqrt{2}} \times 1.113 \times 10^{-2} \times 0.9456 \times 0.82 \times \left(1 + \frac{0.03}{\sqrt{(0.9456 \times 1.113 \times 10^{-2})}}\right) \ln \frac{6.0}{1.113 \times 10^{-2}}$$

= 148.8 KV per phase.

7.4 CORONA LOSS

Formation of corona is always accompanied by energy loss which is dissipated in the form of light, heat, sound and chemical action. Peek determined the following empirical formula for calculating the Corona loss under fair weather conditions :

$$P = 243.5\,\frac{(f+25)}{\delta}\sqrt{\frac{r}{D}}\,(V_p - V_d)^2 \times 10^{-5} \text{ KW/km/phase}$$

where f is the frequency of supply

δ is the air density correction factor,

V_p is the operating voltage in KV,

V_d is critical disruptive voltage.

Peek's formula for corona loss is valid when the ratio $\left(\frac{V_p}{V_d}\right)$ is greater than 1.8.

When this ratio is less than 1.8, Peterson's formula holds good and is given by :

$$P_C = \frac{21\times10^{-6} f V_p^2 F}{\left[\log_{10}\left(\frac{D}{r}\right)\right]^2} \text{ KW/phase/km.}$$

where, F = Corona factor determined by test.

Corona losses under bad weather conditions may be many times more than that under fair weather conditions. The corona loss under foul weather condition is calculated by taking V_d as 80% of its fair weather value.

Example 7.3. *A 3-phase equilateral transmission line has a total corona loss of 53 KW at 106 KV and a loss of 98 KW at 110.9 KV. What is the disruptive critical voltage ? What is the corona loss at 113 KV ?*

Solution. Power loss due to corona is :

$$P = 243.5\times\frac{(f+25)}{\delta}\sqrt{\frac{r}{D}}\,(V_p - V_d)^2 \times 10^{-5} \text{ KW/km/phase.}$$

For 3-phase, power loss due to corona is

$$P = 3\times243.5\times\frac{(f+25)}{\delta}\sqrt{\frac{r}{D}}\,(V_p - V_d)^2 \times 10^{-5} \text{ KW/km.}$$

As f, δ, r and d are same for two cases :

$\therefore$ $P \propto (V_p - V_d)^2$

1. For first case, $P = 53 \text{ KW},\ V_p = \frac{106}{\sqrt{3}} = 61.2 \text{ KV}$

For second case, $P = 98 \text{ KW},\ V_p = \frac{110.9}{\sqrt{3}} = 64 \text{ KV}$

$\therefore$ $53 \propto (61.2 - V_d)^2$ ...(1)

and $98 \propto (64 - V_d)^2$

Dividing both above equations,

$$\frac{53}{98} = \frac{(61.2 - V_d)^2}{(64 - V_d)^2}$$

or $V_d = 54 \text{ KV}.$

2. Let P_C be power loss due to corona at 113 KV,

$$\therefore \qquad P_C \propto \left(\frac{113}{\sqrt{3}} - 54\right)^2 \qquad ...(2)$$

Dividing equation (2) by equation (1), we get

$$\frac{P_C}{53} = \frac{(65.2-54)^2}{(61.2-54)^2}$$

or

$$P_C = \left(\frac{11.2}{7.2}\right)^2 \times 53 = 128 \text{ KW}.$$

7.5 ADVANTAGES OF CORONA

1. Corona causes the attenuation of travelling waves due to lightning. It reduces the magnitude of high voltage steep fronted wave due to lightning or switching by partially dissipating as a corona loss. In this way, corona acts as a safety valve.
2. Due to corona formation, the air surrounding the conductor becomes conducting and hence virtual diameter of the conductor is increased. The increased diameter reduces the electro-static stresses between the conductors.

7.6 DISADVANTAGES OF CORONA

1. Corona is accompanied by a loss of energy. This affects the transmission efficiency of the line.
2. Ozone is produced by corona and may cause corrosion of the conductor due to chemical action.
3. The leakage current due to corona takes place only during peak periods of voltage waves and that current is non-sinusoidal. This non-sinusoidal current causes a non-sinusoidal voltage and large third harmonics are produced. These triple frequency currents and voltages interfere with the communication circuits due to electromagnetic and electrostatic induction effects.

7.7 FACTORS AFFECTING CORONA

Following are the factors that affect corona loss on overhead transmission lines :

1. **Effect of Frequency :** From Peek's formula, it is evident that corona loss is a function of frequency. Thus, higher the frequency of supply, the higher are corona losses. It shows that D.C. corona loss is less as compared with A.C. corona.
2. **Spacing between Conductors :** From Peek's formula, it is clear that :

$$P \propto \sqrt{\frac{r}{D}}$$

where D is spacing between the conductors.

If the spacing between the conductors is made very large as compared to their diameters, there may not be any corona effect. The larger distance between conductors reduces the electro-static stresses at conductor surface, thus avoiding corona formation.

3. **Effect of Density of Air :** From Peek's formula, it is clear that corona loss is inversely proportional to δ. The lower the value of δ, higher the loss. Thus, corona loss increases with decrease in density of air. It is for this reason that the corona loss is more on hilly areas than on plain areas.
4. **Effect of System Voltage :** The system voltage greatly affects corona. If it is low, there is no change in the condition of air surrounding the conductors and hence no corona is formed. On the other hand, if the voltage has such a value that electrostatic stresses developed at the conductor surface make the air around the conductor conducting then corona is formed. Greater the potential difference, greater is the electric field and greater is the corona loss.
5. **Effect of Rain, Dust etc. :** Rain and dust increase the corona loss since rain drops and dust particles settling on the conductor surface establish protruding points which become the regions of local corona discharges.
6. **Surface Conditions of the Conductors :** The corona effect depends upon the shape and conditions of the conductors. The rough and irregular surface will give rise to more corona because unevenness of the surface decreases the value of breakdown voltage. If the conductor surface would have dirt deposited on it, it will lower the disruptive voltage and increase corona loss.
7. **Effect of Conductivity of Air :** The conductivity of the air depends on the number of ions per unit volume of air, the size and charge per ion. Both these factors vary with altitude and atmospheric conditions. Thunderstorms increase the ion content and make the atmosphere more conducting. Higher conductivity leads to greater corona loss.
8. **Effect of Heating of the Conductor by Load Current :** The heating of the conductor by the load current has an indirect reducing effect on the corona loss. Without such heating, tiny dew drop collect on the conductor surface during fog and humid weather and cause larger corona loss. The heating of the conductor due to load current prevents such condensation and reduces corona loss.

7.8 METHODS OF REDUCING CORONA EFFECT

These losses can be reduced by using :

(*a*) Bundled conductors

(*b*) Large dia conductors and

(*c*) Hollow conductors.

7.9 RADIO INTERFERENCE

Radio interference is the adverse effect introduced by corona on wireless broadcasting. The corona discharges emit radiation which may introduce noise signals in the communication lines, radio and television receivers. There is gradual increase in Radio Interference level till the voltage is such that measurable corona loss takes place. Above this voltage, the increase in radio interference is very rapid. The rate of increase in radio interference is affected by conductor surface and its diameter, which is higher in case of smooth and large diameter conductors.

The amplitude of Radio interference level varies inversely as the frequency at which the interference is measured. Thus, services in the higher frequency band *e.g.*, television, frequency modulated broadcasting, radar, microwave relay are less likely to be affected.

Radio interference is considered as a field measured in microvolts per metre at any distance from the transmission line and assumes importance for lines above 200 KV. Thus, radio interference is one of very important factors while designing a transmission line.

SELF TEST

Fill in the blanks by inserting appropriate words / figures :

1. If the spacing between the conductors is increased, then corona effect is __________.
2. When an insulator breaks down by puncture, it is __________ damaged.
3. Corona effect is __________ pronounced in stormy weather as compared to fair weather.
4. If the conductor size is increased, the corona effect is __________.
5. The effect of dirt on the surface of the conductor is to __________ irregularity and thereby __________ the break-down voltage.
6. Due to corona, the probability of flashover is __________.
7. In case of ac system, the current due to corona is __________.
8. The corona effect __________ with the decrease in spacing between the conductors.
9. In stormy weather corona occurs at __________ voltage as compared with fair weather.
10. The corona __________ with the increase in diameter of the conductor.
11. Corona __________ the probability of flashover and __________ the system performance.

Answers to the Self Test

1. reduced
2. permanently
3. more
4. decreased
5. increase, reduce
6. decreased
7. non-sinusoidal
8. increases
9. lower
10. decreases
11. reduces, improves.

OBJECTIVE QUESTIONS

1. The dieletric strength of air under normal conditions is around :
   (*a*) 30 KV/cm (*b*) 100 KV/cm
   (*c*) 150 KV/cm (*d*) 200 KV/cm.
2. The chances of occurrence of corona are mximum during :
   (*a*) Humid weather (*b*) Dry weather
   (*c*) Hot summer (*d*) Winter.
3. The effect of corona is :
   (*a*) Increased reactance (*b*) Increased power loss
   (*c*) Increased inductance (*d*) All of the above.
4. Presence of ozone owing to corona :
   (*a*) reduces the p.f. (*b*) improves voltage regulation
   (*c*) improves the p.f. (*d*) corrodes the material.
5. Corona is likely to occur maximum in case of :
   (*a*) Domestic wiring (*b*) Service lines
   (*c*) Transmission lines (*d*) Distribution lines.
6. Visual critical voltage is :
   (*a*) lower than disruptive critical voltage (*b*) higher than disruptive critical voltage
   (*c*) equal to critical voltage (*d*) None of the above.

7. As the spacing between conductors decreases, the disruptive critical voltage :
   (*a*) increases (*b*) decreases
   (*c*) remains the same.
8. Corona loss increases with:
   (*a*) Decrease in supply frequency and conductor size
   (*b*) Decrease in supply frequency but increase in conductor size
   (*c*) Increase in supply frequency and conductor size
   (*d*) Increase in supply frequency but reduction in conductor size.
9. As the conductor radius increases the disruptive critical voltage
   (*a*) increases (*b*) decreases
   (*c*) remains the same.
10. As pressure decreases the corona loss
    (*a*) increases (*b*) decreases
    (*c*) remains the same.
11. The corona loss on a particular system at 50 Hz is 1 KW/phase per km. The corona loss on the same system with supply frequency 25 Hz will be :
    (*a*) 0.5 KW/phase/km (*b*) 0.667 KW/phase/km
    (*c*) 1 KW/phase/km (*d*) none of the above.
12. Corona loss is less when the shape of the conductor is :
    (*a*) oval (*b*) circular
    (*c*) flat (*d*) independent of shape.
13. As pressure decreases the disruptive critical voltage :
    (*a*) increases (*b*) decreases
    (*c*) remains the same (*d*) any of above.
14. As temperature increases the disruptive critical voltage
    (*a*) increases (*b*) decreases
    (*c*) remains the same (*d*) any of above.
15. Corona loss can be reduced by the use of hollow conductor, because:
    (*a*) of better ventilation in the conductor
    (*b*) the eddy-current in the conductor is eliminated
    (*c*) the curent density is reduced
    (*d*) for a given cross-section, the radius of the conductor is increased.
16. As compared to plains the corona loss in hilly areas is :
    (*a*) more (*b*) less
    (*c*) the same.
17. As compared to the corona loss in a copper conductor line, the corona loss in a line with ACSR conductor and the same current carrying capacity, is:
    (*a*) more (*b*) less
    (*c*) the same.
18. Compared with a solid conductor of the same radius, corona appears on a stranded conductor at a lower voltage, because stranding :
    (*a*) makes the current flow spirally about the axis of the conductor
    (*b*) produces surfaces of smaller radius

(c) produces oblique sections to a plane perpendicular to the axis of the conductor
(d) assists ionisation.

19. The charging current in a transmission line increases due to corona effect because corona increases :
(a) power loss in lines
(b) effective line voltage
(c) line current
(d) the effective conductor diameter.

20. The corona loss in a 50 Hz system is 0.2 KW per phase per km. At the frequency of 60 Hz, the corona loss would be :
(a) 0.17 KW/phase/km
(b) 0.20 KW/phase/km
(c) 0.22 KW/phase/km
(d) 0.24 KW/phase/km.

21. Hollow conductors are used in transmission lines to :
(a) improve stability
(b) reduce corona
(c) reduce weight of copper
(d) increase power transmission capacity.

22. Power loss due to corona is directly proportional to :
(a) radius of conductor
(b) supply frequency
(c) spacing between conductors
(d) none of the above.

23. Which of the following statements is true regarding corona ?
(a) Corona takes place at a voltage higher than breakdown voltage
(b) Corona is a current phenomenon
(c) Corona takes place at a voltage lower than breakdown voltage
(d) Corona increases the transmission line efficiency.

24. Corona loss in a transmission line is dependent on:
(a) material of the conductor
(b) height of the conductor
(c) diameter of the conductor
(d) none of the above.

25. The only advantage of corona is that it:
(a) makes line current sinusoidal
(b) works as a safety valve for surges
(c) ozone gas is produced
(d) produces a pleasing luminous glow.

26. Corona can be reduced by:
(a) reducing the spacing between conductors
(b) increasing the effective conductor diameter
(c) increasing the operating voltage
(d) none of the above.

27. In humid weather, the corona occurs at a voltage :
(a) much less than that needed in fair weather
(b) much higher than that needed in fair weather
(c) equal to that needed in fair weather
(d) none of the above.

28. The current drawn by the line owing to corona loss is :
(a) dc
(b) square
(c) sinusoidal
(d) non-sinusoidal

29. Corona occurs between two transmission conductors when they :
(a) are closely spaced
(b) have high potential difference
(c) carry dc power
(d) both (a) and (b).

30. To reduce corona effect usually :
   (*a*) the conductor diameter is reduced
   (*b*) the distance between the conductors is reduced
   (*c*) bundled conductors are used
   (*d*) stranded conductors are used.
31. Critical voltage limit of a transmission line is increased by :
   (*a*) increasing the spacing between conductors
   (*b*) increasing the radius of the conductors
   (*c*) reducing the spacing between conductors
   (*d*) reducing the radius of conductors.
32. Which of the following affects the corona least ?
   (*a*) Atmospheric temperature (*b*) Size and charge per ion
   (*c*) Number of ions (*d*) Mean free length.
33. In context of corona which statement is not true ?
   (*a*) corona is accompanied with power loss
   (*b*) corona attenuates lightning surges
   (*c*) corona is voltage effect
   (*d*) corona takes place on short transmission lines.

Answers

1. (*a*)	**2.** (*a*)	**3.** (*c*)	**4.** (*d*)	**5.** (*c*)
6. (*b*)	**7.** (*b*)	**8.** (*d*)	**9.** (*a*)	**10.** (*a*)
11. (*b*)	**12.** (*b*)	**13.** (*b*)	**14.** (*b*)	**15.** (*d*)
16. (*a*)	**17.** (*b*)	**18.** (*c*)	**19.** (*d*)	**20.** (*d*)
21. (*b*)	**22.** (*b*)	**23.** (*c*)	**24.** (*c*)	**25.** (*b*)
26. (*b*)	**27.** (*a*)	**28.** (*d*)	**29.** (*d*)	**30.** (*c*)
31. (*b*)	**32.** (*a*)	**33.** (*d*).		

EXERCISE

1. What do you understand by "disruptive critical voltage" and "visual critical voltage"?
2. Explain corona and discuss its effects on performance of transmission lines.
3. Discuss the advantages and disadvantages of corona.
4. Discuss the phenomenon of 'corona'.
5. A 3-phase, 220 KV, 50 Hz transmission line consists of 1.2 cm radius conductors spaced 2 m at the corners of an equilateral triangle. Calculate the disruptive critical voltage between the lines. Irregularity factor = 0.96, temperature 20°C. Barometric pressure 72.2 cm of mercury. Dielectric strength of air = 21.1 KV (rms)/cm. **[Ans:** 208 KV (rms)**]**
6. A single phase overhead line has two conductors of diameter 1 cm with a spacing of 1 m between centres. If the disruptive critical voltage for air is 21.1 KV /cm, for what value of the line voltage will corona commence? **[Ans:** 111.2 KV**]**

8

OVERHEAD LINE INSULATORS

The overhead line conductors should be supported on the poles or towers in such a way that currents from conductors do not flow to earth through supports *i.e.*, line conductors must be properly insulated from supports. This is achieved by securing line conductors to supports with the help of insulators. The insulators are used to separate line conductors from each other and from the supporting structures electrically.

8.1 DESIGN OF AN INSULATOR

While designing an insulator, the following considerations are made :

1. It should possess high mechanical strength in order to withstand conductor load, wind load etc.
2. It should have high electrical resistance in order to avoid leakage currents to earth.
3. The dielectric strength of insulating material should be high so that it can withstand high electrical stresses. The insulator should be able to withstand the over voltages due to lightning, switching or other causes under severe weather conditions in addition to the normal working voltage.
4. The insulator material should be non-porous, free from impurities and cracks which may lower the permittivity and hence dielectric strength.
5. The ratio of puncture voltage to flashover voltage called the factor of safety is kept as high as possible.

The electrical failure of insulators occurs either by puncture or flashover. In case of a puncture, the arc passes through the body of the insulator. Flashover is caused by an arc discharge between the conductor and earth through air surrounding the insulator. It is either due to line surges or due to the formation of wet conducting layer over the insulator surface. Normally, the insulator is not damaged by a flashover but it becomes useless after the puncture.

8.2 INSULATOR MATERIALS

The two most suitable materials for insulators are porcelain and toughened glass.

Porcelain is made from a fine powdered mixture of wet plastic clay, silica and feldspar, which is shaped, covered with glaze and then fired in a kiln. Glazing of insulator is very necessary to keep it relatively free from dust and moisture. Any sealed impurities in this insulator decrease the dielectric strength. It is therefore essential that porcelain must be completely air free and impervious to the entry of gases and liquids.

It is difficult to manufacture porcelain in the thickness required for some types of insulators and it is then necessary to adopt a two or more piece construction, the various pieces being fired and glazed separately and then cemented together. Porcelain is mechanically stronger than glass, gives less trouble from leakage and is less affected by changes of temperature. Porcelain is the most commonly used material for manufacture of insulators.

Toughened or prestressed glass insulators have their surface layers in state of high compression due to which their resistance to withstand mechanical and thermal stresses is greater. The toughening process consists of heating the glass uniformly to a temperature above its strain temperature. It is allowed to cool usually by blowing air on its surface. Toughened glass can withstand greater tension than annealed glass. Due to its transparency, cracks and defects within the insulator material can be detected easily by visual inspection. The glass insulator has the disadvantage that moisture readily condenses on its surface giving rise to high surface leakage. Another disadvantage is that in high tension insulators, the heavy mass of the insulator may result in internal strain of it. In view of this glass insulators are used very rarely and that too for lines below 25 KV.

8.2.1 Polymer Insulator

A synthetic suspension insulator has been developed by General Electric under the trade name 'Gepol'. It is made up of combination of fibre glass and an epoxy polymer instead of porcelain. The polymer insulator is 70% lighter than its equivalent porcelain unit. It is puncture-proof with extremely high mechanical strength. It has high thermal shock resistance to reduce damage to flashover, excellent radio interference voltage performance and better performance in polluted atmosphere.

8.3 TYPES OF INSULATORS

There are 3 main types of insulators used for overhead lines :

1. Pin type insulator.
2. Suspension type insulator or disc or string insulators.
3. Strain or tension type insulator.

8.3.1 Pin Type Insulator

A pin type insulator is small, simple in construction and cheap. It is used on lines upto and including 33 KV lines because for higher voltages they tend to be very heavy and more costly than suspension insulators. The pin insulator is supported on a forged steel pin or bolt which is secured to the cross-arm of the supporting structure. To avoid a direct contact between the porcelain and metal pin, a lead thimble is used. An adequate length of leakage path is obtained by providing the insulator with two or three rain sheds. These are so designed that even when the outer surface of the insulator is wet due to rain, sufficient leakage resistance is still given by the inner dry surfaces. Figure 8.1 hows the pin type insulator.

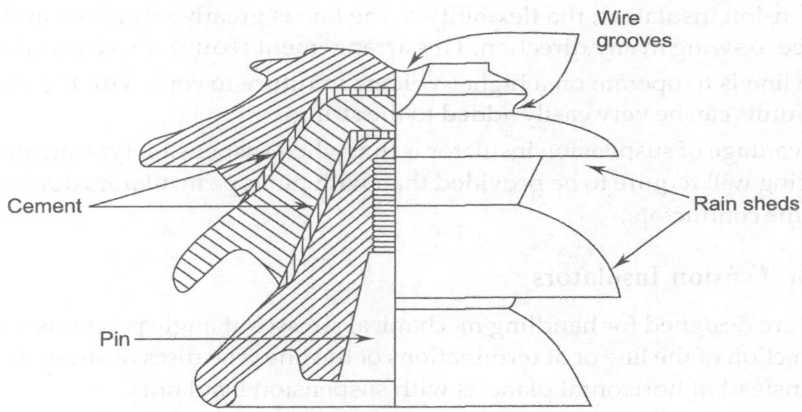

Fig. 8.1 *Pin type insulator.*

8.3.2 Suspension Insulators

The cost of pin insulator increases very rapidly with increase in line voltage. Therefore, suspension insulators are used for lines above 33 KV.

A suspension insulator consists of a number of separate insulator units connected with each other by metal links to form a flexible chain or a string. The insulator string is suspended from the cross-arm of the support. The conductor is attached to the lowest unit. An insulator assembly is shown in Fig. 8.2.

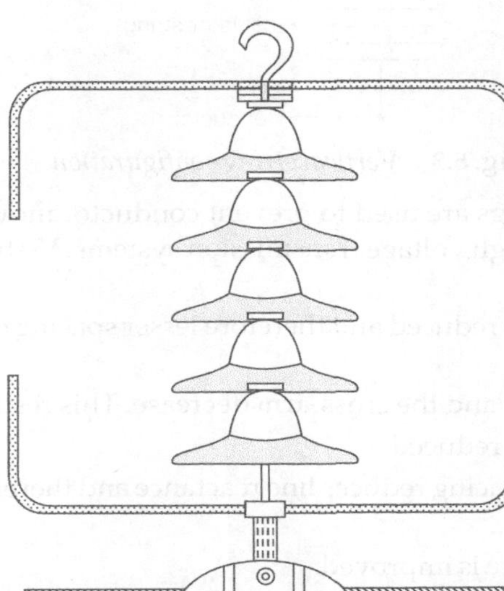

Fig. 8.2 *Suspension insulator string.*

The suspension insulator has many advantages over the pin insulator. These are :

1. Each unit is designed for operating voltage of about 11 KV so that insulation for any working voltage may be obtained by using a string of suitable number of such insulator units.
2. In the event of failure of an insulator unit, only that unit in the string instead of the entire string will require to be replaced.
3. With suspension insulators, the flexibility of the line is greatly improved and the insulator string is free to swing in any direction. This arrangement reduces mechanical stresses.
4. In case the line is to operate on a higher voltage in future to cope with the increasing load, additional units can be very easily added to the string.

But one disadvantage of suspension insulator is that when suspension type insulators are used, small larger spacing will require to be provided than with pin type insulators due to larger amplitude of swing of the conductors.

8.3.3 Strain or Tension Insulators

These insulators are designed for handling mechanical stresses at angle positions where there is a change in the direction of the line or at terminations of the line. The discs of strain insulators are in a vertical plane instead of horizontal plane as with suspension insulators.

8.4 CONFIGURATION OF INSULATOR STRINGS FOR EHV, A.C. AND HVDC TRANSMISSION SYSTEM

(*a*) **Vertical String (Tangent Configuration) :** In this configuration, a single string of insulator follows the conductor and sways like a pendulum in a strong side wind. It is shown in Fig. 8.3.

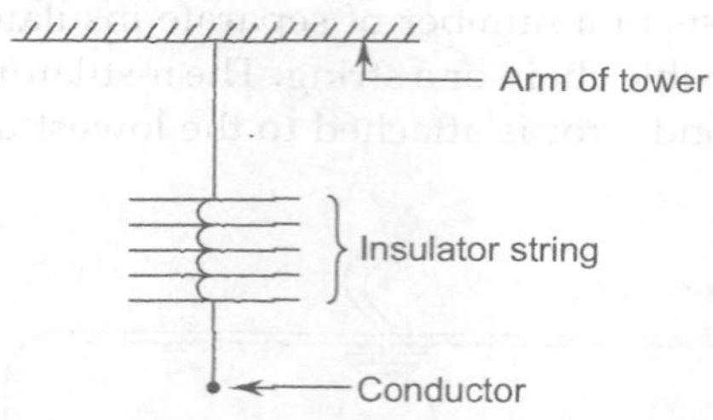

Fig. 8.3 *Vertical string configuration.*

(*b*) **V-Configuration :** V-strings are used to prevent conductor movement at towers. They find increased application in high voltage transmission systems. V-string configuration offers the following advantages :

1. The insulator swing is reduced and therefore lesser spacing can be provided between the conductors.
2. The sizes of the tower and the cross-arm decrease. This results in saving in cost.
3. Right of way width is reduced.
4. Reduction in phase spacing reduces line reactance and therefore the system power limits are higher.
5. Lightning performance is improved.

This configuration is shown in Fig. 8.4.

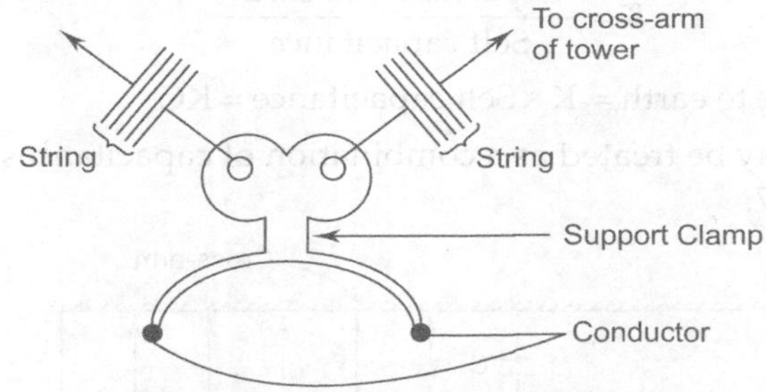

Fig. 8.4 *V-string configuration.*

(*c*) **Horizontal String Configuration:** It is shown in Fig. 8.5

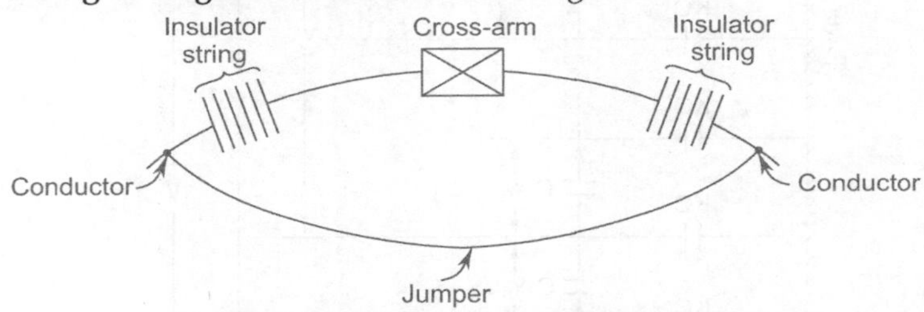

Fig. 8.5 *Horizontal string configuration.*

(*d*) **Special Configuration:** This configuration is used for extra high voltage requirements and it is shown in Fig. 8.6.

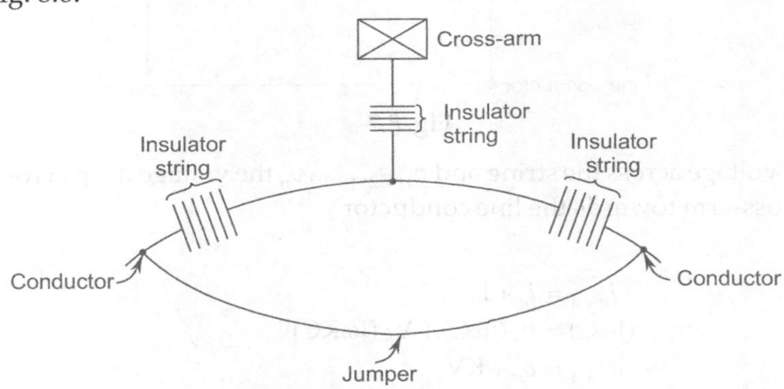

Fig. 8.6 *Special configuration.*

8.5 VOLTAGE DISTRIBUTION OVER INSULATOR STRING

When a voltage is applied across a suspension insulator string, the total system voltage is not equally distributed over the units of the string. The unit nearest to the conductor carries the maximum percentage of voltage and therefore most highly stressed. The determination of voltage distribution on an insulator string is helpful in determining the flashover voltage and the voltage at which localized corona and radio interference are produced.

The insulator material of each insulator being between two metallic pins forms a capacitor of capacitance C called the self capacitance. Also the air between each pin and the tower which is at earth potential forms other set of capacitors. These capacitances are called capacitances to earth.

Let $$K = \frac{\text{Capacitance to earth}}{\text{Self capacitance}}$$

$$\therefore \quad \text{Capacitance to earth} = K \times \text{Self capacitance} = KC$$

A string of insulators may be treated as a combination of capacitances connected in series and parallel as shown in Fig. 8.7.

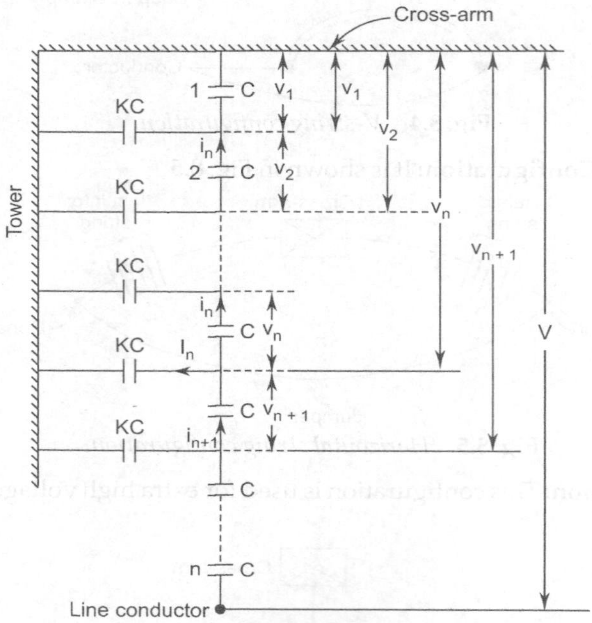

Fig. 8.7

Let V be the total voltage across the string and $v_1, v_2, \ldots\ldots, v_n$ the voltage drop across units 1, 2, ..., n starting from the cross-arm towards the line conductor

From figure,

$$i_{n+1} = i_n + I_n$$

$$v_{n+1}\,(J\omega C) = v_n\,(J\omega C) + V_n\,(J\omega KC)$$

or $$v_{n+1} = v_n + KV_n \qquad \ldots(1)$$

Also $$v_1 + v_2 + \ldots\ldots + v_n = V \qquad \ldots(2)$$

Above both equations (1) and (2) may be used to calculate the voltage across each unit of the insulator string. It is seen that the voltage across the insulator unit nearest to the line conductor is maximum and this unit is most highly stressed. The voltage goes on decreasing as the unit nearest the tower or cross-arm is approached. Consequently, the insulator string as a whole is being inefficiently used. The string efficiency is defined as :

$$\text{String efficiency} = \frac{\text{Voltage across the whole string}}{n \times (\text{Voltage across the unit adjacent to line conductor})}$$

or $$\eta = \frac{V}{nV_l}$$

where, V = voltage across the whole string
n = number of insulator disc in string
and V_1 = Voltage across the lowest unit connected to the line conductor.

If the unit adjacent to the line conductor is about to flashover, then the whole string is about to flashover. Therefore, string efficiency can also be expressed as :

$$\text{String efficiency} = \frac{\text{flashover voltage of string}}{n \times \text{flashover voltage of one unit}}$$

The string efficiency decreases with an increase in the number of units.

Example 8.1. *A string of suspension insulators consists of 4-units. The capacitance between each link pin and earth is 10% of self-capacitance of a unit. The voltage between the line conductor and earth is 100 KV. Find (a) the voltage distribution across each unit and (b) string efficiency.*

Solution. The arrangement is shown in Fig. 8.8.

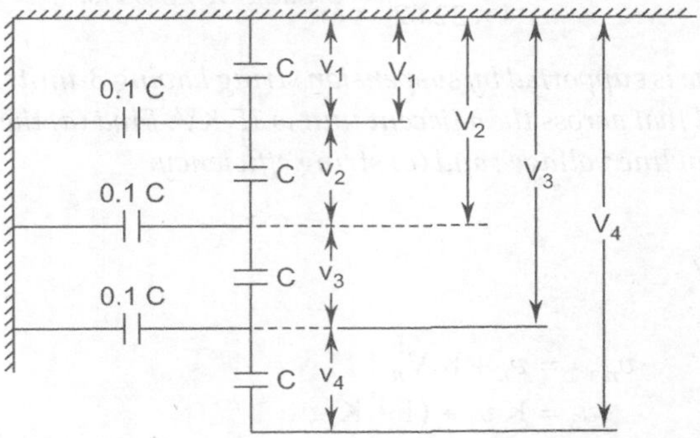

Fig. 8.8

From equation (1),

$$v_{n+1} = v_n + K V_n$$

Now, we will calculate v_2, v_3 and v_4 in terms of v_1

Put $n = 1$,

$$v_{1+1} = v_1 + K V_1$$

But $V_1 = v_1$ [From Fig. (8.8)]

$\therefore$ $v_2 = v_1 + K v_1 = (1 + K) v_1 = (1 + 0.1) v_1 = 1.1 v_1$

Put $n = 2$,

$$v_{2+1} = v_2 + K V_2$$

But $v_2 = v_1 + v_2$ [From Fig. (8.8)]

$\therefore$ $v_3 = v_2 + K (v_1 + v_2) = (1 + K) v_2 + K v_1$

$= (1 + 0.1) (1.1 v_1) + 0.1 v_1 = 1.31 v_1$

Put $n = 3$,

$$v_{3+1} = v_3 + K V_3$$

But $V_3 = v_1 + v_2 + v_3$

$\therefore$ $v_4 = v_3 + K (v_1 + v_2 + v_3) = 0.1 v_1 + 0.1 v_2 + 1.1 v_3$

$$= 0.1\,v_1 + (0.1)\,(1.1\,v_1) + (1.1)\,(1.31\,v_1)$$
$$= 1.651\,v_1$$

Now,
$$V = v_1 + v_2 + v_3 + v_4$$
$$100 = v_1 + 1.1\,v_1 + 1.31\,v_1 + 1.651\,v_1 = 5.061\,v_1$$

$\therefore$
$$v_1 = \frac{100}{5.061} = 19.758 \text{ KV}$$
$$v_2 = 1.1\,v_1 = 1.1 \times 19.758 = 21.73 \text{ KV}$$
$$v_3 = 1.31\,v_1 = 1.31 \times 19.758 = 25.883 \text{ KV}$$
$$v_4 = 1.651\,v_1 = 1.651 \times 19.758 = 32.62 \text{ KV}.$$

$$\text{String efficiency} = \frac{\text{Voltage across the string}}{n \times \text{Voltage across the lowest unit}}$$

$$= \frac{100}{4 \times 32.62} = 0.7664 \text{ or } 76.64\%.$$

Example 8.2. *A 3-phase line is supported by suspension string having 3-units. The voltage across the unit nearest to the line is 20 KV and that across the adjacent unit is 15 KV. Find (a) the ratio of capacitance of Joint to capacitance of disc (b) system line voltage ; and (c) string efficiency.*

Solution.

(*a*) $v_3 = 20$ KV, $v_2 = 15$ KV

From equation (1),

$$v_{n+1} = v_n + \text{K}\,V_n$$
$$v_3 = \text{K}\,v_1 + (1 + \text{K})\,v_2 \quad ...(3)$$

and
$$v_2 = (1 + \text{K})\,v_1 \quad ...(4)$$

Put values of v_2 and v_3 in equations (3) and (4)

$$20 = \text{K}\,v_1 + (1 + \text{K})\,15$$

and
$$15 = (1 + \text{K})\,v_1$$

On solving these equations: K = 0.18 or – 1.847

Since K cannot be negative, K = 0.18

(*b*) From equation (4),

$$v_1 = \frac{v_2}{(\text{K}+1)} = \frac{15}{(1+0.18)} = 12.7 \text{ KV}$$

line to neutral voltage $= 20 + 15 + 12.7 = 47.7$ KV

line to line voltage $= \sqrt{3} \times 47.7 = 82.6$ KV

(*c*) String efficiency $= \dfrac{47.7}{3 \times 20} = 0.795 \text{ or } 79.5\%.$

Example 8.3. *A suspension string has 3-units. Each unit can withstand a maximum voltage of 11 KV. The capacitance of each joint and metal work is 20% of capacitance of each disc. Find (a) maximum line voltage for which the string can be used (b) string efficiency.*

Solution. The arrangement is shown in Fig. 8.9

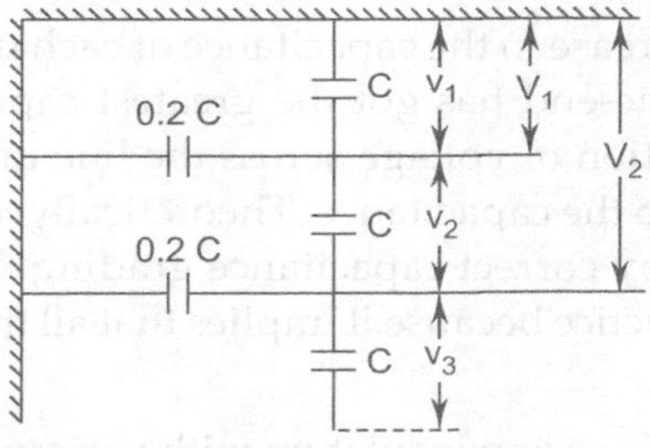

Fig. 8.9

From equation (1),

$$v_{n+1} = v_n + \mathrm{K}\,\mathrm{V}_n$$

Now, $v_2 = v_1 + \mathrm{K}\,\mathrm{V}_1$

$= v_1 + \mathrm{K}\,v_1 = 1.2\,v_1$ [As $\mathrm{V}_1 = v_1$]

Now, $v_3 = v_2 + \mathrm{K}\,\mathrm{V}_2$

$= v_2 + \mathrm{K}\,(v_1 + v_2)$ [As $\mathrm{V}_2 = v_1 + v_2$]

$= 1.2\,v_1 + 0.2\,(v_1 + 1.2\,v_1) = 1.64\,v_1$

Since, $v_3 = 11\,\mathrm{KV}$

$\therefore$ $11 = 1.64\,v_1$

or $v_1 = \dfrac{11}{1.64} = 6.7\,\mathrm{KV}$

$v_2 = 1.2\,v_1 = 1.2 \times 6.7 = 8.04\,\mathrm{KV}$

Maximum line to neutral voltage $= v_1 + v_2 + v_3 = 25.74\,\mathrm{KV}$

Maximum line to line voltage $= \sqrt{3} \times 25.74 = 44.58\,\mathrm{KV}$

String efficiency $= \dfrac{25.74}{3 \times 11} = 0.78$ or 78%.

8.6 IMPROVEMENT OF STRING EFFICIENCY OR EQUALIZING POTENTIAL OVER UNITS OF THE STRING

For satisfactory performance it is essential that the voltage distribution across the units of the string should be uniform. Some methods to improve string efficiency are :

1. Use of long cross-arm.
2. Capacitance grading or grading of units.
3. Use of grading rings or static shielding.

8.6.1 Use of Long Cross-arm

It is obvious that non-uniformity arises mainly due to stray capacitance of each unit to ground. For uniform potential distribution, the ratio of capacitance to earth/capacitance per insulator should be as small as possible. This can be achieved by using a longer cross-arm. But the limitations of cost and strength of towers does not permit the use of very long cross-arms.

8.6.2 Capacitance Grading

Capacitance grading means an increase in the capacitance of each unit from the tower end towards the line end. It means that the line-end has got the greatest capacitance while the top unit the smallest. This will result in reduction of voltage across the line unit since the voltage for a given current is inversely proportional to the capacitance. Theoretically, the voltages across the different units can be made exactly equal by correct capacitance grading. But this method of capacitance grading is very inconvenient is practice because it implies that all the insulator discs in the string to be different from one another.

Figure 8.10 shows a string a suspension insulators with various capacitance :

Let v = voltage across each unit of the string

C_n = self-capacitance of nth unit from the top

C_{n+1} = self-capacitance of $(n+1)^{th}$ unit from the top.

Apply KCL at node P in Fig. 8.10

$$i_{n+1} = i_n + I_n$$

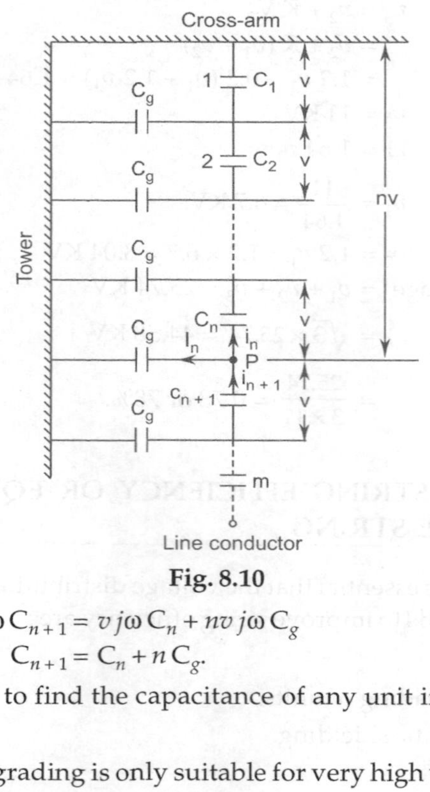

Fig. 8.10

$$v\, j\omega\, C_{n+1} = v\, j\omega\, C_n + nv\, j\omega\, C_g$$

or
$$C_{n+1} = C_n + n\, C_g.$$

This equation may be used to find the capacitance of any unit if the capacitance of one unit is known.

The method of capacitance grading is only suitable for very high voltage systems.

8.6.3 Static Shielding

This method uses a large metal ring surrounding the bottom unit and connected to the line. This ring known as grading ring or guard ring introduces capacitances between different joints and line. This has the effect of increasing the effective capacitance of bottom units. The idea, thus, is to cancel exactly the pin to tower charging currents so that the same current flows through the units of identical capacities to produce equal voltage drops across each unit as shown in Fig. 8.11.

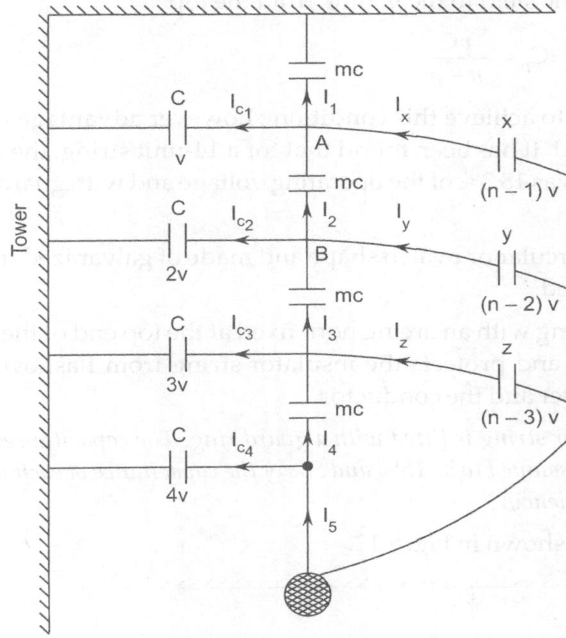

Fig. 8.11

Denoting the capacities to the shield by x, y, z, ..., we can write :

$$\text{At node A} \rightarrow I_2 = I_1 + I_{C_1} - I_x \quad \text{...(5)}$$

$$\text{At node B} \rightarrow I_3 = I_2 + I_{C_2} - I_y \text{ etc.} \quad \text{...(6)}$$

If the voltage is v across each unit and all n units are identical, the currents I_1, I_2, I_3 etc., must be equal. Hence, from equations (5) and (6).

By putting I_1, I_2, I_3 equal

$$I_x = I_{C_1} \quad \text{...(7)}$$

$$I_y = I_{C_2} \quad \text{...(8)}$$

$$I_z = I_{C_3} \quad \text{...(9)}$$

From equation (7) $\quad v\,C\,\omega = (nv - v)\,x\omega$

or $\quad v\,C\,\omega = (n-1)\,v\,x\omega$

or $\quad x = \dfrac{C}{n-1}$

From equation (8) $\quad 2\,v\,C\,\omega = (nv - 2v)\,y\omega$

or $\quad y = \dfrac{2C}{n-2}$

From equation (9) $\quad 3\,v\,C\,\omega = (nv - 3v)\,z\omega$

or $$z = \frac{3C}{n-3}$$

In general, capacitance from the sield to the pth link from the top :

$$C_p = \frac{pC}{n-p}$$

In practice, it is very difficult to achieve this condition ; however advantage is gained partially and this method is normally used. It has been found that for a 14-unit string, the voltage across the bottom unit without guard ring was 18.3% of the operating voltage and with guard ring it was 11.8% which is a great improvement.

Grading rings are generally circular or oval in shape and made of galvanized tube or pipe. Other shapes of ring have also been used.

The guard ring when used along with an arcing horn fixed at the top end of the string also serves the purpose of an arcing shield and protects the insulator string from flashover whenever over voltage appears between the tower and the conductor.

Example 8.4. *A 3-unit insulator string is fitted with a guard ring. The capacitances of the link pins to metal work and guard ring can be assumed to be 15% and 5% of the capacitance of each unit. Determine the voltage distribution and string efficiency.*

Solution. The arrangement is shown in Fig. 8.12.

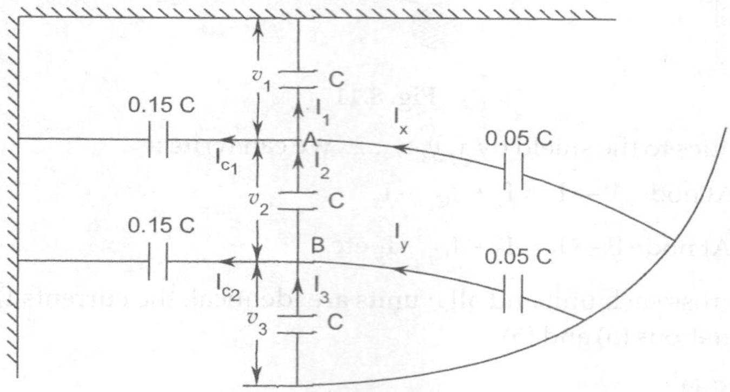

Fig. 8.12

At junction A,

$$I_2 = I_1 + I_{C_1} - I_x$$

$$\omega C\, v_2 = \omega C\, v_1 + 0.15\, C\omega\, v_1 - 0.05\omega C\,(v_2 + v_3)$$

or $$v_2 = v_1 + 0.15\, v_1 - 0.05\,(v_2 + v_3)$$

or $$1.05\, v_2 = 1.15\, v_1 - 0.05\, v_3 \quad \text{...(10)}$$

At junction B,

$$I_3 = I_2 + I_{C_2} - I_y$$

$$\omega C\, v_3 = \omega C\, v_2 + 0.15\,\omega C\,(v_1 + v_2) - 0.05\,\omega C\, v_3$$

or $$v_3 = v_2 + 0.15\,(v_1 + v_2) - 0.05\, v_3$$

or $$1.05\, v_3 = 0.15\, v_1 + 1.15\, v_2 \quad \text{...(11)}$$

Substituting the value of v_3 from equation (11) into equation (10)

$$1.05\ v_2 = 1.15\ v_1 - 0.05\left(\frac{1.15\ v_2}{1.05} + \frac{0.15\ v_1}{1.05}\right)$$

On solving, $v_2 = 1.0345\ v_1$ ...(12)

Substituting the value of v_2 from equation (12) in equation (11)

$$1.05\ v_3 = 0.15\ v_1 + 1.15 \times 1.0345\ v_1$$

or $v_3 = 1.276\ v_1$

Now, $v_1 + v_2 + v_3 = 100$

or $v_1 + 1.0345\ v_1 + 1.276\ v_1 = 100$

or $v_1 = 30.2\%$

$$v_2 = 1.0345\ v_1 = 31.24\%$$

$$v_3 = 1.276\ v_1 = 38.54\%.$$

$$\text{String efficiency} = \frac{100}{3 \times 38.54} = 0.865 \text{ or } 86.5\%.$$

Example 8.5. *Each line of a 3-phase system is suspended by a string of 3-identical insulators of self-capacitance C farad. The shunt capacitance of connecting metal work of each insulator is 0.2 C to earth and 0.1 C to line. Calculate the string efficiency of the system if a guard ring increases the capacitance to the line of metal work of the lowest to 0.3 C.*

Solution. The arrangement is shown in Fig. 8.13.

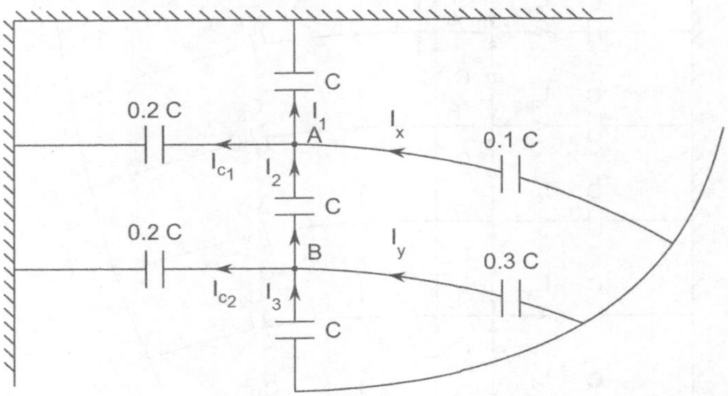

Fig. 8.13

At junction A,

$$I_2 = I_1 + I_{C_1} - I_x$$

or $\omega C\ v_2 = \omega C\ v_1 + 0.2\ \omega C\ v_1 - 0.1\ \omega C\ (v_2 + v_3)$

or $0.1\ v_3 = 1.2\ v_1 - 1.1\ v_2$

or $v_3 = 12\ v_1 - 11\ v_2$ ...(13)

At junction B,

$$I_3 = I_2 + I_{C_2} - I_y$$

$$\omega C\ v_3 = \omega C\ v_2 + 0.2\ \omega C\ (v_1 + v_2) - 0.3\ \omega C\ v_3$$

or $$1.3\, v_3 = 0.2\, v_1 + 1.2\, v_2 \quad ...(14)$$

Put the value of v_3 from equation (13) into equation (14)

$$1.3\,(12\, v_1 - 11\, v_2) = 0.2\, v_1 + 1.2\, v_2$$

or $$v_2 = 0.993\, v_1 \quad ...(15)$$

Substitute the value of v_2 from equation (15) into equation (13)

$$v_3 = 12\, v_1 - 11 \times 0.993\, v_1$$
$$= 1.077\, v_1.$$

Voltage between conductor and earth $= V_1 + V_2 + V_3$

$$= V_1 + 0.993\, v_1 + 1.077\, v_1 = 3.07\, v_1$$

$$\text{String efficiency} = \frac{3.07\, v_1}{3 \times 1.077\, V_1} \times 100 = 95\%$$

Example 8.6. *It is required to grade a string having 7-suspension insulators. If the pin to earth capacitance are all equal to C, determine the line to pin capacitance that would give the same voltage across each insulator of the string.*

Solution. The arrangement is shown in Fig. 8.14. As voltage across each insulator has to be same, therefore,

$$I_1 = I_2 = I_3 = I_4 = I_5 = I_6 = I_7$$

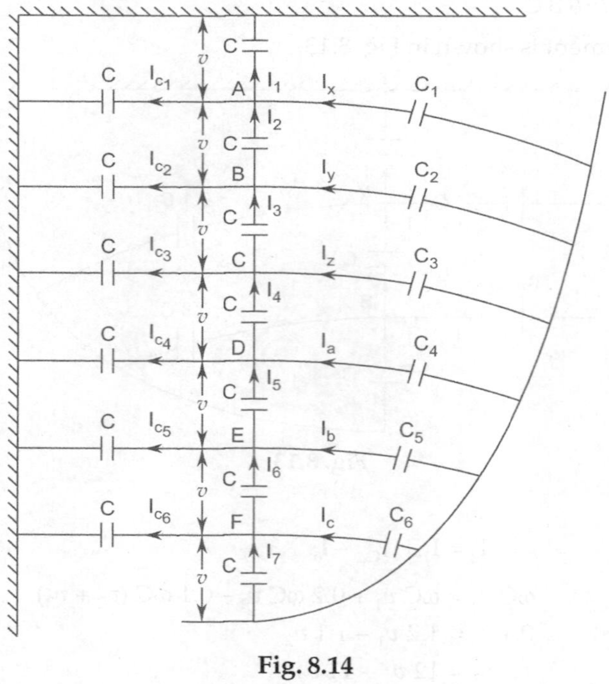

Fig. 8.14

At junction A,

$$I_2 = I_1 + I_{C_1} - I_x$$

or $$I_{C_1} = I_x \quad [\because \; I_1 = I_2]$$

$$\omega C v = 6 v \omega C_1$$

or $$C_1 = \frac{C}{6} = 0.167\,C$$

At junction B,

$$I_3 = I_2 + I_{C_2} - I_y$$

or $$I_{C_2} = I_y \qquad [\because \; I_2 = I_3]$$

$$2 v \omega C = 5 v \omega C_2$$

or $$C_2 = \frac{2C}{5} = 0.4\,C$$

At junction C,

$$I_4 = I_3 + I_{C_3} - I_z$$

or $$I_{C_3} = I_z \qquad [\because \; I_3 = I_4]$$

$$3 v \omega C = 4 v \omega C_3$$

or $$C_3 = \frac{3C}{4} = 0.75\,C.$$

Similarly, $$C_4 = 1.33\,C$$

$$C_5 = 2.5\,C$$

$$C_6 = 6C.$$

8.7 SELECTION OF INSULATION

The insulation level affects the cost and reliability of high voltage system considerably. Normal operating and transient voltages are the primary considerations in selecting the insulation for a line. The transient voltages include switching and lightning surges. It is found that when sufficient insulation is provided for switching and lightning surges, power frequency voltage requirements are usually satisfied. The magnitude of switching surges is 2.5 to 3 times normal voltage. Switching surges assume greater importance above 230 KV. Insulation requirements on lines upto 230 KV lightning is the major factor to decide the amount of insulation. The lightning insulation requirements depend on other factors also like tower footing resistance, number of ground wires, conductor configuration etc., in addition to the number of insulators. Final choice of the number of insulators for a line is also dictated by the atmospheric conditions, temperature and altitude of the place.

Atmospheric conditions affect considerably the flashover voltages. Wet conditions cause flashovers at lower voltages. Dirty and wet surfaces of the insulators allow the leakage currents to flow over them resulting in flashovers. The flashover voltage is proportional to the length of leakage path. The leakage distances or creepage paths vary depending upon the quality and quantity of deposition on insulators. Usually about 25 to 28 mm of leakage distance per KV of line-to-ground voltage is provided. In heavily polluted areas either antifog type of units are used or the number of units is increased in the string to increase the leakage path. In such cases the insulators should be washed periodically.

8.8 INSULATOR FAILURE

The main causes of an insulator failure are abnormal stress, cracks, porosity, puncture and flashover.

The failure due to mechanial stresses is very rare because defective pieces are weeded out in the routine factory tests.

Cracking is due to continued action of alternate and sometimes rapidly varying heat and cold, dryness and dampness which produce high stresses on the insulator and lead to failure.

Porosity which may be due to underfiring or other causes leads to failure because porous material absorbs water from atmosphere and its electrical resistance decreases. A porosity test is a necessary routine factory test for insulators.

Flow of leakage current through a portion of the insulator may heat up the localized portion and may cause its failure. The failure of an insulator by puncture is quite rare because the electrical characteristics of insulator are chosen to be the best.

The most common cause of insulator failure is due to flashover. The flashover voltage is the voltage which will cause an arc through the air surrounding the insulator. Switching and lightning surge are also responsible for insulator flashover. The arc produced by the flashover ionizes the surrounding medium causing insulator failure and subsequent short circuit of the conductor. For this reason insulators are often fitted with arcing horns to keep this power arc away from the insulator.

Pollution, rain drop, dirt, mist etc., may cause deposition of materials on insulator. This reduces the dielectric strength of the insulator units and increases surface leakage current.

8.9 TESTING OF INSULATORS

The following tests are performed on the insulators :

8.9.1 Mechanical Tests

The mechanical tests are performed to confirm capabilities of insulators to withstand the corresponding stresses. The important mechanical tests on insulators and their fittings include the following :

(*a*) Tensile strength
(*b*) Compression test
(*c*) Torsional test
(*d*) Bending minimum test
(*e*) Mechanical vibration test.

8.9.2 Porosity Test

This is a destructive test. The specimen insulator is broken into pieces and pieces are immersed in a 1% solution of dye in alcohol under a pressure of 15×10^6 N/m^2. After sufficient time, the pieces are removed from the container and examined. A deep penetration of dye indicates porosity in insulator.

8.9.3 Power Frequency Withstand Test

Normal power frequency voltage is continuously applied to the insulators. The effect of this voltage is to align the dust particles on the surface and cause leakage currents over the surface. To avoid flashover, the insulators are provided with sufficient creepage distance. During the power frequency voltage withstand test, voltage of order of twice the rated voltage is applied for a period of a minute. There should be no flashover or puncture. Power frequency withstand voltage is a type test as well as a routine test. Power frequency test is conducted in two categories :

(*a*) 1 minute power frequency withstand test-Dry.

(*b*) 1 minute power frequency withstand test-Wet.

The wet test is applicable only to the outdoor insulators.

8.9.4 Impulse Voltage Withstand Test

During this test, a standard impulse voltage wave is applied to the insulator. The standard lightning impulse wave is of 1.2 μ-sec wavefront and 50 μ-sec wave tail. This wave is called $\frac{1.2}{50}$ wave. The standard switching impulse wave is $\frac{250}{2500}$ μ-sec. Switching impulse tests are applicable for rated voltage above 220 KV.

8.9.5 Puncture Voltage Test

This is a destructive test. Here, the insulator with the pin and conductor in proper position is immersed in an insulating oil and a steadily increasing power frequency voltage is applied. The voltage at which conduction starts is called the puncture voltage. In air, flashover occurs at a voltage lower than the puncture voltage, so the test is conducted in oil at room temperature.

For systematizing the tests, the following categories are defined in the standard :

1. **Type tests :** These are mechanical and electrical tests to prove the design and assigned ratings of insulators of particular type, shape and design.
2. **Sample tests :** These are mechanical tests to prove the quality, material and mechanical properties.
3. **Routine tests :** These are performed on every insulator.

8.10 NUMBER OF INSULATOR DISCS FOR VARIOUS LINE VOLTAGES

While designing a transmission line at high voltage it is necessary to find out the number of insulators in a string. The approximate number of insulators used for various transmission voltages is given below :

Line Voltage (KV)	No. of Discs (in clean atmosphere)
66	5
132	10
220	14
400	21
765	35

SELF TEST

Fill in the blanks by inserting appropriate words / figures :

1. The most commonly used material for insulators of overhead lines is __________.
2. By using a guard ring, string efficiency is __________.
3. If the string efficiency is 100%, it means that __________.
4. Suspension type insulators are used for voltages beyond __________.
5. The longer the cross-arm, the __________ the string efficiency.
6. In a string of suspension insulators, the maximum voltage appears across the unit __________ to the conductor.
7. If shunt capacitance is reduced, then string efficiency is __________.
8. In a string of suspension insulators, if the unit nearest to the conductor breaks down, then other units will __________.
9. The discs of the strain insulators are used in __________ plane.
10. A shorter string has __________ string efficiency than a larger one.
11. The potential across the various discs of suspension string is different because of __________ capacitance.
12. The modern practice is not to use pin type insulators beyond __________.
13. Lowest disc in a string of suspension insulators has __________ pd across it.
14. Glass insulators can be used upto __________ KV under ordinary atmospheric conditions and well upto __________ KV in dry atmosphere.
15. The insulators employed in overhead lines must be mechanically __________ and of __________ relative permittivity and __________ insulation resistance.
16. __________ is extensively used material for insulators and is produced by firing at a controlled temperature a mixture of Kaolin, feldspar and quartz.

Answers to Self Test

1. porcelain
2. increased
3. potential across each disc is the same.
4. 33 KV
5. greater
6. nearest
7. increased
8. also breakdown
9. vertical
10. more
11. shunt
12. 33 KV
13. maximum
14. 25, 50
15. very strong, high, high
16. porcelain.

OBJECTIVE QUESTIONS

1. The string efficiency of a high voltage line is around:
   (*a*) 0.1
   (*b*) 0.3
   (*c*) 0.8
   (*d*) 1.0.
2. If the cross-arms of a high voltage line are made longer, the string efficiency:
   (*a*) increases
   (*b*) decreases
   (*c*) remains the same.

3. Strain type insulators are used:
   (a) at intermediate anchor towers (b) at dead ends
   (c) on straight runs (d) any of (a) or (b).
4. The insulators used on 220 KV transmission lines are of:
   (a) pin type (b) shackle type
   (c) suspension type (d) none of the above.
5. A string insulator has 4 units. The voltage across the bottom-most unit is 33.33% of the total voltage. Its string efficiency is:
   (a) 75% (b) 66.66%
   (c) 33.33% (d) 25%.
6. In actual practice the string efficiency is improved by using:
   (a) capacitance grading (b) grading ring
   (c) discs of different sizes (d) none of the above.
7. Insulators used on EHT transmission lines are made of:
   (a) glass (b) porcelain
   (c) PVC (d) stealite.
8. Pin type insulators are generally not used for voltages exceeding:
   (a) 11 KV (b) 25 KV
   (c) 33 KV (d) 66 KV.
9. The 11 KV lines use
   (a) suspension insulators (b) pin insulators
   (c) both pin and suspension insulators (d) none of the above.
10. Whenever the conductors are dead-ended or there is a change in the direction of transmission line, the insulators used are of the:
    (a) shackle type (b) pin type
    (c) suspension type (d) strain type.
11. For a 220 KV line, the number of discs in an insulator string is around :
    (a) 5 (b) 10
    (c) 15 (d) 20.
12. For a 400 KV line, the number of discs in an insulator string is around :
    (a) 10 (b) 16
    (c) 22 (d) 31.
13. In a string of suspension insulators, the voltage distribution across the different units of a string could be made uniform by the use of a grading ring, because it:
    (a) forms capacitances which help to cancel the charging current from link-pins.
    (b) forms capacitances with link-pins to cancel the charging current from link-pins.
    (c) decreases the capacitances of upper insulator units to cause equal voltage drop.
    (d) increases the capacitances of lower insulator units to cause equal voltage drop.
14. Which of the following materials is not a constituent of material used in making porcelain insulators ?
    (a) kaolin (b) felspar
    (c) silica (d) quartz.

15. String efficiency can be improved by :

(*a*) using a guard ring (*b*) using long cross-arms

(*c*) grading the insulators (*d*) any of the above.

16. The insulators may fail due to:

(*a*) short-circuits (*b*) flashover

(*c*) deposition of dust (*d*) any of above.

17. The non-uniform distribution of voltage across the units in a string of suspension type insulators is due to :

(*a*) non-uniform distance of separation of the units from the tower body.

(*b*) the existence of stray capacitance between the metallic junctions of the units and the tower body.

(*c*) unequal self-capacitance of the units.

(*d*) non-uniform distance between the cross-arm and the units.

18. The purpose of guard ring in transmission lines is to :

(*a*) increase the earth capacitance of the lowest unit.

(*b*) reduce the earth capacitance of the lowest unit.

(*c*) reduce the transmission line losses.

(*d*) none of the above.

19. If the frequency of a transmission system is changed from 50 Hz to 100 Hz, the string efficiency :

(*a*) will increase

(*b*) will decrease

(*c*) remain unchanged

(*d*) may increase or decrease depending on the line parameters.

20. The ratio of puncture voltage to the flashover voltage of a line insulator is :

(*a*) lower than 1 (*b*) much greater than 1

(*c*) equal to 1 (*d*) other than 1.

21. A 66 KV system has string insulator having five discs and the earth to disc capacitance ratio of 0.10. The string efficiency will be :

(*a*) 55% (*b*) 67%

(*c*) 75% (*d*) 85%.

22. The string efficiency of insulators can be increased by :

(*a*) correct grading of insulators of various capacitances

(*b*) increasing the number of strings

(*c*) reducing the number of strings

(*d*) changing the orientation of strings.

23. Consider the following statements :

In case of suspension-type insulators, the string efficiency can be improved by :

1. using a longer cross-arm 2. using a guard ring

3. grading the insulator discs 4. reducing the cross-arm length.

Of these statements :

(*a*) 1, 2 and 3 are correct (*b*) 2, 3 and 4 are correct

(*c*) 2 and 4 are correct (*d*) 1 and 3 are correct.

Answers to Objective Questions

1. (*c*)	**2.** (*a*)	**3.** (*d*)	**4.** (*c*)
5. (*a*)	**6.** (*b*)	**7.** (*b*)	**8.** (*c*)
9. (*b*)	**10.** (*d*)	**11.** (*c*)	**12.** (*c*)
13. (*a*)	**14.** (*c*)	**15.** (*d*)	**16.** (*d*)
17. (*b*)	**18.** (*b*)	**19.** (*c*)	**20.** (*b*)
21. (*b*)	**22.** (*a*)	**23.** (*a*).	

EXERCISE

1. A 3-phase 66 kV transmission line is carried by string of 5 suspension insulators. The capacity of each unit insulator to the capacity relative to earth is 4:1. Calculate the potential across each unit and the string efficiency. Assume that there is no leakage. [**Ans:** 3.32 kV, 4.16 kV, 6.16 kV, 9.36 kV and 15.1 kV; 50.4%]

2. Why voltage distribution across the insulator discs in an insulator string is not uniform?

3. Name the different types of insulators used in transmission and distribution systems.

4. Show that in a string of suspension insulators, the disc nearest to the conductor has the highest voltage across it.

5. What are the two functions that are performed by grading ring?

6. Discuss the advantages and disadvantages of (*i*) pin-type insulators (*ii*) suspension type insulators.

7. A 3-phase overhead transmission line is being supported by 3-disc suspensions insulators, the potential across the first and second insulator are 8 and 11kV respectively. Calculate (*a*) the line voltage (*b*) the ratio of capacitance between pin and earth to self-capacitance of each unit, (*c*) the string efficiency. [**Ans:** (*a*) 64.28 kV, (*b*) 0.375 (*c*) 68.28%]

8. What are the causes of failure of insulators?

9. What is meant by string efficiency and how can it be improved?

10. Name the various materials that are commonly used for overhead line insulators. Why glass insulators cannot be used above 50 kV?

11. Explain why the potential distribution is not, in general, uniform over the string in a suspension type of insulator.

12. Explain the methods used for improving the voltage distribution along the string of insulators in overhead lines.

13. A string of 6 suspension insulators is to be fitted with a guard ring. The pin-to-earth capacitances are all equal to C. What should be the value of line-to-pin capacitances so as to have uniform voltage distribution over the string? [**Ans:** $C_1 = \frac{1}{5}C$, $C_2 = \frac{1}{2}C$, $C_3 = C$, $C_4 = 2C$, $C_5 = 5C$]

14. Each line of a 3-phase system is supported by a string of 3 similar insulators. If the voltage across the line unit is 17.5 kV, calculate the line-to-neutral voltage. Assume that the shunt capacitance between each insulator and earth is $\frac{1}{8}^{th}$ of the capacitance of the insulator itselt. Also find the string efficiency. [**Ans:** 44.24 kV, 84.26%]

9

PERFORMANCE OF TRANSMISSION LINES

A transmission line is a set of conductors that run from one place to another with support on transmission towers. These lines have four distributed parameters — series resistance and inductance, shunt capacitance and conductance. The electrical power is transmitted over the overhead lines at approximately the speed of light.

9.1 CLASSIFICATION OF OVERHEAD TRANSMISSION LINES

1. **Short Transmission Lines :** When the length of an overhead transmission line is upto about 80 km and line voltage is comparatively low (< 20 KV), it is considered as a short transmission line. Due to smaller length and lower voltage, the capacitance effects can be neglected.
2. **Medium Transmission Lines :** When the length of an overhead transmission line is about 80–160 km and line voltage is moderately high (> 20 KV and < 100 KV), it is considered as a medium transmission line. Due to sufficient length and voltage of the line, the capacitance effects are taken into account. For medium length lines, the shunt capacitance can be assumed to be lumped at the middle of line or half of the shunt capacitance may be considered to be lumped at each end of the line. These two representations are termed as nominal-T and nominal-π respectively.
3. **Long Transmission Lines :** When the length of an overhead transmission line is more than 160 km and line voltage is very high (> 100 KV), it is considered as a long transmission line. For such a line, the line constants are considered uniformly distributed over the whole length of the line.

9.2 SOME BASIC TERMS

(*a*) **Voltage Regulation :** When a transmission line is carrying current, there is a voltage drop in the line due to resistance and inductance of the line. Thus, the receiving and voltage (V_R) of the line is less than the sending end voltage (V_S). This voltage drop in the line ($V_S - V_R$) expressed as a percentage of V_R is called voltage regulation.

$$\% \text{age Voltage Regulation} = \frac{V_S - V_R}{V_R} \times 100$$

For better performance, voltage regulation of the system should be low. An ideal system has zero voltage regulation.

(*b*) **Transmission Efficiency :** The ratio of receiving end power to the sending end power of a transmission line is known as the transmission efficiency of the line *i.e.*,

$$\% \text{ efficiency} = \frac{\text{Receiving end power}}{\text{Sending end power}} \times 100$$

$$= \frac{V_R I_R \cos\phi_R}{V_S I_S \cos\phi_S} \times 100$$

or
$$\% \text{ efficiency} = \frac{\text{Receiving end power}}{\text{Receiving end power + losses}} \times 100$$

Where V_S, V_R, I_S, I_R are sending end and receiving end voltages and the currents.

9.3 PERFORMANCE OF SHORT TRANSMISSION LINES

The equivalent circuit of a short transmission line is shown in Fig. 9.1. As stated earlier, the effects of line capacitances are neglected for a short transmission line. The phasor diagram is also shown in Fig. 9.2 for logging p.f.

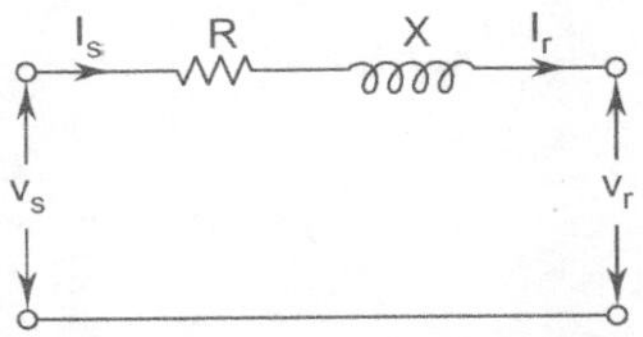

Fig. 9.1

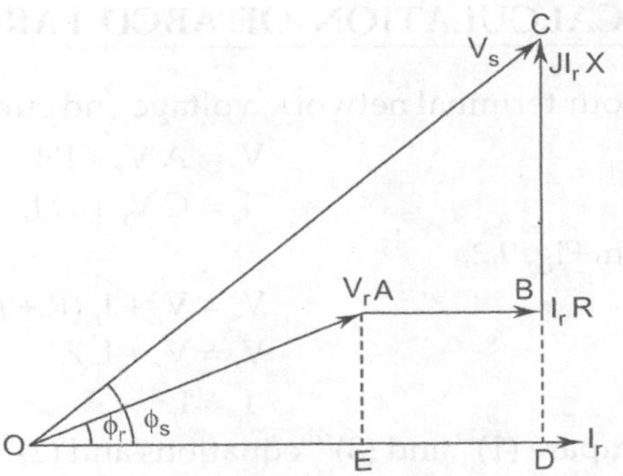

Fig. 9.2

In figure,
V_s = Sending end voltage
V_r = Receiving end voltage
I_s = Sending end current
I_r = Receiving end current.
ϕ_S and ϕ_R = Sending and Receiving end power factor.

From phasor diagram, (for lagging p.f.)

$$(OC)^2 = (OD)^2 + (DC)^2$$

or
$$V_s^2 = (OE + ED)^2 + (DB + BC)^2$$

or $$V_s = \sqrt{(V_r \cos\phi_r + I_r R)^2 + (V_r \sin\phi_r + I_r X)^2}$$

$$= \sqrt{V_r^2 \cos^2\phi_r + I_r^2 R^2 + 2I_r R V_r \cos\phi_r + V_r^2 \sin^2\phi_r + I_r^2 X^2 + 2I_r X V_r \sin\phi_r}$$

$$= \sqrt{V_r^2 + I_r^2 R^2 + 2I_r R V_r \cos\phi_r + I_r^2 X^2 + 2I_r X V_r \sin\phi_r}$$

or $$V_s = V_r \sqrt{1 + \frac{2I_r R \cos\phi_r}{V_r} + \frac{2I_r X \sin\phi_r}{V_r} + \frac{I_r^2}{V_r^2}(R^2 + X^2)}$$

$$V_s = V_r \left\{1 + \left(\frac{2I_r R}{V_r} \cos\phi_r + \frac{2I_r X}{V_r} \sin\phi_r\right)\right\}^{\frac{1}{2}}$$

Using Binomial Expansion upon second term only

$$V_s = V_r + I_r R \cos\phi_r + I_r X \sin\phi_r$$

As we know,

$$\%\text{ regulation} = \frac{V_s - V_r}{V_r} \times 100$$

$$= \left(\frac{I_r R \cos\phi_r}{V_r} + \frac{I_r}{V_r} X \sin\phi_r\right) \times 100$$

or Regulation per unit = $V_r \cos\phi_r + V_x \sin\phi_r$

Where V_r and V_x are the per unit values of resistance and reactance of the line.

9.4 CALCULATION OF ABCD PARAMETERS OF SHORT LINE

For a four terminal network, voltage and current at sending end are :

$$V_s = A\,V_r + B\,I_r \quad ...(1)$$

$$I_s = C\,V_r + D\,I_r \quad ...(2)$$

From Fig. 9.2,

$$V_s = V_r + I_r (R + jX)$$

or $$V_s = V_r + I_r Z \quad ...(3)$$

and $$I_s = I_r \quad ...(4)$$

Compare (1)st and (3)rd equations and (2)nd and (4)th equations :

$$A = 1, B = Z, C = 0, D = 1.$$

$$\%\,\eta = \frac{\text{Power received at the receiving end}}{\text{Power delivered from the sending end}} \times 100$$

or $$\%\,\eta = \frac{\text{Receiving end power}}{\text{Receiving end power + losses}} \times 100$$

or $$\%\eta = \frac{P}{P + 3I_r^2 R} \times 100$$

where, $P = V_r I_r \cos\phi_r$

and R is resistance per phase of the line.

Example 9.1. *A single phase overhead transmission line delivers 1100 kW at 33 KV at 0.8 p.f. lagging. The total resistance and inductive reactance of the line are 10 Ω and 15 Ω. Determine :*

1. *Sending end voltage.*
2. *Sending end p.f.*
3. *Transmission efficiency.*

Solution. Given
$$\cos \phi_r = 0.8 \text{ (lagging)}$$
$$Z = R + jX_L = 10 + j15$$
$$V_r = 33 \text{ KV}$$

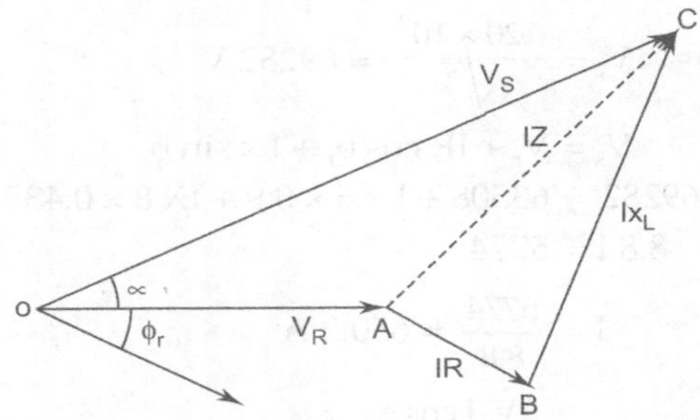

∴ Line current,
$$I = \frac{1100 \times 10^3}{33000 \times 0.8} = 41.67 \text{ A}$$

As $\cos \phi_r = 0.8$

∴ $\sin \phi_r = 0.6$

1.
$$V_s = V_r + I R \cos \phi_r + IX \sin \phi_r$$
$$= 33000 + 41.67 \times 10 \times 0.8 + 41.67 \times 15 \times 0.6$$
$$= 33000 + 333.36 + 375.03$$
$$= 33708.39 \text{ V}$$

Take V as a reference and
$$I = 41.67 (\cos \phi_r - j \sin \phi_r)$$
$$= 33.33 - j25$$

2.
$$V_s = V_r + IZ = 33000 + (33.33 - j25.0)(10 + j15) = 33708.39 + j250.$$

Angle between V_s and V_r
$$\alpha = \tan^{-1} \frac{250}{33708.39} = \tan^{-1} 0.0074 = 0° 25'$$

Sending end p.f.
$$\phi_s = \phi_r + \alpha = 36° 52' + 0° 25' = 37° 17'$$

∴ Sending end p.f. $\cos \phi_s = \cos 37° 17' = 0.7956$ lagging

3. Transmission efficiency
$$= \frac{\text{Receiving end power}}{\text{Receiving end power + losses}} \times 100$$
$$\text{Losses} = I^2 R = (41.67)^2 \times 10 = 17364 \text{ W}$$
$$= 17.364 \text{ KW}$$

∴
$$\% \eta = \frac{1100}{1100 + 17.364} \times 100 = 98.44\%.$$

Example 9.2. *A short 3-ϕ transmission line with an impedance of (6 + j8) Ω per phase has sending and receiving end voltages of 120 KV and 110 KV for receiving end load p.f. of 0.9 lagging. Determine:*

1. *Power output*
2. *Sending end p.f.*

Solution. Receiving end voltage/phase

$$V_r = \frac{110 \times 10^3}{\sqrt{3}} = 63508 \text{ V}$$

Sending end voltage/phase – $V_s = \dfrac{120 \times 10^3}{\sqrt{3}} = 69282$ V

$$V_s = V_r + IR \cos \phi_r + I \times \sin \phi_r$$

$$69282 = 63508 + I \times 6 \times 0.9 + IX\, 8 \times 0.435$$

or $\quad 8.8\, I = 5774$

$$I = \frac{5774}{8.8} = 650.2 \text{ A}$$

1. For 3-ϕ, power output $= \dfrac{3V_r I \cos \phi_r}{1000}$ KW

$$= \frac{3 \times 63508 \times 650.2 \times 0.9}{1000} = 111490 \text{ KW.}$$

2. Sending end p.f.

$$\cos \phi_s = \frac{OD}{OC} = \frac{V_r \cos \phi_r + IR}{V_s}$$

$$= \frac{63508 \times 0.9 + 650.2 \times 6}{69282}$$

$$= 0.88 \text{ lag.}$$

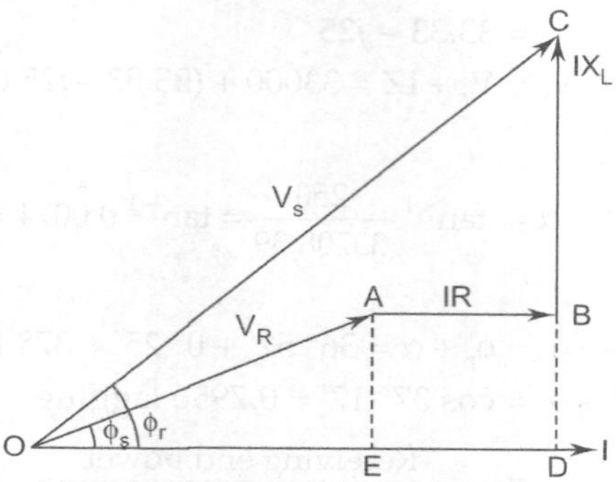

Example 9.3. *What is maximum length in km for a 1-ϕ transmission line having copper conductor of 0.775 cm^2 cross-section over which 200 KW at unity p.f. and at 3300 V are to be delivered. The efficiency of transmission is 90%. Take specific resistance as 1.725 mΩcm.*

Solution. Receiving end power $= 200$ kW $= 200000$ W

$\therefore$ Sending end power $= \dfrac{200000}{0.9} = 222222$ W

$$\text{Line losses} = 222222 - 200000 = 22222 \text{ W}$$

Now line current, $$I = \frac{200 \times 10^3}{3300 \times 1} = 60.6 \text{ A}$$

$$\text{line losses} = 2I^2 R$$

or $$22222 = 2 \times (60.6)^2 \times R$$

$\therefore$ $$R = \frac{22222}{2 \times (60.6)^2} = 3.025\ \Omega$$

Now, $$R = \rho \frac{l}{A}$$

$$3.025 = 1.725 \times 10^{-6} \times \frac{l}{0.775}$$

or $$l = 1.36 \times 10^6 \text{ cm}$$
$$= 13.6 \text{ km.}$$

9.5 MEDIUM TRANSMISSION LINES

Since the medium transmission lines have sufficient length (80–160 km) and operate at voltages greater than 20 KV, the effect of capacitances cannot be neglected. There are two configuration depending upon whether the shunt capacitance is assumed to be concentrated at the middle of line or half of the total capacitance is concentrated at each end of the line. They are :

1. Nominal-T Method.
2. Nominal-π Method.

9.5.1 Nominal-T Method

In this method, the whole line capacitance is assumed to be concentrated at the middle point of the line and half the line resistance and reactance are lumped on its either side as shown in Fig. 9.3.

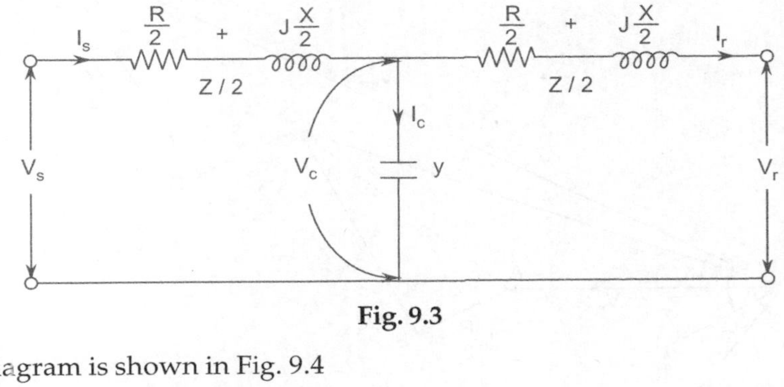

Fig. 9.3

The phasor diagram is shown in Fig. 9.4

We will calculate A, B, C, D constants :

From Fig. 9.3 $$V_s = V_c + I_s \frac{Z}{2} \quad ...(5)$$

and $$I_s = I_c + I_r \quad ...(6)$$

Now, $$I_c = V_c Y = \left(V_r + I_r \frac{Z}{2}\right) Y \quad ...(7)$$

Put the value of I_c from equation (7) into equation (6),

$$I_s = \left(V_r + I_r \frac{Z}{2}\right) Y + I_r$$

$$= YV_r + I_r \left(1 + \frac{YZ}{2}\right)$$

Put the value of I_s in equation (5)

$$V_s = \left(V_r + I_r \frac{Z}{2}\right) + \left[YV_r + I_r \left(1 + \frac{YZ}{2}\right)\right] \frac{Z}{2}$$

$$= V_r \left(1 + \frac{YZ}{2}\right) + I_r \left[\frac{Z}{2} + \frac{Z}{2} + \frac{YZ^2}{4}\right]$$

$$= V_r \left(1 + \frac{YZ}{2}\right) + I_r Z \left(1 + \frac{YZ}{4}\right)$$

Compare with the 2-port equations :

$$V_s = A V_r + B I_r$$

and $$I_s = C V_r + D I_r$$

Hence, $$A = D = \left(1 + \frac{YZ}{2}\right)$$

$$B = Z \left(1 + \frac{YZ}{4}\right)$$

$$C = Y.$$

The phasor diagram of nominal-T is shown in Fig. 9.4.

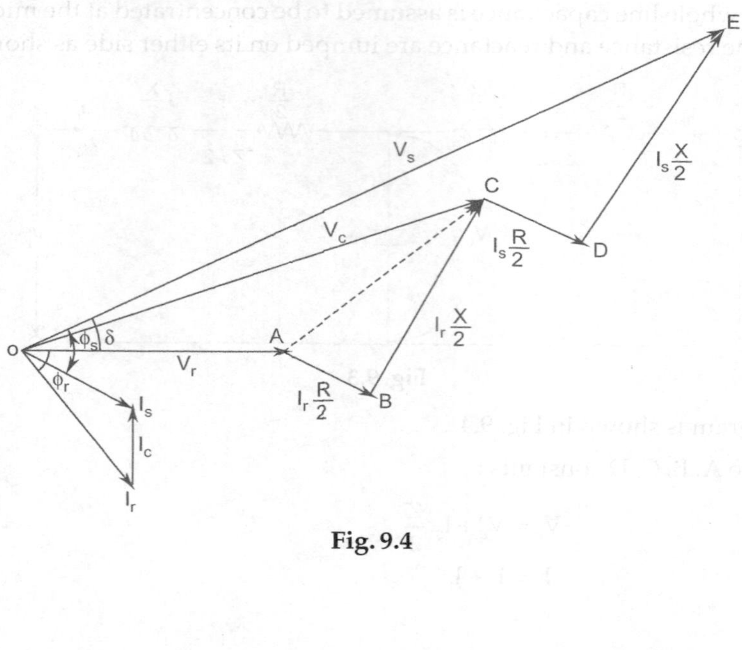

Fig. 9.4

Example 9.4. *A 3-phase 11 KV overhead line has series impedance of 5.2 + j6.4 Ω/ph and shunt susceptance of 950 μS/ph. The total 3-phase load is 1.7 MVA at 0.9 p.f. lag. Calculate : 1. Sending-end voltage ; 2. Sending-end current ; 3. Sending-end p.f. ; and 4. Power angle δ using nominal T-equivalent.*

Solution. Given :

$$V_r = \frac{11000}{\sqrt{3}} = 6351 \angle 0° \text{ V (Taking } V_r \text{ as a reference)}$$

$$I_r = \frac{1.7 \times 10^3}{\sqrt{3} \times 11 \times \angle -\cos^{-1} 0.9}$$

$$= 89.23 \angle -25.84° \text{ A}$$

$$= 80.33 - J\,38.84 \text{ A.}$$

From Fig. 9.3,

$$V_c = V_r + \frac{1}{2} I_r Z$$

Given :

$$Z = 5.2 + j6.4$$

or

$$\frac{1}{2} Z = \frac{1}{2}(5.2 + j6.4) = 4.123 \angle 50.9° \ \Omega/\text{ph.}$$

or

$$\frac{1}{2} I_r Z = 89.23 \angle -25.84° \times 4.123 \angle 50.9°$$

$$= 367.9 \angle 25.1°$$

$$= 333.2 + j156.1 \text{ V}$$

∴

$$V_c = 6351 + 333.2 + j156.1$$

or

$$V_c = 6684 + j156.1$$

$$= 6686 \angle 1.34° \text{ V}$$

From Fig. 9.3,

$$I_s = I_r + V_c Y$$

Given,

$$Y = 950 \ \mu\text{S/ph.}$$

or

$$Y = 950 \times 10^{-6} \angle 90° \text{ S/ph.}$$

or

$$V_c Y = 6686 \angle 1.34° \times 950 \times 10^{-6} \angle 90°$$

$$= 6.352 \angle 91.34°$$

$$= -0.149 + j6.35 \text{ A}$$

∴

$$I_s = 80.33 - j38.84 - 0.149 + j6.35$$

$$= 86.51 \angle -22.06° \text{ A}$$

From Fig. 9.3,

$$V_s = V_c + \frac{1}{2} I_s Z$$

$$\frac{1}{2} I_s Z = 86.51 \angle -22.06° \times 4.123 \angle 50.9°$$

$$= 312.4 + j172.1 \text{ V}$$

∴

$$V_s = 6684 + j156.1 + 312.4 + j172.1$$

$$= 6696 + j328.2$$

$$= 7004 \angle 2.81° \text{ V/ph.}$$

Sending end p.f. ϕ_s is angle between V_s and I_s.

i.e.,
$$\phi_s = 22.06° + 2.81° \text{ lag}$$
$$= 24.87° \text{ lag.}$$

i.e., sending-end power factor is 0.907 lag

$$\text{load-angle } \delta = \angle V_s - \angle V_r$$
$$= 2.81 - 0 = 2.81°$$

9.5.2 Nominal-π Method

In this method, capacitance of each conductor is divided into two halves, one half being lumped at the sending end and the other half at the receiving end as shown in Fig. 9.5.

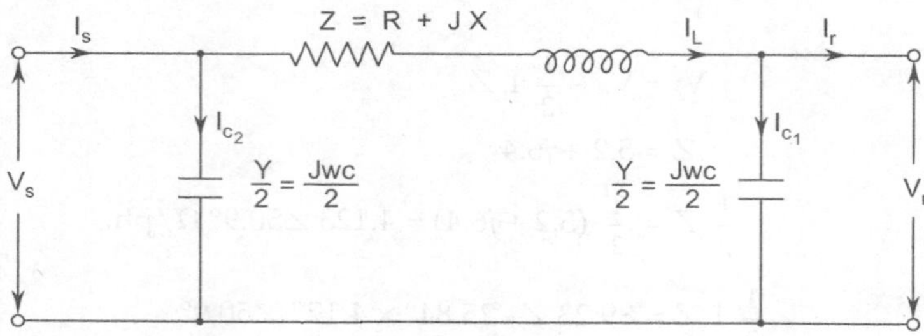

Fig. 9.5

The phasor diagram is shown in Fig. 9.6.

We will calculate A, B, C, D constants as :

From Fig. 9.6,

$$I_s = I_L + I_{C_2}$$

$$= I_r + \frac{1}{2} V_r Y + \frac{1}{2} V_s Y \qquad ...(8)$$

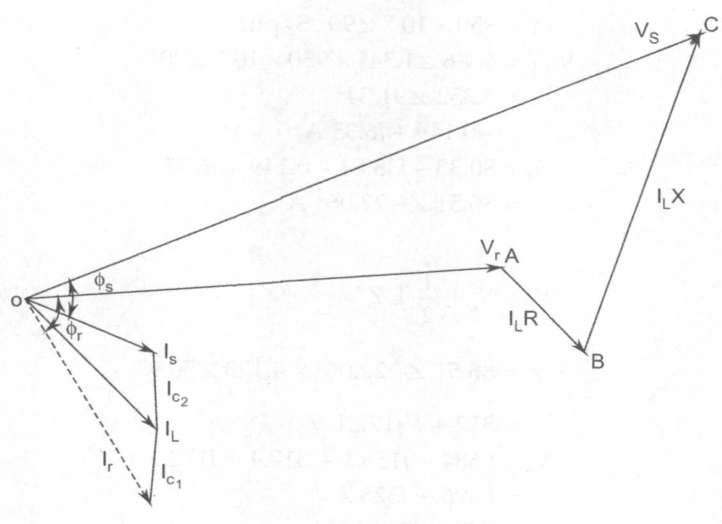

Fig. 9.6

$$V_s = V_r + I_L Z$$

$$= V_r + \left(I_r + \frac{1}{2} V_r Y\right) Z$$

$$= V_r \left(1 + \frac{YZ}{2}\right) + I_r Z \qquad ...(9)$$

Put value of V_s in equation (8) from equation (9)

$$I_s = I_r + \frac{1}{2} V_r Y + \frac{1}{2} Y \left[V_r \left(1 + \frac{YZ}{2}\right) + I_r Z\right]$$

$$= V_r Y \left(1 + \frac{1}{4} YZ\right) + I_r \left(1 + \frac{YZ}{2}\right)$$

Compare with the 2-port equations :

$$V_s = A V_r + B I_r$$

and

$$I_s = C V_r + D I_r$$

Hence,

$$A = D = \left(1 + \frac{YZ}{2}\right)$$

$$B = Z$$

$$C = Y \left(1 + \frac{YZ}{4}\right).$$

Example 9.5. *A 3-phase, 11 KV line has following constants :*

$R = 5.2$ *Ω/ph*

$X = 6.4$ *Ω/ph*

$Y = 950$ *μS/ph.*

Total 3-phase load is 170 MVA at 0.9 p.f. lag. Calculate :

1. *Sending-end voltage*
2. *Sending-end current.*

Solution.

$$V_r = \frac{11000}{\sqrt{3}} = 6351 \angle 0^\circ \text{ V} \qquad \text{[reference]}$$

$$I_r = \frac{1.7 \times 10^3}{\sqrt{3} \times 11 \times \angle -\cos^{-1} 0.9} = 89.23 \angle -25.84^\circ \text{ A}$$

$$= 80.33 - j38.84 \text{ A}$$

$$Z = 5.2 + j6.4 = 8.246 \angle 50.9^\circ \ \Omega/\text{ph}.$$

$$\frac{1}{2} Y = \frac{1}{2} (950 \times 10^{-6}) \angle 90^\circ$$

$$= 425 \times 10^{-6} \angle 90^\circ \text{ S/ph}.$$

$$\frac{1}{2} YV_r = 425 \times 10^{-6} \angle 90^\circ \times 6351 \angle 0^\circ$$

$$= 2.70 \angle 90^\circ \text{ A}.$$

From Fig. 9.5,

$$I_L = I_r + \frac{1}{2} Y V_r$$
$$= 80.33 - j38.84 + j2.70$$
$$= 80.33 - j36.14$$
$$= 88.09 \angle -24.22° \text{ A}$$
$$I_L Z = 88.09 \angle -24.22° \times 8.246 \angle 50.9°$$
$$= 726.4 \angle 26.68°$$
$$= 649.1 + j326.2 \text{ V}$$

From Fig. 9.5,

$$V_s = V_r + I_L Z$$
$$= 6351 + 649 + j326.2$$
$$= 7008 \angle 2.67° \text{ V/ph.}$$

From Fig. 9.5,

$$I_s = I_L + \frac{1}{2} Y V_s$$

$$\frac{1}{2} Y V_s = 425 \times 10^{-6} \angle 90° \times 7008 \angle 2.67°$$
$$= 2.978 \angle 92.67°$$
$$= -0.139 + j2.975 \text{ A}$$

$$\therefore \quad I_s = 80.33 - j36.14 - 0.139 + j2.975$$
$$= 86.78 \angle -22.5° \text{ A.}$$

Example 9.6. *A 100 km long 3-phase, 50 Hz transmission line has following line constants :*

Resistance/phase/km = 0.1 Ω

Reactance/phase/km = 0.5 Ω

Susceptance/phase/km = 10×10^{-6} ℧

If the line supplies load of 20 MW at 0.9 p.f. lagging at 66 KV at the receiving end. Calculate voltage regulation by nominal π-method.

Solution. Total resistance/phase, $R = 0.1 \times 100 = 10\ \Omega$

Total reactance/phase, $x = 0.5 \times 100 = 50\ \Omega$

Susceptance/phase, $Y = 10 \times 10^{-6} \times 100 = 10 \times 10^{-4}$ ℧

$$V_r = \frac{66 \times 10^3}{\sqrt{3}} = 38105 \angle 0° \text{ V} \qquad \text{(reference)}$$

$$I_r = \frac{20 \times 10^6}{\sqrt{3} \times 66 \times 10^3 \times 0.9} = 195 \text{ A}$$

$$\cos \phi_r = 0.9 \text{ and } \sin \phi_r = 0.435$$

$$\text{load current } I_r = I_r (\cos \phi_r - j \sin \phi_r)$$

$$= 195\,(0.9 - j0.435)$$
$$= 176 - j85 \text{ A.}$$

From Fig. 9.5,

$$I_{C_1} = \frac{1}{2}\,Y V_r$$
$$= \frac{1}{2} \times 10 \times 10^{-4}\ \angle 90° \times 38105\ \angle 0°$$
$$= 19\ \angle 90°$$
$$= j19 \text{ A.}$$

$$\therefore \quad I_L = I_{C_1} + I_r$$
$$= j19 + 176 - j85$$
$$= 176 - j66 \text{ A}$$

From Fig. 9.5,

$$V_s = V_r + I_L Z$$
$$= V_r + I_L (R + jX)$$
$$= 38105 + (176 - j66)\,(10 + j50)$$
$$= 43165 + j8140$$
$$= 43925\ \angle 10.65° \text{ V}$$

Voltage Regulation (%)
$$= \frac{V_s - V_r}{V_r} \times 100$$
$$= \frac{43925 - 38105}{38105} \times 100$$
$$= 15.27\%.$$

9.6 LONG TRANSMISSION LINES

So far electrically short transmission lines less than 160 km in length have been considered wherein the parameters are assumed to be lumped. In case the lines are more than 160 km long, in order to obtain fair degree of accuracy in the performance calculations of long lines, the line constants are considered as uniformly distributed throughout the length of the line. As a result of which the voltages and currents will vary from point to point on the line. Figure 9.7 shows the equivalent circuit of a 3-phase long transmission line on a phase-neutral basis. The whole line is divided into n sections, each section having line constants $\frac{1}{n}$ th of those for the whole line. Here, the resistance and inductive reactance are the series elements. The leakage susceptance (B) and leakage conductane (G) are shunt elements. Admittance = $\sqrt{G^2 + B^2}$. The leakage current through shunt admittance is maximum at the sending end of the line and decreases continuously as the receiving end of the circuit is approached at which point its value is zero.

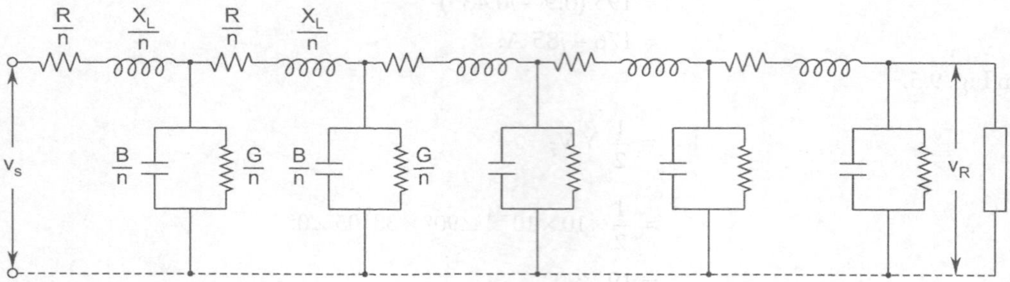

Fig. 9.7

9.7 ANALYSIS OF LONG TRANSMISSION LINE (RIGOROUS METHOD)

Figure 9.8 shows one-phase and neutral connection of a 3-phase line with impedance and shunt admittance of the line uniformly distributed.

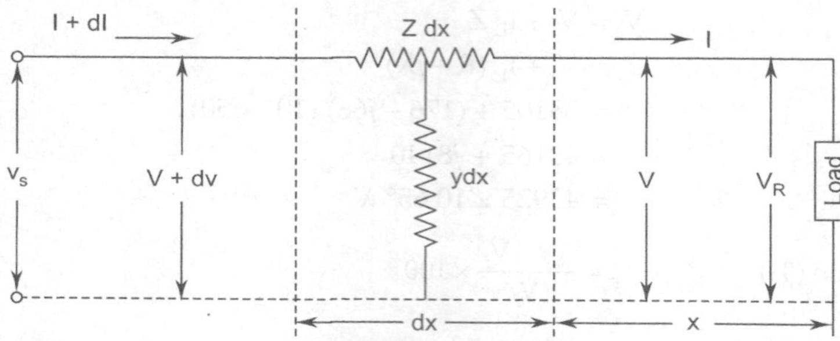

Fig. 9.8

Let us consider a very small element in the line and calculate the difference in voltage and difference in current between the ends of the element. Let x be the distance measured from the receiving end of the line to the small element of line and let the length of the element be dx.

In Fig. 9.8,

z = series impedance of the line per unit length

y = shunt admittance of the line per unit length

V = voltage at the end of element towards receiving end

V + dV = voltage at the end of element towards sending end

I + dI = current entering the element dx

I = current leaving the element dx.

For small element dx,

$$\text{Series impedance} = z\,dx$$

$$\text{Shunt admittance} = y\,dx.$$

The rise in voltage in the direction of increasing x is product of the current in the element flowing opposite to the direction of increasing x and the impedance of the element. Thus,

$$dV = I\,z\,dx$$

or $$\frac{dV}{dx} = I\,z \qquad ...(10)$$

Similarly, the current entering the element is I + dI whereas the current leaving the element is I. The difference in the current flows through shunt admittance of the element *i.e.*,

$$dI = V\,y\,dx$$

or $$\frac{dI}{dx} = V\,y \qquad ...(11)$$

Differentiate equations (10) and (11) w.r.t. x, we obtain

$$\frac{d^2V}{dx^2} = z\,\frac{dI}{dx} \qquad ...(12)$$

and $$\frac{d^2I}{dx^2} = y\,\frac{dV}{dx} \qquad ...(13)$$

Substitute value of $\frac{dI}{dx}$ from equation (11) in equation (12) and value of $\frac{dV}{dx}$ from equation (10) in equation (13), we get

$$\frac{d^2V}{dx^2} = yz\,V \qquad ...(14)$$

and $$\frac{d^2I}{dx^2} = yz\,I \qquad ...(15)$$

Equations (14) and (15) are linear differential equations of second order whose general solution can be written as :

$$V = A\,e^{\sqrt{yz}\,x} + B\,e^{-\sqrt{yz}\,x}$$

or $$V = A\,e^{\gamma x} + B\,e^{-\gamma x} \qquad ...(16)$$

where, $$\gamma = \sqrt{yz} = \text{Propagation constant} = \alpha + j\beta$$

$$I = \frac{V}{Z_c} = \frac{1}{Z_c}\left[A\,e^{\gamma x} - B\,e^{-\gamma x}\right] \qquad ...(17)$$

where, $$Z_c = \sqrt{\frac{z}{y}} = \text{Characteristic impedance} \qquad \left[\text{From equation (10)}\ \frac{dv}{dx} = IZ\right]$$

The values of A and B can be found by applying end conditions as under :

At $x = 0$,

$$V = V_r$$

and $$I = I_r$$

Putting these values in equations (16) and (17)

$$V_r = A + B$$

and $$I_r = \frac{1}{Z_c}(A - B)$$

Thus, $$A = \frac{V_r + I_r Z_c}{2} \text{ and } B = \frac{V_r - I_r Z_c}{2}.$$

Substitute the values of A and B in equations (16) and (17)

$$V = \frac{V_r + I_r Z_c}{2} e^{\gamma x} + \frac{V_r - I_r Z_c}{2} e^{-\gamma x} \quad ...(18)$$

and $$I = \frac{1}{Z_c}\left[\frac{V_r + I_r Z_c}{2} e^{\gamma x} - \frac{V_r - I_r Z_c}{2} e^{-\gamma x}\right] \quad ...(19)$$

Putting $\gamma = \alpha + j\beta$, the real part is known as attenuation constant and the quadrature component, β is phase constant.

$$V = \frac{V_r + I_r Z_c}{2} e^{\alpha x} \cdot e^{j\beta x} + \frac{V_r - I_r Z_c}{2} e^{-\alpha x} \cdot e^{-j\beta x} \quad ...(20)$$

First term in above expression is called the incident voltage wave as its value increases as x is increased. The second part is called the reflected voltage. At any point along the line, voltage is the sum of these two components *i.e.*, sums of incident and reflected voltages.

By rearranging equations (18) and (19),

$$V = V_r \frac{e^{\gamma x} + e^{-\gamma x}}{2} + I_r Z_c \frac{e^{\gamma x} - e^{-\gamma x}}{2}$$

$$= V_r \cosh \gamma x + I_r Z_c \sinh \gamma x$$

and $$I = \frac{1}{Z_c}\left[V_r \frac{e^{\gamma x} - e^{-\gamma x}}{2} + I_r Z_c \frac{e^{\gamma x} + e^{-\gamma x}}{2}\right]$$

$$= \frac{V_r}{Z_c} \sinh \gamma x + I_r \cosh \gamma x$$

Letting $x = l$ to obtain the voltage and current at the sending end.

$$V_s = V_r \cosh \gamma l + I_r Z_c \sinh \gamma l \quad ...(21)$$

and $$I_s = V_r \frac{\sinh \gamma l}{Z_c} + I_r \cosh \gamma l \quad ...(22)$$

Compare with the 2-port equations :

$$V_s = A V_r + B I_r$$

$$I_s = C V_r + D I_r$$

Hence, $$A = D = \cosh \gamma l$$

$$B = Z_c \sinh \gamma l$$

$$C = \frac{\sinh \gamma l}{Z_c}$$

$$AD - BC = \cosh^2 \gamma l - Z_c \sinh \gamma l \cdot \frac{\sinh \gamma l}{Z_c}$$

$$= 1.$$

9.8 SURGE IMPEDANCE OF THE LINE

In power system work, characteristic impedance is sometimes called surge impedance. The term surge impedance, however, is usually reserved for the special case of a lossless line.

$$Z_c = \sqrt{\frac{z}{y}} = \text{Characteristic Impedance}$$

$$= \sqrt{\frac{r + j\omega L}{g + j\omega C}}.$$

For a lossless line, $r = 0$ and $g = 0$

$$Z_c = \sqrt{\frac{L}{C}}$$

a pure resistance and this is known as surge impedance of the line. When dealing with high frequencies or with surges due to lightning, losses are often neglected and surge impedance becomes important Surge Impedance Loading (SIL) of a line is the power delivered by a line to a purely resistive load equal to its surge impedance. When so loaded, the line supplies a current of

$$|I_L| = \frac{|V_L|}{\sqrt{3} \times \sqrt{\frac{L}{C}}} \text{ A}$$

Where $|V_L|$ is the line-to-line voltage at the load. Since the load is pure resistance.

$$\text{SIL} = \sqrt{3}\,|V_L| \frac{|V_L|}{\sqrt{3} \times \sqrt{\frac{L}{C}}} \text{ W}$$

or with $|V_L|$ in kilovolts,

$$\text{SIL} = \frac{|V_L|^2}{\sqrt{\frac{L}{C}}} \text{ MW.}$$

Power system engineers sometimes find it convenient to express the power transmitted by a line in terms of per unit of SIL, that is, the ratio of the power transmitted to the surge-impedance loading. For instance, the permissible loading of a transmission line may be expressed as a fraction of its SIL and SIL provides a comparison of load-carrying capabilities of lines.

The approximate value of surge impedance for overhead lines is 400 ohms and that for cables is about 40 ohms. The phase angle of Z_c for transmission lines is usually between 0° and – 15°. The lower value of surge impedance in case of cables is due to the relatively large capacitance and low inductance of the cables.

If a line is terminated in its characteristic impedance Z_c, the receiving-end voltage V_r is equal to $I_r Z_c$ and there is no reflected wave of either voltage or current. A line terminated in its characteristic impedance is called a flat line or an infinite line. Usually power lines are not terminated in their characteristic impedance but communication lines are frequently so terminated in order to eliminate the reflected wave.

9.9 PROPAGATION CONSTANT (γ)

$$\gamma = \sqrt{zy} = \alpha + j\beta$$

where the real part α is known as attenuation constant and the quadrature component β is known as phase constant and is measured in radians per unit length.

z = Series impedance of the line per unit length, Ω/m

y = Shunt admittance of the line per unit length, mho/m.

A wavelength λ is the distance along a line between two points of a wave which differ in phase by 360° or 2π rad. If β is the phase shift in radians per meter, the wavelength in meter is

$$\lambda = \frac{2\pi}{\beta}.$$

The velocity of propagation of a wave is the product of the wavelength and frequency (Hz) or

$$\text{Velocity} = f\lambda.$$

Example 9.7. *AC line is 150 km long is rated voltage at 138 KV at 50 Hz. The line has following parameters - series resistance r = 0.1858 Ω/km. Series inductance per phase L = 3.12 mH/km, shunt capacitance per phase C = 0.0144 μF/km. A load of 50 MVA at p.f. 0.85 lag at voltage 138 KV is prevailing at receiving end. Calculate A, B, C, D constants of the line by rigorous solution.*

Solution.

$$Z = 0.1858 + j2\pi \times 50 \times 3.12 \times 10^{-3}$$
$$= 0.1858 + j0.9802$$
$$= 0.9977 \angle 79.27° \ \Omega/\text{km}.$$
$$y = j2\pi \times 50 \times 0.0144 \times 10^{-6}$$
$$= 4.5239 \times 10^{-6} \angle 90° \ \text{S/km}.$$

Propagation constant (γ) of the line is

$$\gamma = \sqrt{yz}$$
$$= [(4.5239 \times 10^{-6} \angle 90°)(0.9977 \angle 79.27°)]^{1/2}$$
$$= [4.5135 \times 10^{-6}]^{1/2} \angle \tfrac{1}{2}(90° + 79.27°)$$
$$= 0.002144 \angle 84.63°$$
$$\gamma = \alpha + j\beta$$
$$= 0.002144 \cos 84.63° + j0.002144 \sin 84.63°$$

Characteristic impedance (Z_c) of the line is

$$Z_c = \sqrt{\frac{z}{y}} = \left[\frac{0.9977 \angle 72.27°}{4.5239 \times 10^{-6} \angle 90°}\right]^{\frac{1}{2}}$$
$$= \left(\frac{0.9977 \times 10^{6}}{4.5239}\right)^{\frac{1}{2}} \angle \tfrac{1}{2}(79.27° - 90°)$$
$$= 469.62 \angle -5.37° \ \Omega$$
$$\gamma = \alpha + j\beta = 0.002144 \angle 84.63°$$
$$= 0.002144(\cos 84.63° + j \sin 84.63°)$$

$$= 0.00206 + j0.0021345$$

$$\gamma l = \gamma \times 150 = 0.301 + j0.3202$$

$$= \alpha l + j\,\beta l$$

$$A = \cosh \gamma l = \cosh (\alpha + j\beta)\, l$$

$$= \frac{1}{2}\left(e^{\alpha l}\, e^{j\beta l} + e^{-\alpha l}\, e^{-j\beta l}\right)$$

or

$$A = \frac{1}{2}\left(e^{0.301}\, e^{j0.3202} + e^{-0.301}\, e^{-j0.3202}\right)$$

$$= \frac{1}{2}\left(e^{0.301}\ \angle 18.35° + e^{-0.301}\ \angle -18.35°\right)$$

$$= 0.9496 + j0.0095$$

$$= 0.9497\ \angle 0.57°.$$

$$B = Z_c \sinh \gamma l = Z_c \sinh (a + j\beta)\, l$$

$$= Z_c\left[\frac{1}{2}\left(e^{\alpha l}\, e^{J\beta l} - e^{-\alpha l - J\beta l}\right)\right]$$

$$= \frac{1}{2}\left[469.69\ \angle -5.37°\right]\left[e^{0.301}\, e^{j0.3202} - e^{-0.301}\, e^{-j0.3202}\right]$$

$$= 234.81\ \angle -5.37°\ [1.0306\ \angle 18.35° - 0.9703\ \angle -18.35°]$$

$$= (234.81\ \angle -5.37°)\ [0.0572 - j0.6300]$$

$$= (234.81\ \angle -5.37°)\ (0.6326\ \angle 84.81°)$$

$$= 148.54\ \angle 79.44°\ \Omega$$

$$C = Y_c \sinh \gamma l$$

$$= \frac{1}{Z_c} \sinh \gamma l$$

$$= \frac{1}{469.62\ \angle -5.37°} \times 0.3163\ \angle 84.81°$$

$$= 0.00067\ \angle 90.18°\ S$$

$$D = A = \cosh \gamma l = 0.9497\ \angle 0.57°.$$

Example 9.8. *A 3ϕ transmission line 200 km long has the following constants :*

Resistance / phase / km = 0.16 Ω

Reactance / phase / km = 0.25 Ω

Shunt admittance / phase / km = 1.5×10^{-6} ℧.

Calculate by rigorous method the sending end voltage and current when the line is delivering a load of 20 MW at 0.8 p.f. lagging. The receiving end voltage is kept constant at 110 KV.

Solution. As we know that

$$V_s = V_r \cosh \gamma l + I_r Z_c \sinh \gamma l$$

and

$$I_s = V_r \frac{\sinh \gamma l}{Z_c} + I_r \cosh \gamma l$$

Putting the values of Z_c and γ in above equations :

$$V_s = V_r \cosh\left(l\sqrt{yz}\right) + \sqrt{\frac{z}{y}}\, I_r \sinh\left(l\sqrt{yz}\right)$$

$$I_s = \sqrt{\frac{y}{z}}\, V_r \sinh\left(l\sqrt{yz}\right) + I_r \cosh\left(l\sqrt{yz}\right)$$

Now, $$l\sqrt{yz} = \sqrt{ly \cdot lz} = \sqrt{YZ}$$

and, $$\sqrt{\frac{y}{z}} = \sqrt{\frac{yl}{yz}} = \sqrt{\frac{Y}{Z}}$$

where, Y = total shunt admittance of the line

Z = total series impedance of the line

Therefore, expressions for V_s and I_s become :

$$V_s = V_r \cosh\sqrt{YZ} + I_r\sqrt{\frac{Z}{Y}} \sinh\sqrt{YZ}$$

$$I_s = I_r\sqrt{\frac{Y}{Z}} \sinh\sqrt{YZ} + I_r \cosh\sqrt{YZ}$$

where, $$\cosh\sqrt{YZ} = \left(1 + \frac{ZY}{2} + \frac{Z^2Y^2}{24} + \cdots\cdots\right)$$

and $$\sinh\sqrt{YZ} = \left(\sqrt{YZ} + \frac{(YZ)^{\frac{3}{2}}}{6} + \cdots\cdots\right)$$

Given: Total resistance/phase, $R = 0.16 \times 200 = 32\ \Omega$

Total reactance/phase, $\chi_L = 0.25 \times 200 = 50\ \Omega$

Total shunt admittance/phase, $Y = j1.5 \times 10^{-6} \times 200$

$= 0.0003\ \angle 90°$

Series impedance/phase, $Z = R + j\chi_L$

$= 32 + j50° = 59.4\ \angle 58°$

$\therefore$ $$V_s = V_r \cosh\sqrt{YZ} + I_r\sqrt{\frac{Z}{Y}} \sinh\sqrt{ZY}$$

Now, $$\sqrt{ZY} = \sqrt{59.4\angle 58° \times 0.0003\angle 90°} = 0.133\ \angle 74°$$

$$ZY = 0.0178\ \angle 148°$$

$$Z^2Y^2 = 0.00032\ \angle 296°$$

$$\sqrt{\frac{Z}{Y}} = \sqrt{\frac{59.4\ \angle 58°}{0.0003\ \angle 90°}} = 445\ \angle -16°$$

$$\sqrt{\frac{Y}{Z}} = 0.00224\ \angle 16°$$

$$\cosh\sqrt{YZ} = 1+\frac{ZY}{2}+\frac{Z^2Y^2}{24} \text{ approx.}$$

$$= 1+\frac{0.0178}{2}\angle 148° + \frac{0.00032}{24}\angle 296°$$

$$= 1 + 0.0089 \angle 148° + 0.0000133 \angle 296°$$

$$= 1 + 0.0089 (-0.848 + j0.529) + 0.0000133 (0.438 - j0.9)$$

$$= 0.992 + j0.00469$$

$$= 0.992 \angle 0.26°$$

$$\sinh\sqrt{YZ} = \sqrt{YZ}+\frac{(YZ)^{\frac{3}{2}}}{6} \text{ approx.}$$

$$= 0.133 \angle 74° + \frac{0.0024}{6}\angle 222°$$

$$= 0.133 (0.275 + j0.961) + 0.0004 (-0.743 - j0.67)$$

$$= 0.0362 + j0.1275$$

$$= 0.1324 \angle 74.6°$$

Given : $$V_r = 110\times\frac{10^3}{\sqrt{3}} = 63508 \text{ V}$$

$$I_r = \frac{20\times 10^6}{\sqrt{3}\times 110\times 10^3\times 0.8} = 131 \text{ A}$$

$\therefore$ $$V_s = 63508\times 0.992 \angle 0.26° + 131\times 445 \angle -16° \times 0.1325 \angle 74.6°$$

$$= 63000 \angle 0.26° + 7724 \angle 58.6°$$

$$= 63000 (0.999 + j0.0045) + 7724 (0.5284 + j0.8489)$$

$$= 67{,}366 \angle 5.50°$$

Sending end voltage (line-to-line)

$$= 67366\times\sqrt{3} = 116.67 \text{ KV.}$$

Now, $$I_s = V_r\sqrt{\frac{Y}{Z}}\sinh\sqrt{YZ} + I_r\cosh\sqrt{YZ}$$

or $$I_s = 63508\times 0.00224 \angle 16° \times 0.1325 \angle 74.6° + 131\times 0.992 \angle 0.26°$$

$$= 18.85 \angle 90.6° + 130 \angle 0.26°$$

$$= 129.83 + j18.24$$

$$= 131.1 \angle 8°$$

Sending-end current $= 131.1$ A.

9.10 THE EQUIVALENT CIRCUIT OF A LONG LINE

The nominal-T and nominal-π circuits do not represent a transmission line exactly because they do not account for the parameters of the line being uniformly distributed. The discrepancy between the nominal T and π and the actual line becomes larger as the length of line increases. It is possible, however, to find the equivalent circuit of a long transmission line and to represent the line accurately, in so far as measurements at the ends of the line are concerned by a network of lumped parameters. Let us assume that a π circuit similar to that of Fig. 9.5 is the equivalent circuit of a long line but let us call the series arm of our equivalent-π circuit Z′ and the shunt arms $\frac{Y'}{2}$ to distinguish them from the arms of the nominal-π circuit. Equation (9) gives the sending-end voltage of a symmetrical-π circuit in terms of its series and shunt arms and the voltage and current at the receiving end. By substituting Z′ and $\frac{Y'}{2}$ for Z and $\frac{Y}{2}$ in equation (9), we obtain the sending-end voltage of our equivalent circuit in terms of its series and shunt arms and the voltage and current at the receiving end.

$$V_s = V_r\left(1+\frac{Y'Z'}{2}\right)+I_r Z' \qquad ...(23)$$

For our circuit to be equivalent to the long transmission line, the coefficients of V_r and I_r in equation (23) must be identical respectively to the coefficients of V_r and I_r in equation (21). Equating the coefficients of I_r in the two equations yields.

$$Z' = Z_c \sinh \gamma l$$

$$Z' = \sqrt{\frac{z}{y}}\sinh \gamma l = zl\frac{\sinh \gamma l}{\sqrt{zy}\, l}$$

$$Z' = Z\frac{\sinh \gamma l}{\gamma l}$$

where $Z = zl$ = Total series impedance of the line. The term $\frac{(\sinh \gamma l)}{\gamma l}$ is the factor by which the series impedance of the nominal π must be multiplied to convert the nominal-π to the equivalent-π. For small values of γl, $\sinh \gamma l$ and γl are almost identical and this fact shows that the nominal-π represents the medium-length transmission line quite accurately, in so far as the series arm is concerned.

To investigate the shunt arms of the equivalent-π circuit, we equate the coefficients of V_r in equations (21) and (23).

$$\frac{Z'Y'}{2}+1 = \cosh \gamma l$$

Substituting, $$Z' = Z_c \sinh \gamma l$$

$$\frac{Y' Z_c \sinh \gamma l}{2}+1 = \cosh \gamma l$$

and $$\frac{Y'}{2} = \frac{1}{Z_c}\frac{\cosh \gamma l - 1}{\sinh \gamma l}$$

or $$\frac{Y'}{2} = \frac{1}{Z_c}\tanh\frac{\gamma l}{2} = \sqrt{\frac{z}{y}}\tanh\frac{\gamma l}{2}$$

$$= \frac{\gamma l}{2}\left(\frac{\tanh \gamma l/2}{\gamma l/2}\right)$$

or

$$\frac{Y'}{2} = \frac{Y}{2}\left(\frac{\tanh \gamma l/2}{\gamma l/2}\right).$$

Thus, $\left(\frac{\tanh \gamma l/2}{\gamma l/2}\right)$ is the factor by which the shunt admittance arm of the nominal-π must be multiplied to obtain the shunt parameter $\left(\frac{Y'}{2}\right)$ of the equivalent-π. The equivalent-π circuit is shown in Fig. 9.9

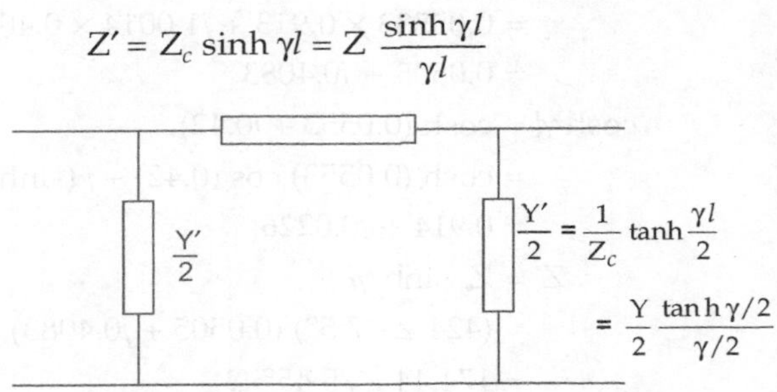

Fig. 9.9

Similarly, equivalent-T network parameters of a transmission line are obtained on similar lines and shown in Fig. 9.10.

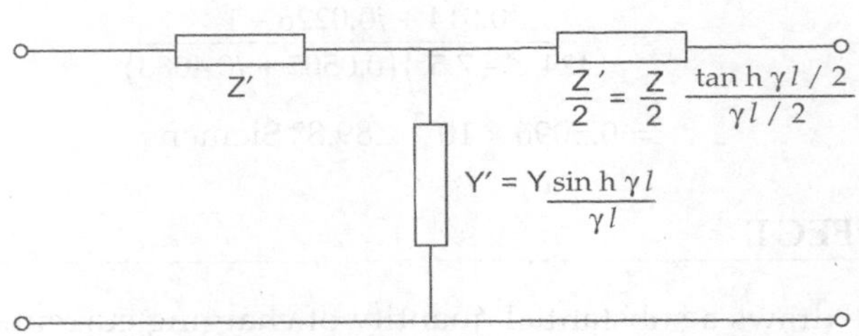

Fig. 9.10

Example 9.9. *A 350 km long line as total series impedance of 180 ∠75° ohms per phase and a total shunt admittance to neutral per phase of 1×10^{-3} ∠90° Siemens. Find the constants of the equivalent-π circuit.*

Solution.

$$Z = \frac{180}{350}\angle 75° = \frac{18}{35}\angle 75° \text{ ohms per km}$$

$$y = \frac{1\times 10^{-3}}{350}\angle 90° = \frac{10^{-4}}{35}\angle 90° \text{ siemens per km.}$$

$$\gamma = \sqrt{yz} = \sqrt{\frac{18}{35} \times \frac{10^{-4}}{35}} \angle \frac{(90° + 75°)}{2}$$

$$= \frac{4.24 \times 10^{-2}}{35} \angle 82.5°$$

and $$\gamma l = \frac{4.24 \times 10^{-2}}{35} \angle 82.5° \times 350 = 0.424 \angle 82.5° = 0.0553 + j0.42$$

$$Z_c = \sqrt{\frac{z}{y}} = \left[\left(\frac{18}{35} \angle 75°\right) \Big/ \left(\frac{10^{-4}}{35} \angle 90°\right)\right]^{\frac{1}{2}}$$

$$= 424 \angle -7.5°$$

$$\sinh \gamma l = \sinh (0.0553 + j0.42)$$

$$= \sinh (0.0553) \cos (0.42) + j \cosh (0.0553) \sin (0.42)$$

$$= 0.05533 \times 0.913 + j1.0014 \times 0.40776$$

$$= 0.0505 + j0.4083$$

$$\cosh \gamma l = \cosh (0.0553 + j0.42)$$

$$= \cosh (0.0553) \cos (0.42) + j (\sinh 0.0553) \sin (0.42)$$

$$= 0.914 + j0.0226$$

$$Z' = Z_c \sinh \gamma l$$

$$= (424 \angle -7.5°)(0.0505 + j0.4083)$$

$$= 174.44 \angle 75.45° \ \Omega$$

$$\frac{Y'}{2} = \frac{1}{Z_c}\left[\frac{\cosh \gamma l - 1}{\sinh \gamma l}\right]$$

$$= \frac{0.914 + j0.0226 - 1}{(424 \angle -7.5°)(0.0505 + j0.4083)}$$

$$= 0.5096 \times 10^{-3} \angle 89.8° \text{ Siemens.}$$

9.11 FERRANTI EFFECT

A long transmission line draws a substantial quantity of charging current. If such a line is open circuited or very lightly loaded at the receiving end, the voltage at the receiving end may become higher than the voltage at the sending end. This is known as Ferranti effect. The effect becomes more pronounced as the line length increases. As the capacitance (and charging current) is negligible in short line but significant in medium length lines and appreciable in long lines, therefore, this phenomenon occurs only in medium and long lines.

A simple explanation of the ferranti effect on an approximate basis can be advanced by lumping the inductance and capacitance parameters of the line. As shown in Fig. 9.11, the capacitance is lumped at the receiving-end of the line.

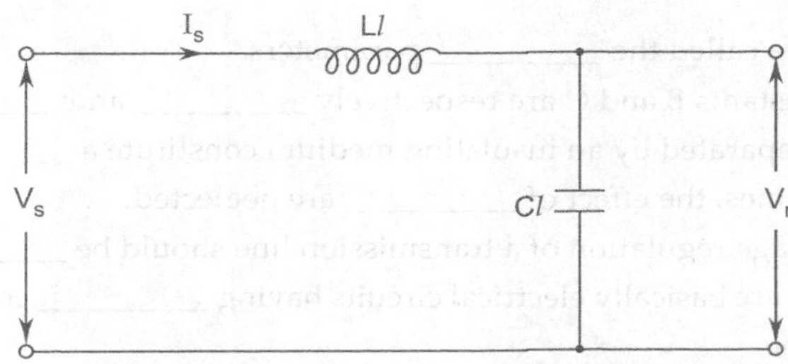

Fig. 9.11

Here, $$I_s = \frac{V_s}{\left(\frac{1}{j\omega Cl} + j\omega Ll\right)}$$

Since C is small compared to L, ωLl can be neglected in comparison to $\frac{1}{\omega Cl}$. Thus,

$$I_s = j\,V_s\,\omega\,Cl$$

Now,
$$\begin{aligned} V_r &= V_s - I_s\,(j\omega\,Ll) \\ &= V_s + V_s\,\omega^2\,CLl^2 \\ &= V_s\,(1 + \omega^2\,CLl^2) \end{aligned}$$

Magnitude of voltage rise
$$\begin{aligned} &= |\,V_s\,|\,\omega^2\,CLl^2 \\ &= |\,V_s\,|\,\frac{\omega^2 l^2}{v^2}. \end{aligned}$$

Where $v = \frac{1}{\sqrt{LC}}$ is constant for all overhead lines and is the velocity of propagation of electro-magnetic waves on the line, which is nearly equal to the velocity of light *i.e.*, 3×10^8 m/sec.

The phasor diagram is shown in Fig. 9.12 for this operating condition.

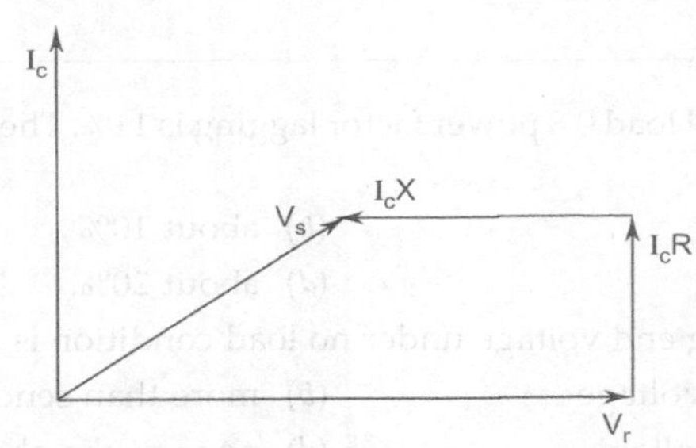

Fig. 9.12

SELF TEST

Fill in the blanks by inserting appropriate words/figures :

1. In a transmission line, generalised constants __________ and __________ are equal.
2. The line constants of a transmission line are __________.

3. The line parameters are called the __________ parameters.
4. The dimensions of constants B and C are respectively __________ and __________.
5. Any two conductors separated by an insulating medium constitute a __________.
6. In short transmission lines, the effect of __________ are neglected.
7. It is desirable that voltage regulation of a transmission line should be __________.
8. The transmission lines are basically electrical circuits having __________ constants.
9. In any transmission line,

$$AD - BC = \_\_\_\_\_\_\_\_\_\_$$

10. The generalised constants A and D of the transmission line have __________.
11. An electric transmission line has four parameters viz., resistance, inductance, capacitance and __________.
12. When the regulation is positive, then receiving end voltage (V_R) is __________ than sending end voltage (V_S).
13. In medium transmission lines, effect of __________ are taken into account.
14. __________ of transmission lines, is the most important cause of power loss in the line.
15. In the analysis of 3-phase transmission line, only __________ is considered.

Answers to Self Test

1. A and D
2. uniformly distributed
3. distributed
4. ohm, mho
5. capacitor
6. capacitance
7. low
8. distributed
9. 1
10. no dimension
11. shunt conductance
12. less
13. capacitance
14. resistance
15. one phase.

OBJECTIVE QUESTIONS

1. The regulation of a line at full load 0.8 power factor lagging is 11%. The regulation at full load 0.8 power factor leading can be :
   (*a*) about 3% (*b*) about 10%
   (*c*) about 15% (*d*) about 20%.
2. For a long line, the receiving end voltage under no load condition is :
   (*a*) less than sending end voltage (*b*) more than sending end voltage
   (*c*) equal to sending end voltage (*d*) none of the above.
3. Charging current in transmission line:
   (*a*) unpredictable (*b*) will not affect the line losses
   (*c*) reduces the line losses (*d*) increases the line losses.
4. For an A.C transmission line of length not exceeding 80 km, it is usual to lump the line capacitance at :
   (*a*) the mid point (*b*) the sending end
   (*c*) the receiving end (*d*) any convenient point.

5. Ferranti effect on long overhead lines is experienced when it is :

(*a*) on full load at 0.8 p.f. lag (*b*) on full load at unity p.f.

(*c*) lightly loaded (*d*) in all these cases.

6. Ferranti effect in transmission lines is due to the presence of :

(*a*) line inductance (*b*) line capacitance

(*c*) line inductance and capacitance (*d*) line resistance and capacitance.

7. A 160 km, 110 KV transmission line falls under the category of :

(*a*) long transmission line (*b*) short transmission line

(*c*) medium transmission line (*d*) ultra high voltage line.

8. 120 km long transmission line is considered as a:

(*a*) short line (*b*) medium line

(*c*) long line (*d*) either (*a*) or (*b*).

9. For 11 KV transmission line the capacitance per km will be about:

(*a*) 0.1 μF (*b*) 0.01 μF

(*c*) 0.1 F (*d*) 0.01 F.

10. For 11 KV transmission line the inductance per km will be about :

(*a*) 0.1 H (*b*) 1 H

(*c*) 0.1 mH (*d*) 1 mH.

11. A transmission line of 210 km length has certain values of parameters A, B, C, D. If the length is made 100 km thus the parameter :

(*a*) A increases B decreases (*c*) A decreases B increases

(*c*) A and B both increase (*d*) A decreases B also decreases.

12. In the above problem, the parameters :

(*a*) C increases D decreases (*b*) C decreases D increases

(*c*) C and D both increase (*d*) C and D both decrease.

13. For a transmission line :

$$V_s = A\,V_r + B\,I_r$$

$$I_s = C\,V_r + D\,I_r$$

Then I_r equals

(*a*) $A\,V_s + D\,I_s$ (*b*) $D\,V_s + A\,I_s$

(*c*) $-C\,V_s + A\,I_s$ (*d*) $D\,V_s - B\,I_s$.

14. In the above problem (Q. No. 13), V_r equals :

(*a*) $C\,V_s + A\,I_s$ (*b*) $D\,V_s + B\,I_s$

(*c*) $D\,V_s - C\,I_s$ (*d*) $D\,V_s + B\,I_s$.

15. A 25 km 33 KV transmission line is considered to be :

(*a*) long transmission line (*b*) short transmission line

(*c*) medium transmission line (*d*) high power line.

16. In a transmission line of negligible resistance, the surge impedance will be :

(*a*) $\sqrt{\dfrac{C}{L}}$ (*b*) $\sqrt{\dfrac{L}{C}}$

(*c*) $\sqrt{LC}$ (*d*) $\dfrac{1}{\sqrt{LC}}$.

17. The square root of the ratio of line impedance and shunt admittance is known as the :

(*a*) line surge impedance
(*b*) line regulation
(*c*) line susceptance
(*d*) line conductance.

18. A transmission line of 80 km length is operating at 400 Hz, it can be classified as :

(*a*) short length line
(*b*) medium length line
(*c*) long length line
(*d*) all of above.

19. As compared to the sending end voltage (V_S), the receiving end voltage (V_R) of a short line under no load condition is :

(*a*) higher
(*b*) lower
(*c*) the same.

20. As compared to regulation at unity power factor, the regulation at 0.8 leading power factor is :

(*a*) higher
(*b*) lower
(*c*) the same.

21. The critical value of surge impedance of a long transmission line is :

(*a*) 50 Ω
(*b*) 250 Ω
(*c*) 400 Ω
(*d*) 500 Ω.

22. The propagation constant of a transmission line is given as :

(*a*) $j\omega \sqrt{LC}$
(*b*) $j \sqrt{LC}$
(*c*) $j \sqrt{\frac{C}{L}}$
(*d*) $j \sqrt{\frac{L}{C}}$.

23. If, in as short transmission line, resistance and inductive reactance are found to be equal and regulation appears to be zero, then the load will :

(*a*) have zero power factor
(*b*) have unity power factor
(*c*) be 0.707 lagging
(*d*) be 0.707 leading.

24. Transmission efficiency of a transmission line increases with the :

(*a*) increase in power factor but decrease in voltage.
(*b*) decrease in power factor but increase in voltage.
(*c*) increase in power factor and voltage both.
(*d*) decrease in power factor and voltage.

25. If L, C and Y are the inductance, capacitance and shunt admittance of a line per unit length, then for length *l*:

(*a*) the inductance is $\frac{L}{l}$
(*b*) the shunt admittance is Y *l*
(*c*) the capacitance is $\frac{C}{l}$
(*d*) the shunt admittance is $\frac{Y}{l}$.

26. For a short line with $\frac{R}{X}$ ratio 1.0, the regulation will be zero when the load power factor is :

(*a*) 0.5
(*b*) 0.707 leading
(*c*) 0 leading
(*d*) unity.

27. Under no load conditions, the current in a transmission line is because of :

(*a*) corona effect
(*b*) proximity effect
(*c*) capacitance effect
(*d*) back flow from earth.

28. Percentage regulation of a transmission line is given by the expression

(*a*) $\frac{V_R - V_S}{V_S} \times 100$ (*b*) $\frac{V_R - V_S}{V_R} \times 100$ (*c*) $\frac{V_S - V_R}{V_S} \times 100$ (*d*) $\frac{V_S - V_R}{V_R} \times 100$.

29. For a transmission line with resistance R, reactance X, and negligible capacitance, the transmission constant A is :

(*a*) 1 (*b*) 0 (*c*) R + X (*d*) R + *j*X.

30. The ABCD constants of a 3-phase transposed transmission line with linear and passive elements :

(*a*) are always equal (*b*) never equal
(*c*) A and D are equal (*d*) B and C are equal.

31. For a distortionless transmission line (where G is the shunt conductance between two conductors) :

(*a*) RG = LC (*b*) RL = GC (*c*) RC = GL (*d*) RLGC = 0.

32. The phenomenon of rise in receiving end voltage of the open circuited or lightly load line is called the :

(*a*) skin effect (*b*) proximity effect
(*c*) seebeck effect (*d*) Ferranti effect.

33. When the load on a transmission line is equal to the surge impedance loading :

(*a*) the sending end voltage is less than the receiving end voltage
(*b*) the receiving end voltage is less than the sending end voltage
(*c*) the receiving end voltage is equal to sending end voltage
(*d*) none of these.

34. In a long transmission line under no-load condition :

(*a*) the receiving end voltage is less than the sending end voltage
(*b*) the sending end voltage is less than the receiving end voltage
(*c*) the sending end voltage is equal to the receiving end voltage
(*d*) none of these.

35. In terms of constants A, B, C and D for short transmission lines, which of the following relation (*s*) is/are valid ?

(*a*) B = Z (*b*) C = 0
(*c*) A = D = 1 (*d*) all of the above.

Answers to Objective Questions

1. (*a*)	2. (*b*)	3. (*d*)	4. (*c*)	5. (*c*)
6. (*c*)	7. (*a*)	8. (*b*)	9. (*d*)	10. (*d*)
11. (*a*)	12. (*b*)	13. (*c*)	14. (*d*)	15. (*b*)
16. (*b*)	17. (*a*)	18. (*c*)	19. (*c*)	20. (*b*)
21. (*c*)	22. (*a*)	23. (*d*)	24. (*c*)	25. (*b*)
26. (*b*)	27. (*c*)	28. (*d*)	29. (*a*)	30. (*c*)
31. (*c*)	32. (*d*)	33. (*c*)	34. (*b*)	35. (*d*).

EXERCISE

1. Discuss the terms voltage regulation and transmission efficiency as applied to transmission line.
2. What do you understand by long transmission line? How capacitance effects are taken into account in such lines?

3. Explain clearly the 'Ferranti effect' with a phasor diagram.
4. Explain the short, medium and long transmission lines.
5. Explain the classification of lines based on their length of transmission.
6. Obtain an expression for the rise in voltage at receiving end due to Ferranti effect.
7. Draw the phasor diagram of a short transmission line and derive an expression for voltage regulation.
8. Define A, B, C, D constants of transmission lines. What are their values in short lines?
9. Derive the A, B, C, D constants of a medium length transmission line and draw the phasor diagram assuming a π configuration.
10. What are the factors which govern the performance of a transmission line?
11. Deduce expression for the magnitudes of incident and reflected voltages at any point on a transmission line in terms of voltage and current at receiving and characteristic impedance and propagation constant of the line show how these expressions are useful in determining voltage and current distribution along the line.
12. Determine A, B, C, D constants for a 3-phase 50 Hz transmission line 200 km long having the following distributed parameters,

    $L = 1.20 \times 10^{-3}$ H/km

    $C = 8 \times 10^{-9}$ F/km

    $R = 0.15\ \Omega$/km

    $G = 0$. **[Ans:** $A = D = 0.981 \angle 0.44°$, $B = 80.59 \angle 68.429°\ \Omega$, $C = 0.49 \times 10^{-3} \angle 90.149°$ mho]
13. A single phase line has an impedance of $5 \angle 60°$ ohm. It is supplying a load of 120 A, 3.3 KV at 0.8 pf, lagging. Calculate its regulation. **[Ans:** 16.94%]
14. Determine the efficiency and regulation of a 3-phase, 100 km, 50 Hz transmission line delivering 20 MW at a p.f of 0.8 lagging and 66 kV to a balanced load. The conductors are of copper, each having resistance 0.1 ohm per km, 1.5 cm outside dia, spaced equilaterally 2 meters between centres. Neglect leakance and use (*i*) nominal-T and (*ii*) nominal-π method. **[Ans:** 93.5%, 17.47%]
15. The parameters of a 132 KV, 50 Hz, 3 – ϕ transmission line are: $R = 12.84\ \Omega$, $L = 0.162$ H, $C = 1.014 \times 10^{-6}$F. Calculate the line constants A, B, C and D for (*i*) nominal T-circuit and (*ii*) nominal π-circuit.

    [Ans: (*i*) $A = D = 0.992 \angle 0.1°$, $B = 52.2 \angle 75.95°\ \Omega$, $C = 0.00032 \angle 90°$ S

    (*ii*) $A = D = 0.992 \angle 0.1°$, $B = 52.4 \angle 75.9°\ \Omega$, $C = 0.000318 \angle 90.05°$ S]
16. A 3-phase overhead transmission line has the following constants. Resistance/phase = 10 Ω.

    Inductive Reactance/Phase = 35 Ω.

    Capacitive admittance/Phase = 3×10^{-4} mho

    If the line supplied a balanced load of 40,000 KVA at 110 KV and 0.8 p.f lagging, calculate

    (*i*) sending end power factor

    (*ii*) percentage regulation

    (*iii*) transmission efficiency **[Ans:** (*i*) 0.798 lag (*ii*) 10% (*iii*) 96.38%]
17. A 1000 km long, 3-phase 50 Hz transmission line has resistance per phase per km = 0.1 Ω, reactance per phase per km = 0.5 Ω, susceptance per phase per km = 10×10^{-6} ℧. If the line supplies a load of 20 MW at 0.9 p.f lagging at 65 KV at the receiving end, calculate by the normal π-method the regulation and efficiency of the line. Neglect leakage **[Ans:** 1515%, 94.56%]
18. A 132 kV overhead transmission line has the following line constants: $A = D = 0.99 \angle 0.3°$, $B = 70 \angle 69°\ \Omega$, $C = 4 \times 10^{-4} \angle 90°$ ℧ calculate the sending end voltage of transmission line if the load delivered at the receiving end is 45 MVA at 0.8 p.f lagging. **[Ans:** 167.215 kV]

10

MECHANICAL DESIGN OF TRANSMISSION LINES

The design of a transmission line has to be satisfactory from electrical as well as mechanical considerations. As far as electrical aspects are concerned, the line should have sufficient current carrying capacity and the line losses should be small. As far as mechanical aspects are concerned, the line conductors, supports and cross-arms should have sufficient mechanical strength to cope with the worst probable weather conditions. The line conductors, supports and cross-arms must be strong enough to give satisfactory service over a long period of time without the necessity of too much maintenance.

10.1 LINE SUPPORTS

The supporting structures for overhead line conductors are various types of poles and towers called line supports. The line supports should have following properties :

(*a*) High mechanical strength to withstand the weight of conductors and wind loads etc.
(*b*) Light in weight without the loss of mechanical strength
(*c*) Easy accessibility of conductors for maintenance
(*d*) Cheap in cost and economical to maintain
(*e*) Longer life

A line support has to perform the following functions :

(*a*) To keep the proper spacing between the conductors.
(*b*) To maintain the specified ground clearance.
(*c*) To keep the conductors at prescribed distances from its grounded parts.

The various types of supports include :

1. Wooden poles
2. Reinforced concrete poles

3. Tubular steel poles
4. Latticed steel towers.

10.1.1 Wooden Poles

Of the various types of supports, wood pole is the cheapest. These poles are suitable for line where spans are short and tensions are low. Wooden poles are useful for voltages up to 132 KV and for spans up to about 60 metres. These poles are used for distribution lines especially in areas where ample supplies of good quality wood are available. They are very economical but susceptible to decay. Wooden poles are protected by metallic cap at the top to protect it from the decay. In order to obtain greater strength. Double pole structures of 'A' and 'H' type are used as shown in Figs. 10.1 and 10.2. The strength of these types of constructions varies from two to four times the strength of the single pole.

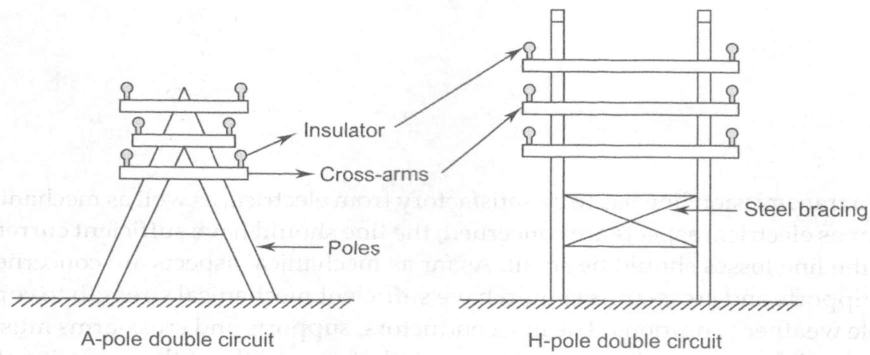

Fig. 10.1 **Fig. 10.2**

Wood has got a natural insulating property and therefore lesser flash-overs are likely to take place due to lightning.

The main objections to wooden supports are :

(*a*) Tendency to rot below the ground level.
(*b*) Less mechanical strength.
(*c*) Comparatively smaller life.
(*d*) Require periodic inspection.

10.1.2 Reinforced Concrete (RCC) poles

Poles made of reinforced cement concrete are stronger but more costly than wood poles. RCC poles has a longer life than that of wood poles because of little deterioration. The maintenance cost is also low.

The main difficulty with the use of these poles is the high cost of transport owing to their heavy weight. Handling and transportation difficulties are overcome by the use of pre-stressed concrete supports which can be manufactured in pieces and then assmebled at the job sites. The weight of pre-stressed concrete pole is considerably less than that of reinforced concrete pole.

Figures 10.5 and 10.4 shows R.C.C. poles for single and double circuit.

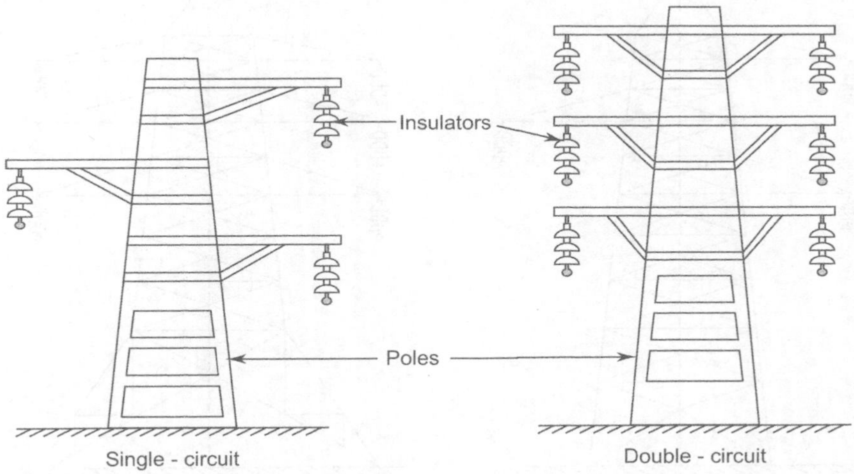

Fig. 10.3 Fig. 10.4

10.1.3 Tubular Steel Poles

The steel poles are often used as a substitute for wooden poles. They possess greater mechanical strength, longer life and permit longer spans to be used. The use of tubular steel poles is favoured for low and medium voltage distribution work.

These poles may be stepped or swaged type. A stepped pole is manufactured from a single tube, the diameter being reduced in parallel steps by passing the tube through a series of dies. A swaged pole is manufactured from poles of different diameters which are swaged together when hot.

This type of support needs to be galvanised which prolongs its life. They are costlier than wooden or RCC-poles and their maintenance cost is also more. Hence, they are widely used for lines up to 33 KV.

10.1.4 Steel Towers

Lines of 66 KV and above with long distance transmission lines are supported on steel towers. They are fabricated from painted or galvanized angle sections which can be transported separately and the erection done on site. Steel towers have greater mechanical strength, longer life, can withstand most severe climatic conditions, have a high degree of reliability and permit the use of longer spans. The risk of service interruptions due to damage of insulators is greatly reduced since large spans can be used. The height of the tower depends on the line voltage and span length. The legs of the towers are set in special concrete foundations. The forces which have to be taken into account in the tower design include the vertical loads of conductor, insulators, fittings and tower itself, wind pressure on conductor and wind pressure on tower itself.

Steel towers are mostly suitable for double-circuit lines. Figures 10.5 and 10.6 shows single and double circuit steel towers. The double circuit has advantage that it ensures continuity of supply. In case there is breakdown of one circuit, the continuity of supply can be maintained by the other circuit.

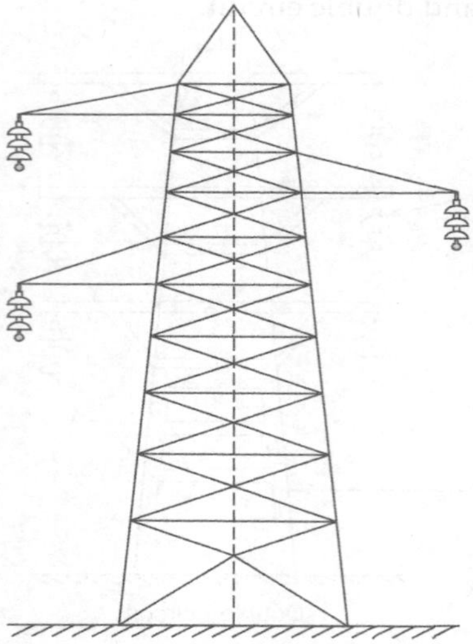

Single-circuit tower

Fig. 10.5

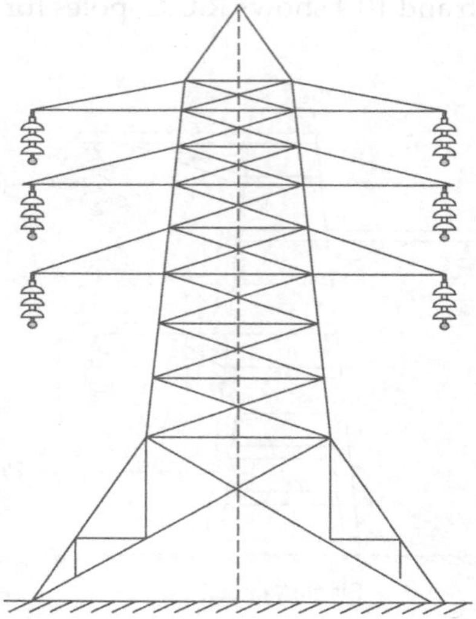

Double-circuit tower

Fig. 10.6

10.2 TYPES OF STEEL TOWERS

Steel towers can be broadly classified as :

(*a*) **Tangent Towers :** They are used for straight runs of the line and up to 2° line deviation from the straight run. The line is straight or along the tangent to line route. In these towers, the stress is due to the weight of the line, ice and wind loads. The base of the tower may be square or rectangular. Suspension type insulators are used with these towers.

(*a*) **Deviation Towers :** They are used at points where the transmission line changes direction. In addition to forces to which the tangent towers are subjected, these towers have also to withstand the resultant forces due to change in direction. They have broader base, stronger members and are costlier as compared to tangent towers. Strain insulators are used with these towers.

Deviation towers are further classified as :

1. **Small angle towers :** Used for 2° to 15° change in direction.
2. **Medium angle towers :** Used for 15° to 30° change in direction.
3. **Large angle towers :** Used for 30° to 60° change in direction.

10.3 SPAN

For every proposed line there is a definite length of span which will give the minimum cost of the line. From mechanical consideration there is a maximum value of span for each conductor size. A longer span means a small number of towers but the towers are taller and more costly. The higher the operating voltage, the greater should be the span to reduce the high cost of insulators. Modern high voltage lines have spans between 200 and 400 m. For river and ravine crossings exceptionally long spans up to 800 m or so have been satisfactory employed.

10.4 CONDUCTOR CONFIGURATION

In most of the cases, flat horizontal or vertical configurations are used from mechanical considerations. A flat horizontal configuration means a lesser tower height but a wider right of way. A vertical configuration means a taller tower and increased lightning hazards.

10.5 SPACING BETWEEN CONDUCTORS

There must be adequate spacing between conductors so that they do not come within sparking distance of each other even while swinging due to the wind. An empirical formula commonly used for determining the spacing of aluminium conductor line is

$$\text{Spacing} = \sqrt{S} + \frac{V}{150} \text{ metres}$$

where, S = Sag in metres

and V = Line voltage in KV.

10.6 SAG IN OVERHEAD LINES

The difference in level between points of supports and the lowest point on the conductor is called sag.

Figure 10.7 shows a conductor suspended between two equi-level supports A and B. The conductor is not fully stretched but is allowed to have a dip. The lowest point on the conductor is O and the sag is S.

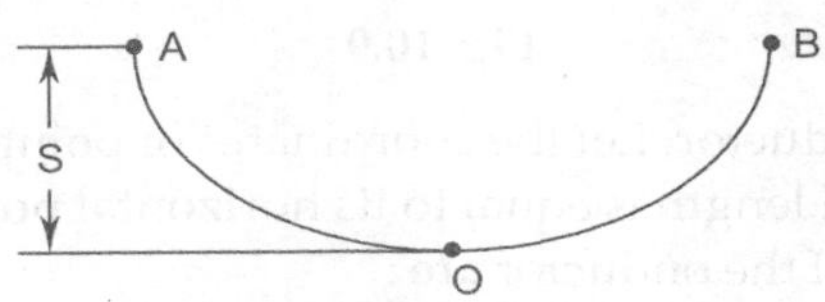

Fig. 10.7

The following points may be noted :

(*a*) When the conductor is suspended between two supports at the same level, it takes the shape of catenary. However, if the sag is very small compared with the span, then sag-span curve is like a parabola.

(*b*) The tension at any point on the conductor acts tangentially. Thus, tension T_0 at the lowest point O acts horizontally as shown in Fig. 10.8.

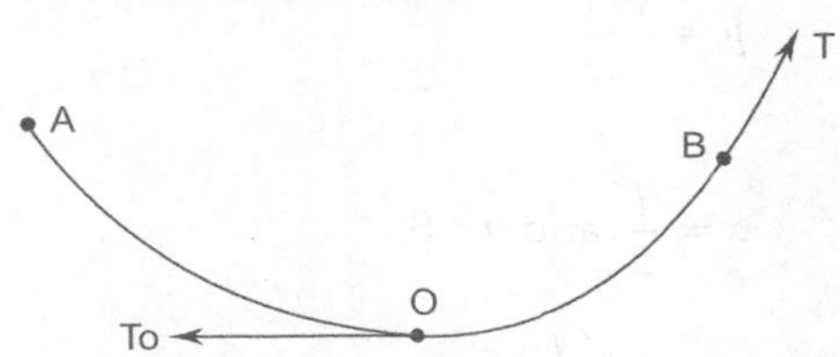

Fig. 10.8

(c) The tension at supports is approximately equal to the horizontal tension acting at any point on the wire. Thus, if T is a tension at the support B, then $T = T_0$.

(d) The conductor sag should be kept to a minimum in order to reduce the conductor material required and to avoid extra pole height for sufficient clearance above ground level.

10.7 CALCULATION OF SAG

10.7.1 Supports at Same Level

Figure 10.9 shown an overhead line suspended between two supports A and B at te same level. Mathematically, lowest point will be at the mid-span.

Let l = length of span *i.e.*, horizontal distance between supports in metres.

W = weight per unit length of conductor, N/m

T = Tension in the conductor in Newtons.

S = Sag at mid span in metres.

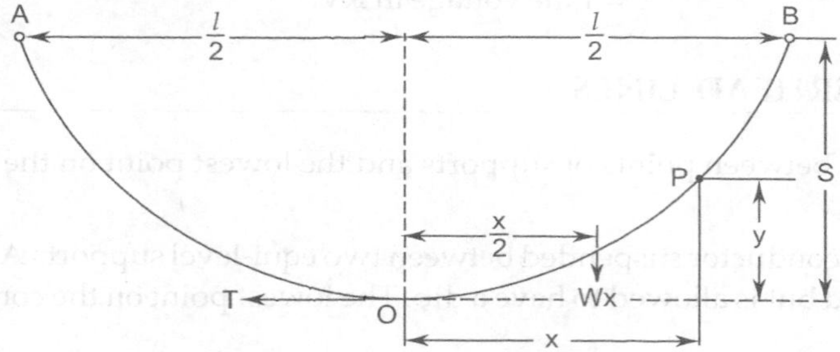

Fig. 10.9

Consider a point P on the conductor. Let the coordinates of point be x and y. Assuming that the curvature is so small that curved length is equal to its horizontal portion, *i.e.*, OP = x, then the two forces acting on the portion OP of the onductor are :

1. The weight Wx of conductor acting at a distance $\frac{x}{2}$ from O.

2. The tension T acting at O.

Equating the moments of above two forces about point O :

$$T \cdot y = Wx \cdot \frac{x}{2}$$

or

$$y = \frac{Wx^2}{2T}$$

At support A,

$$x = \frac{l}{2} \text{ and } y = S$$

$$\therefore \quad \text{Sag (S)} = \frac{W\left(\frac{l}{2}\right)^2}{2T}$$

$$= \frac{Wl^2}{8T}.$$

10.7.2 Supports at Different Level

In hilly areas, the two supports of a span may be at different levels. Figure 10.10 shows a conductor suspended between two supports A and B which are at different levels.

Let l = span length.

h = Difference in levels between two supports.

x_1 = Distance of support at lower level *i.e.*, A from O.

x_2 = Distance of support at higher level *i.e.*, B from O.

T = Tension in the conductor.

W = Weight per unit length of the conductor.

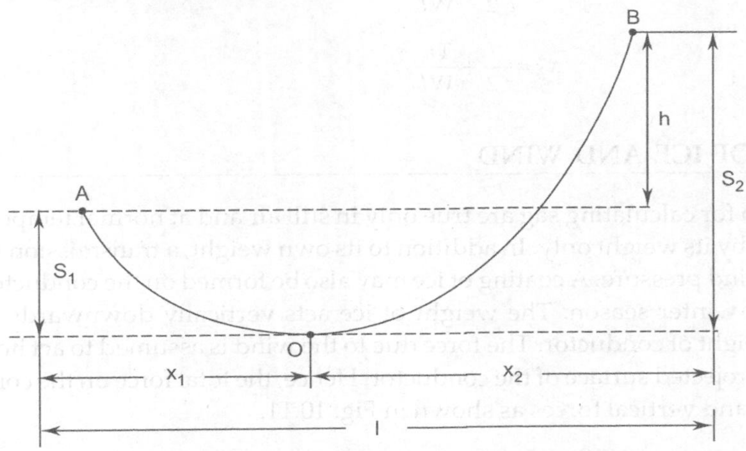

Fig. 10.10

As we have already proved that

$$y = \frac{Wx^2}{2T}$$

In Fig. 10.10, at support A

$$x = x_1 \text{ and } y = S_1$$

$\therefore$

$$\text{Sag } S_1 = \frac{Wx_1^2}{2T}$$

and

$$\text{sag } S_2 = \frac{Wx_2^2}{2T}$$

$$x_1 + x_2 = l \qquad ...(1)$$

Now,

$$S_2 - S_1 = \frac{Wx_2^2}{2T} - \frac{Wx_1^2}{2T}$$

$$= \frac{W}{2T}\left(x_2^2 - x_1^2\right)$$

$$= \frac{W}{2T}(x_2 + x_1)(x_2 - x_1)$$

$$= \frac{Wl}{2T}(x_2 - x_1) \qquad [\text{As } x_1 + x_2 = l]$$

From Fig. 10.10, $S_2 - S_1 = h$

$$\therefore \quad h = \frac{Wl}{2T}(x_2 - x_1)$$

or

$$x_2 - x_1 = \frac{2Th}{Wl} \qquad \text{...(2)}$$

Solving equations (1) and (2),

$$x_1 = \frac{l}{2} - \frac{Th}{Wl}$$

and

$$x_2 = \frac{l}{2} + \frac{Th}{Wl}.$$

10.8 EFFECT OF ICE AND WIND

The above formula for calculating sag are true only in still air and at normal temperature when the conductor is acted by its weight only. In addition to its own weight, a transmission line conductor is also subjected to wind pressure. A coating of ice may also be formed on the conductors of the lines in hilly areas during winter season. The weight of ice acts vertically downwards *i.e.,* in the same direction as the weight of conductor. The force due to the wind is assumed to act horizontally *i.e.,* at right angle to the projected surface of the conductor. Hence, the total force on the conductor is vector sum of horizontal and vertical forces as shown in Fig. 10.11.

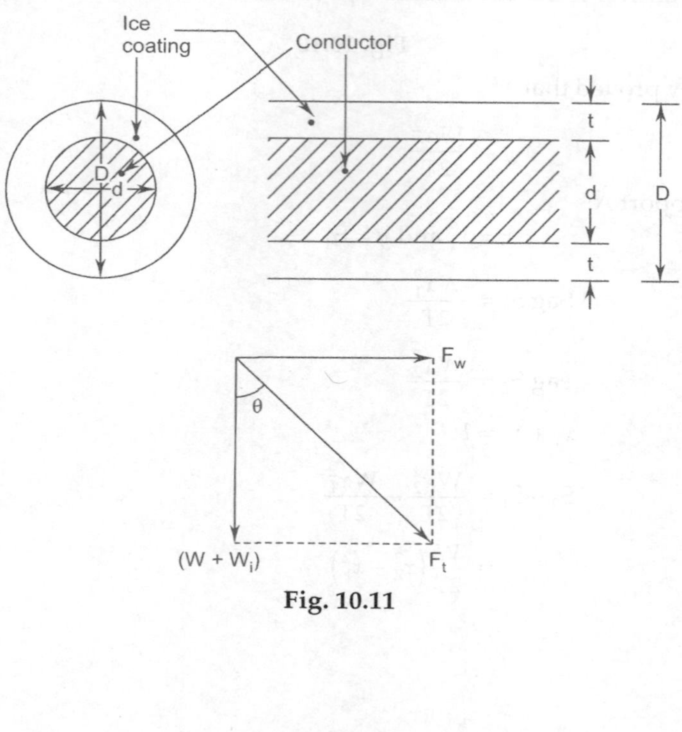

Fig. 10.11

The total force F_t acting on the conductor per metre length is given by :

$$F_t = \left[(W + W_i)^2 + F_w^2\right]^{\frac{1}{2}} \text{ N/m.}$$

where W = Weight of conductor per unit length

= Conductor material density × volume per unit length.

W_i = Weight of ice per unit length

= Density of ice × Volume of ice per unit length

$= \text{Density of ice} \times \frac{\pi}{4}\left[(d+2t)^2 - d^2\right]$

$= \text{Density of ice} \times \frac{\pi}{4}\left[d^2 + 4t^2 + 4dt - d^2\right]$

$= \text{Density of ice} \times \pi t\,(d+t)$

As density of ice ≈ 8920 N/m^3.

∴ $W_i = 8920 \times 3.14t\,(d+t)$ N/m.

$= 2.8 \times 10^4\, t\,(d+t)$ N/m.

and F_w = Wind force per unit length

= Wind pressure per unit area × projected area per unit length

= Wind pressure × $[(d + 2t) \times 1]$

= pD N/m.

where D is overall diameter of ice covered conductor and is equal to $(d + 2t)$

The conductor sets itself in a plane at an angle θ to the vertical where,

$$\tan\theta = \frac{F_w}{W + W_i}$$

The sag in the conductor is given by

$$S = \frac{F_t l^2}{8T} \quad ...(3)$$

The vertical sag is S cos θ.

Example 10.1. *A 132 KV transmission line has following data :*

Weight of the conductor = 680 kg/km

Length of span = 260 m

Ultimate strength = 3100 kg

Safety factor = 2.

Calculate the height above ground at which the conductor should be supported. Ground clearance required is 12 metres.

Solution. Weight of conductor/metre run

$$W = \frac{680}{1000} = 0.68 \text{ kg}$$

Tension, $$T = \frac{\text{Ultimate strength}}{\text{Safety factor}}$$

$$= \frac{3100}{2} = 1550 \text{ kg}$$

Span length, $l = 260$ m

$$\therefore \quad S = \frac{Wl^2}{8T} = \frac{0.68 \times (260)^2}{8 \times 1550} = 3.7 \text{ m.}$$

Hence, conductor should be supported at a height of 12 + 3.7 = 15.7 m.

Example 10.2. *A transmission line has a span of 150 m between level supports. The conductor has a cross-sectional area of 2 cm^2. The tension in the conductor is 2000 kg. If the specific gravity of the conductor material is 9.9 gm/cm^3 and wind pressure is 1.5 kg/m length. Calculate the sag. Determine the vertical sag also.*

Solution. Tension = 2000 kg

Span length, $l = 150$ m

Wind force/*m* length of conductor = F_w = 1.5 kg

Weight of conductor/*m* length

$$W = \text{specific gravity} \times \text{volume of 1 m conductor}$$
$$= 9.9 \times 2 \times 100 = 1980 \text{ gm}$$
$$= 1.98 \text{ kg.}$$

Total weight of conductor per metre length is

$$F_t = \sqrt{W^2 + F_w^2}$$
$$= \sqrt{(1.98)^2 + (1.5)^2} = 2.48 \text{ kg.}$$
$$S = \frac{F_t l^2}{8T}$$
$$= \frac{2.48 \times (150^2)}{8 \times 2000} = 3.48 \text{ m.}$$

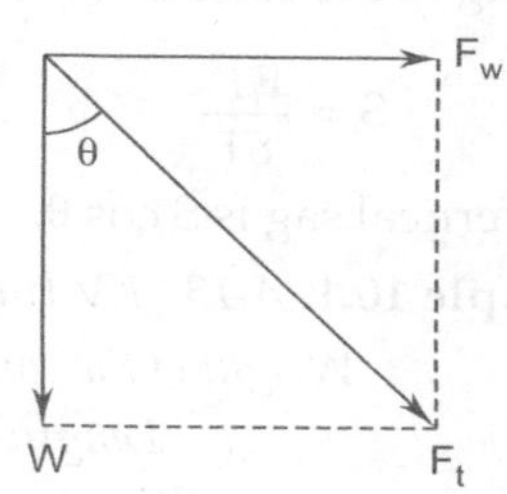

$$\text{Vertical sag} = S \cos \theta$$
$$= S \times \frac{W}{F_t}$$
$$= 3.48 \times \frac{1.98}{2.48}$$
$$= 2.77 \text{ m.}$$

Example 10.3. *An overhead line has following data :*

Span length = 160 m, conductor diameter = 0.95 cm. Weight per unit length of conductor = 0.65 kg/metre. Ultimate stress = 4250 kg/cm^2, wind pressure is 40 kg/cm^2 of projected area, factor of safety = 5. Calculate the sag.

Solution. Ultimate stress = 4250 kg/cm^2

Factor of safety = 5

$$\text{Working Stress} = \frac{\text{Ultimate Stress}}{\text{Safety Factor}}$$

$$= \frac{4250}{5} = 850 \text{ kg/cm}^2.$$

$$\text{Area of conductor} = \frac{\pi}{4} \times d^2$$

$$= \frac{\pi}{4} \times (0.95)^2 = 0.7084 \text{ cm}^2.$$

Working Tension, $T = 850 \times 0.7084 = 602.5$ kg.

Weight of conductor per unit length = 0.65 kg/m.

Wind pressure per unit length of the conductor

$$= 40 \times 0.95 \times 10^{-2} = 0.38 \text{ kg/m}.$$

Total weight of conductor per unit length is

$$F_t = \sqrt{(0.65)^2 + (0.38)^2}$$

$$= \sqrt{0.5669} = 0.7529 \text{ kg/m}.$$

$$\therefore \quad S = \frac{F_t l^2}{8T}$$

$$= \frac{0.7529 \times (160)^2}{8 \times 602.5}$$

$$= 4 \text{ metres}.$$

Example 10.4. *A transmission line has a span of 200 metres between level supports. The conductor has a cross-sectional area of 1.29 cm². Weighs 1170 kg/km and has a breaking stress of 4218 kg/cm². Calculate the sag for a safety factor of 5 allowing a wind pressure of 122 kg per square metre of projected area. What is the vertical sag ?*

Solution. Working stress $= \dfrac{\text{Breaking Stress}}{\text{Safety Factor}}$

$$= \frac{4218}{5} = 843.6 \text{ kg/cm}^2.$$

Working Tension $= 843.6 \times 1.29 = 1088$ kg.

Weight of Conductor = 1170 kg/km.

or Weight of conductor per meter length $= \dfrac{1170}{1000} = 1.17$ kg.

Area of conductor $= \dfrac{\pi}{4} \times d^2$

$\therefore$ Diameter of conductor, $d = \sqrt{\dfrac{4 \times \text{Area}}{\pi}}$

$$= \sqrt{\frac{4 \times 1.29}{3.14}} = 1.28 \text{ cm}.$$

Wind pressure per unit length of the conductor

$$= 122 \times 1.28 \times 10^{-2}$$

$$= 1.56 \text{ kg.}$$

Total weigt of conductor per metre length

$$F_t = \sqrt{(1.17)^2 + (1.56)^2}$$

$$= 1.95 \text{ kg.}$$

Sag,
$$S = \frac{F_t l^2}{8T}$$

$$= \frac{1.95 \times (200)^2}{8 \times 1088} = 8.96 \text{ m.}$$

$$\text{Vertical sag} = S \cos \theta$$

$$= S \times \frac{W}{F_t}$$

$$= 8.96 \times \frac{1.17}{1.95}$$

$$= 5.37 \text{ m.}$$

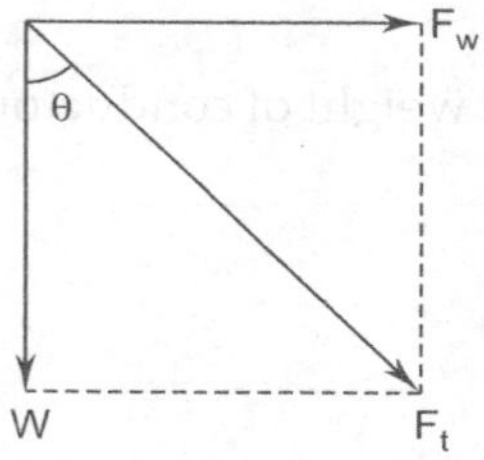

Example 10.5. *A transmission line has a span of 214 metres between level supports. The conductor have a cross-sectional area of 3.225 cm². Calculate the factor of safety under following conditions :*

Vertical sag = 2.35 m.

Wind pressure = 1.5 kg/m run

Breaking stress = 2540 kg/cm²

Weight of conductor = 1.125 kg/m run.

Solution. Wind pressure per metre run = 1.5 kg

Weight of conductor per metre run = 1.125 kg

Total weight per metre length of conductor

$$F_t = \sqrt{(1.5)^2 + (1.125)^2}$$

$$= 1.875 \text{ kg.}$$

Sag,
$$S = \frac{F_t l^2}{8T}$$

or
$$T = \frac{F_t l^2}{8S}$$

$$= \frac{1.875 \times (214)^2}{8 \times S}$$

$$\text{Vertical sag} = S \cos \theta$$

$$= S \times \frac{W}{F_t}$$

or
$$S = \frac{2.35 \times 1.875}{1.125} = 3.92 \text{ m.}$$

$$\therefore \qquad T = \frac{1.875 \times (214)^2}{8 \times 3.92}$$

$$\text{Breaking stress} = 2540 \text{ kg/cm}^2.$$

$$\text{Area} = 3.225 \text{ cm}^2$$

$$\therefore \quad \text{Breaking Tension} = 2540 \times 3.225 \text{ kg}$$

$$\therefore \quad \text{Working Tension,} \qquad T = \frac{2540 \times 3.225}{f} \text{ kg}$$

where f is factor of safety.

$$\frac{2540 \times 3.225}{f} = \frac{1.875 \times (214)^2}{8 \times 3.92}$$

or

$$f = \frac{2540 \times 3.225 \times 8 \times 3.92}{1.875 \times (214)^2}$$

$$= 3$$

Example 10.6. *An overhead transmission line conductor having a parabolic configuration weights 1.925 kg per metre of length. The area of cross-section of the conductor is 2.2 cm² and ultimate strength is 8000 kg/cm². The supports are 500 m apart having 15 m difference. Calculate the sag from the taller of two supports which must be allowed so that the factor of safety should be 5. Assume that ice-load is 1 kg per metre run and there is no wind pressure.*

Solution. Ultimate strength $= 8000 \text{ kg/cm}^2$

Area of cross-section of conductor $= 2.2 \text{ cm}^2$

Factor of safety $= 5$

$$\therefore \qquad T = \frac{8000 \times 2.2}{5} = 3520 \text{ kg.}$$

Figure 10.12 shows the conductor suspended between two supports at A and B at different levels with O as the lowest point of the conductor.

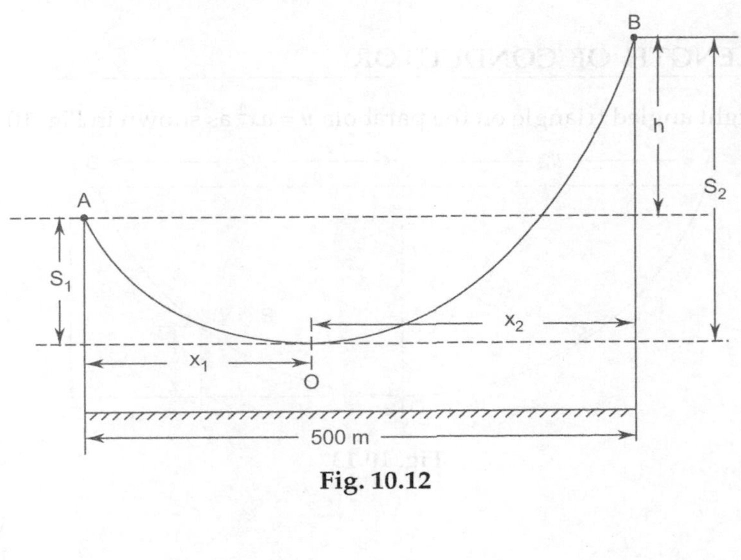

Fig. 10.12

Here, $l = 500$ m, $h = 15$ m

$W_i = 1$ kg and $W = 1.925$ kg

Total weight of 1 m length of conductor

$$W_t = W + W_i = 1 + 1.925$$
$$= 2.925 \text{ kg.}$$

Suppose x_1 is distance of lowest point O of the conductor from the support at lower level (*i.e.*, A) and x_2 is from the support at higher level (*i.e.*, B). It is clear from Fig. 10.12, that

$$x_1 + x_2 = 500 \text{ m} \quad ...(4)$$

$$h = s_2 - s_1 = \frac{W_t x_2^2}{2T} - \frac{W_t x_1^2}{2T}$$

$$15 = \frac{W_t}{2T}\left(x_2^2 - x_1^2\right)$$

or
$$15 = \frac{2.925}{2 \times 3520}\left(x_2 + x_1\right)\left(x_2 - x_1\right)$$

or
$$x_2 - x_1 = \frac{2 \times 3520 \times 15}{2.925 \times 500} = 72 \text{ m} \quad ...(5)$$

Solving equations (4) and (5)

$$x_2 = 286 \text{ m}$$

and
$$x_1 = 214 \text{ m.}$$

Sag from the taller of two towers

$$s_2 = \frac{W_t x_2^2}{2T}$$
$$= \frac{2.925 \times (286)^2}{2 \times 3520}$$
$$= 33.94 \text{ m.}$$

10.9 TOTAL LENGTH OF CONDUCTOR

Consider a small right angled triangle on the parabola $y = ax^2$ as shown in Fig. 10.13.

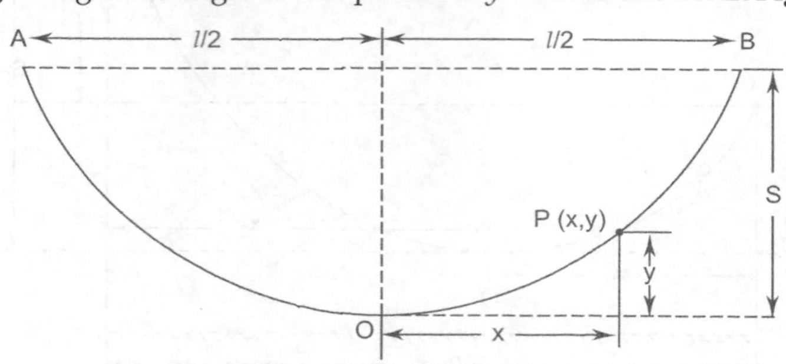

Fig. 10.13

It gives

$$(dS)^2 = (dx)^2 + (dy)^2$$

or

$$\left(\frac{dS}{dx}\right)^2 = 1 + \left(\frac{dy}{dx}\right)^2$$

$$= 1 + 4a^2x^2 \qquad \left[\text{As } \frac{dy}{dx} = \frac{d\left(ax^2\right)}{dx} = 2ax\right]$$

or

$$\frac{dS}{dx} = \left(1 + 4a^2x^2\right)^{\frac{1}{2}}$$

$$= 1 + 2a^2x^2$$

or

$$dS = (1 + 2a^2x^2)\, dx$$

Integrating the above equation :

$$S = x + \frac{2a^2x^3}{3} + k$$

When $x = 0$, $\quad S = 0$

Therefore, $\quad k = 0$

Thus,

$$S = x + \frac{2a^3x^3}{3} \qquad \text{...(6)}$$

From Fig. 10.13, when $\quad x = \frac{l}{2}$ then $y = S$.

$\therefore$

$$S = a\left(\frac{l}{2}\right)^2$$

or

$$a = \frac{4S}{l^2} \qquad \text{...(7)}$$

Let length of line be Z. If we put $x = \frac{l}{2}$ and value of a from equation (7) into equation (6), we get the length of half line *i.e.*, $\frac{Z}{2}$.

Thus,

$$\frac{Z}{2} = \frac{l}{2} + \frac{2 \times 16S^2}{l^4} \frac{l^3}{3 \times 8}$$

$$= \frac{l}{2} + \frac{4}{3} \frac{S^2}{l}$$

or

$$Z = l + \frac{8S^2}{3l}$$

Substitute the value of

$$S = \frac{F_t l^2}{8T}$$

$$\therefore \qquad Z = l\left(1+\frac{F_t^2 l^2}{24T^2}\right) \qquad ...(8)$$

10.10 EQUIVALENT SPAN

A section of an overhead line may consist of a number of spans of different lengths due to changing profile and the desire to locate the towers in such positions as to cause minimum interference with the normal use of land. When successive spans are unequal, changes in load or temperature will bring unequal changes in tension in the different spans. It is very tedious to make calculation of sag and tension for each and every span and then to make adjustment while erecting the transmission line.

The tensions are made equal by defining an equivalent span l_e and basing all calculations on the equivalent span l_e. The two conditions which have to be satisfied are :

(*a*) The sum total of horizontal spans should be equal ; and

(*b*) The sum total of curved lengths should be equal for the actual line and the equivalent line.

If there are *n*-sections, then

$$n\, l_e = \Sigma l$$

Thus, the equation $Z = l\left(1+\dfrac{F_t^2 l^2}{24T^2}\right)$ becomes

$$n\left(l_e + \frac{F_t^2 l_e^3}{24T^2}\right) = (\Sigma l) + \frac{F_t^2\left(\Sigma l^3\right)}{24T^2}$$

Substitute the $n\, l_e = \Sigma l$ in above equation

$$l_e^2 = \frac{\Sigma l^3}{n l_e}$$

$$= \frac{\Sigma l^3}{\Sigma l}$$

or

$$l_e = \left[\frac{\left(l_a^3 + l_b^3 + l_c^3 + \ldots\ldots\right)}{l_a + l_b + l_c}\right]^{\frac{1}{2}}$$

where *a*, *b*, *c* etc., are the sections.

10.11 ERECTION CONDITIONS

The erection conditions should be such that when loading conditions occur, there will be no infringement on the factor of safety.

Suppose Z_1, T_1, θ_1 and F_{t_1} denote the total conductor length, tension, temperature and total load per metre under the worst probable conditions and Z_2, T_2, θ_2, F_{t_2} denote the same quantities under the erection conditions. A relationship between them is derived as under :-

An increase in tension from T_1 to T_2 elongates the conductor by $Z_1 \dfrac{(T_2 - T_1)}{AE}$ where A is cross-sectional area and E is Young's modulus of elasticity. The increase in temperature from θ_1 to θ_2 increases the conductor length by $\alpha Z_1 (\theta_2 - \theta_1)$ where α is coefficient of linear expansion. Thus,

$$Z_2 = Z_1 + Z_1 \frac{(T_2 - T_1)}{A \cdot E} + \alpha Z_1 (\theta_2 - \theta_1) \quad \text{...(9)}$$

Substituting value of Z_1 and Z_2 from equation (8) and using binomial expansion with neglecting second order terms equation (9) becomes

$$\left(\frac{l^2}{24}\right)\left(\frac{F_{t_2}^2}{T_2^2} - \frac{F_{t_1}^2}{T_1^2}\right) = \frac{T_2 - T_1}{AE} + \alpha(\theta_2 - \theta_1)$$

or $$T_2^2 \left[T_2 - \left\{ T_1 - \alpha AE(\theta_2 - \theta_1) - \frac{AE F_{t_1}^2 l^2}{24 T_1^2} \right\} \right] = \frac{AE F_{t_2}^2 l^2}{24} \quad \text{...(10)}$$

The above cubic equation can be solved graphically or analytically.

10.12 STRINGING CHART

The plotting of the graph of tension vs temperature and sag vs temperature for a fixed span is known as stringing chart. It is shown in Fig. 10.14. From this chart the values of sag and tension can be found for any temperature. A stringing chart is very useful in the field for stringing the conductors and adjusting them to the proper sag and tension.

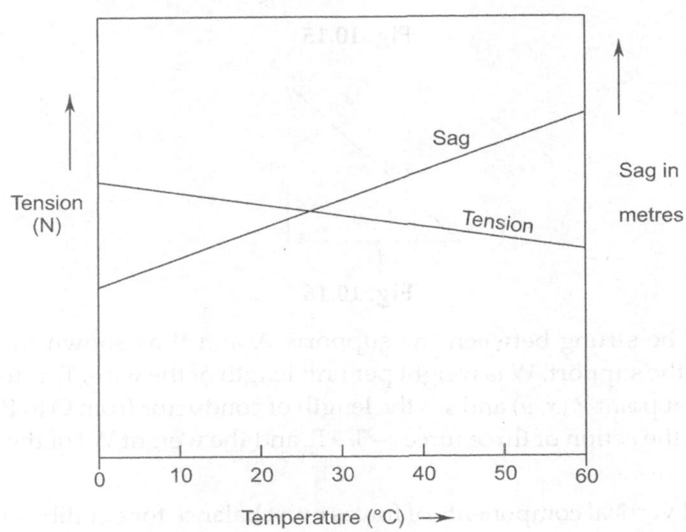

Fig. 10.14

10.13 FACTORS AFFECTING SAG

It can be seen from equation (3) that sag is affected by following factors :

(a) **Weight of Conductor :** Sag is directly proportional to weight per unit length of conductor. Ice and wind loads also increase the sag.

(b) **Span :** Sag is proportional to square of span, thus a longer span causes more sag.

(c) **Conductor Tension :** Sag is inversely proportional to conductor tension, thus an increase in the tension will reduce the sag.

(d) **Temperature :** A decrease in temperature reduces the sag. However, when a decrease in temperature is accompanied by snow and wind, the sag may increase.

10.14 THE CATENARY CURVE

The exact shape of a transmission line is of a catenary. For extra high voltage lines, when sag is greater than 10% of the span, formula based on catenary should be used.

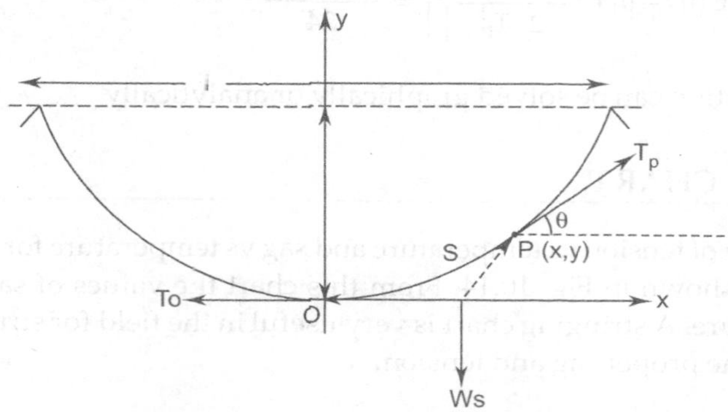

Fig. 10.15

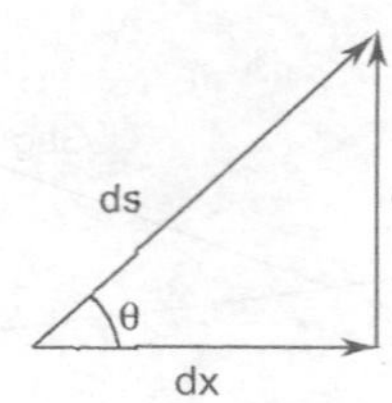

Fig. 10.16

Let the conductor be strung between the supports A and B as shown in Fig. 10.15 and l is the distance between the support, W is weight per unit length of the wire, T_0 is tension at the lowest point O, T_p is tension at point P (x, y) and s is the length of conductor from O to P. The portion OP is in equilibrium under the action of three forces – T_o, T_p and the weight Ws of the conductor portion OP.

The horizontal and vertical components of forces must balance for equilibrium. Thus,

$$T_p \cos \theta = T_o \qquad ...(11)$$

and

$$T_p \sin \theta = Ws \qquad ...(12)$$

or $$\frac{T_p \sin\theta}{T_p \cos\theta} = \frac{Ws}{T_o}$$

or $$\tan\theta = \frac{Ws}{T_o} \qquad \text{...(13)}$$

Consider a small right angled triangle on catenary as shown in Fig. 10.16

$$(ds)^2 = (dx)^2 + (dy)^2$$

or $$\left(\frac{ds}{dx}\right)^2 = 1 + \left(\frac{dy}{dx}\right)^2$$

or $$\frac{ds}{dx} = \sqrt{(1 + \tan^2\theta)}$$

Substituting the value of tan θ from equation (13),

$$\frac{ds}{dx} = \left(1 + \frac{W^2 S^2}{T_o^2}\right)^{\frac{1}{2}}$$

or $$dx = \frac{ds}{\left(1 + \frac{W^2 S^2}{T_o^2}\right)^{0.5}}$$

Integrating both sides of above equation,

$$x = \left(\frac{T_o}{W}\right)\sinh^{-1}\frac{Ws}{T_o} + A,$$

where A is constant of integration.

At the lowest point on the curve, $x = 0$ and $s = 0$.

Thus, $$A = 0$$

Then, $$x = \left(\frac{T_o}{W}\right)\sinh^{-1}\frac{Ws}{T_o}$$

or $$s = \left(\frac{T_o}{W}\right)\sinh\frac{Wx}{T_o} \qquad \text{...(14)}$$

Substituting the value of s into equation (13)

$$\tan\theta = \frac{dy}{dx} = \frac{W}{T_o}\cdot\frac{T_o}{W}\sinh\frac{Wx}{T_o}$$

$$= \sinh\frac{Wx}{T_o}$$

or $$dy = \left(\sinh\frac{Wx}{T_o}\right)\cdot dx$$

Integrating both sides of above equation,

$$y = \frac{T_o}{W}\cosh \cdot \frac{Wx}{T_o} + B$$

where B is the constant of integration.

At the lowest point, $y = 0$ and $x = 0$.

Thus, $$B = -\frac{T_o}{W}$$

Then, $$y = \left(\frac{T_o}{W}\right)\left[\cosh\frac{Wx}{T_o} - 1\right] \quad ...(15)$$

Squaring and adding equations (11) and (12)

$$T_p^2 = T_o^2 + W^2 s^2$$

Substituting the value of s from equation (14)

$$T_p^2 = T_o^2 + W^2\left[\frac{T_o^2}{W^2}\sinh^2\left(\frac{Wx}{T_o}\right)\right]$$

$$= T_o^2 \cosh^2\left(\frac{Wx}{T_o}\right)$$

or $$T_P = T_o \cosh\left(\frac{Wx}{T_o}\right).$$

As x increases, T_p increases for a given value of T_o. The minimum tension is line is T_o. The sag s can be found by putting $x = \frac{l}{2}$ in equation (15),

$$S = \frac{T_o}{W}\left[\cosh\frac{Wl}{2T_o} - 1\right]$$

or $$S = \frac{Wl^2}{8T_o}\left[1 + \left(\frac{l^2}{48}\right)\left(\frac{W}{T_o}\right)^2 + \cdots\cdots\right]$$

The tension at support point A is

$$T_A = T_o \cosh\left(\frac{Wl}{2T_o}\right) = T_o + Ws$$

If ice and wind loads exist, F_t will replace, W in above all equations.

10.15 SAG TEMPLATE

A sag template is a very convenient way of locating the towers in the field. This template is prepared on celluloid or tracing cloth and the curves are drawn to the same horizontal and vertical scales as the profile of the line. The horizontal distance represents the span length and vertical distance represents the sag. The plotting of a sag template on a piece of transparent paper which consists of a set of curves is shown in Fig. 10.17. DEF is the ground clearance line *i.e.*, minimum clearance of the power conductor from the ground. GHI is the tower footing line *i.e.*, this line gives the location of the footing of the tower. Curve ABC is such that with a span length of AC, the maximum sag of the conductor would be O′B and with span Jk, the maximum sag is O″ B. The curve DEF is obtained from ABC by subtracting ordinates from ABC equal to the desired minimum ground clearance. The curve GHI is obtained from ABC by setting off from ABC, a distance representing the height of standard tower from the point of attachment of the lowest conductor to the ground level.

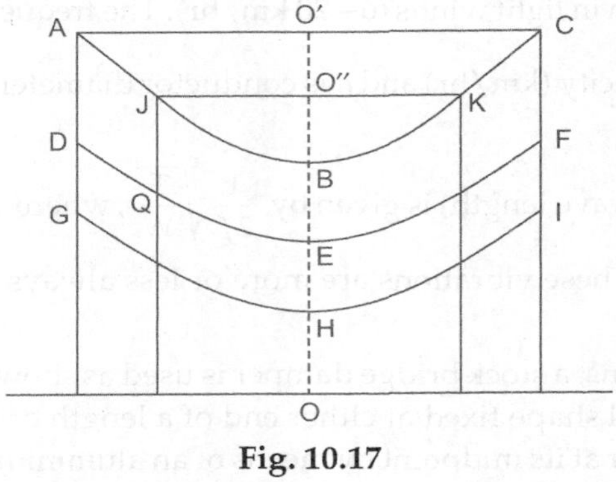

Fig. 10.17

If such a transparent template is placed upon a profile map of the route as indicated such that the ground clearance line just touches the profile at Q and then at points G and I where the support foot line cuts the profile will indicate the position of the towers represented by GA and IC. The curve ABC will represent the actual shape and position of the lowest conductor and JQ represents the correct ground clearane. Hence, the conductor approaches the ground by more than the safe amount.

Example 10.7. *A 500 m span over a river is to have its two supports at the same level. The conductor weighs 12 N/m and ice load is 14 N/m. The tension in the conductor under the above conditions should not exceed 50000 N. Find the sag by (a) representing the line as a parabola (b) representing the line as catenary.*

Solution.

(*a*) Total vertical load $= 12 + 14 = 26$ N/m

Horizontal load $= 0$

$$s = \frac{F_t l^2}{8T_o}$$

$$= \frac{26 \times 500 \times 500}{8 \times 50000} = 16.25 \text{ m.}$$

(*b*)
$$s = \frac{F_t l^2}{8T_o}\left[1 + \frac{l^2}{48}\left(\frac{F_t}{T_o}\right)^2\right]$$

$$= \frac{26 \times 500 \times 500}{8 \times 50000}\left[1 + \frac{500 \times 500}{48}\left(\frac{26}{50000}\right)^2\right]$$

$$= 16.272 \text{ m.}$$

10.16 CONDUCTOR VIBRATION

In addition to normal swinging in wind, overhead line conductor may be subject to vibrations which can be classified as :

(*a*) Resonant or Aeolian Vibrations (*b*) Dancing or Galloping.

10.16.1 Resonant Vibrations

These vibrations are of high frequencies (50 – 100 Hz) and low amplitudes (up to 5 mm). They are caused by vortex phenomenon in light winds (6 – 20 km/hr). The frequency *f* is empirically given by $-50\left(\frac{v}{d}\right)$ where v is wind velocity (km/hr) and d is conductor diameter (mm).

The length of a loop (half-wave length) is given by $\frac{1}{2f}\sqrt{\frac{T}{W}}$, where T is tension (N) and W is the conductor weight (kg f/m). These vibrations are more or less always present and common to all conductors.

To minimize these vibrations, a stock bridge damper is used as shown in Fig. 10.18. It consists of two hollow weights of special shape fixed at either end of a length of flexible steel cable which is itself fastened to the conductor at its midpoint by means of an aluminium clamp. The weights are of galvanized iron or in the smaller sizes of zinc and the flexible cable is protected by enclosing it in a water-tight flexible aluminium tube. The vibration causes movement of the damper and energy is absorbed by the interstrand friction in the steel cable. A typical damper on 132 KV lines for use is about 60 cm long and weighs about 5 kg.

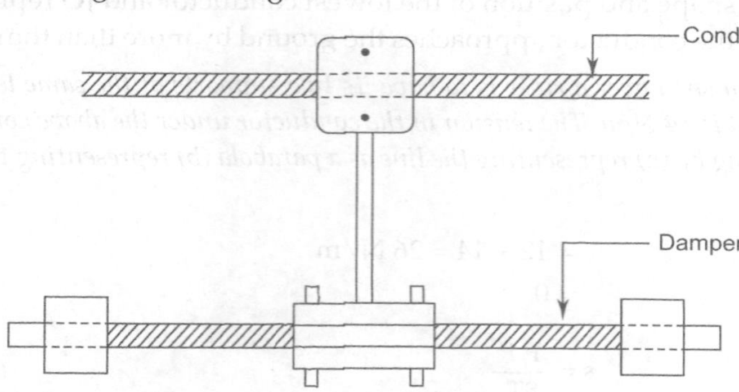

Fig. 10.18

10.16.2 Galloping or Dancing Conductors

These vibrations have high amplitude (up to 3 m) and low frequency (1–2 Hz). These vibrations are of self-excited type and are caused by asymmetrical layers of ice formation. When the ice coated conductor is acted upon by a light drift wind particularly where the ground slopes at right angles to the line, the wind travels up the slope and appears to get underneath the conductor and initiate vibrations. The travel of the conductor follows the path of an ellipse.

In order to prevent these vibrations, the adoption of horizontal conductor configuration reduces the danger due to these vibrations.

SELF TEST

Fill in the blanks by inserting appropriate words/figures :

1. Sag is provided in overhead lines so that __________.
2. RCC poles are __________ in weight and need __________ maintenance.
3. The conductivity of aluminium is __________ times of that of copper.
4. In an overhead system, the number of crossarms carried by a pole depends on the number of __________ it has to carry.
5. If sag in an overhead line increases, tension in the line __________.
6. __________ conductor is the best conductor owing to its high electrical conductivity and greater tensile strength for all types of transmission.
7. The conductor material used for transmission and distribution of electrical power must be of __________ conductivity, __________ tensile strength, __________ specific gravity and __________ cost.
8. __________ poles are generally used for distribution purposes in rural areas, steel tubular poles and __________ poles are usually used for distribution purposes in urban areas to give good apperance and steel rails, or narrow base lattice steel towers are used for transmission at __________ KV or 33 KV and broad base lattice steel towers are used for transmission at __________ KV and above.

Answers to Self Test

1. Safe tension is not exceeded.
2. heavy, little
3. 0.6
4. wires
5. decreases
6. hard down copper
7. high, high, low, low
8. wooden, RCC, 11, 66.

OBJECTIVE QUESTIONS

1. RCC poles usually have the spans of :
   (*a*) 25 – 50 m
   (*b*) 50–80 m
   (*c*) 80 – 150 m
   (*d*) 250–400 m.
2. Between two supports, due to sag the conductor takes the form of :
   (*a*) hyperbola
   (*b*) catenary
   (*c*) semicircle
   (*d*) none of the above.
3. The sag of a transmission line is least affected owing to :
   (*a*) ice deposited on the conductor
   (*b*) weight of the conductor
   (*c*) atmospheric temperature
   (*d*) current through the conductor.
4. The sag of the conductors of a transmission line is 2.5 m when the span is 250 m. Now if the height of supporting tower is increased by 25%, the sag will :
   (*a*) increase by 25%
   (*b*) decrease by 25%
   (*c*) decrease by 12.5%
   (*d*) remain unchanged.
5. Stringing chart is useful for :
   (*a*) Finding the distance between the tower
   (*b*) Finding the sag in the conductor
   (*c*) In the design of tower
   (*d*) In the design of insulator string.

6. The effect of ice deposition on conductor is to increase the :
   (*a*) weight of the conductor (*b*) resistance to flow of current
   (*c*) skin effect (*d*) transmission losses.

7. The effect of wind pressure is more predominant on :
   (*a*) insulators (*b*) supporting towers
   (*c*) transmission lines (*d*) none of above.

8. Maximum permissible span with wooden poles is :
   (*a*) 100 m (*b*) 50 m
   (*c*) 30 m (*d*) 20 m.

9. In India for distribution of electric power we usually use :
   (*a*) steel poles (*b*) RCC poles
   (*c*) wooden poles (*d*) both (*a*) and (*b*).

10. In a transmission line sag depends upon :
   (*a*) span length (*b*) weight of the conductor per unit length
   (*c*) tension in conductors (*d*) all of the above.

11. Which of the following poles are bulky poles ?
   (*a*) Tubular steel poles (*b*) Rail poles
   (*c*) RCC poles (*d*) Wooden poles.

12. In an overhead transmission line with the conductors supported between two supports, the variation in tension T with atmopheric temperature *t* is represented by :
   (*a*) Curve I (*b*) Curve II
   (*c*) Curve III (*d*) Curve IV.

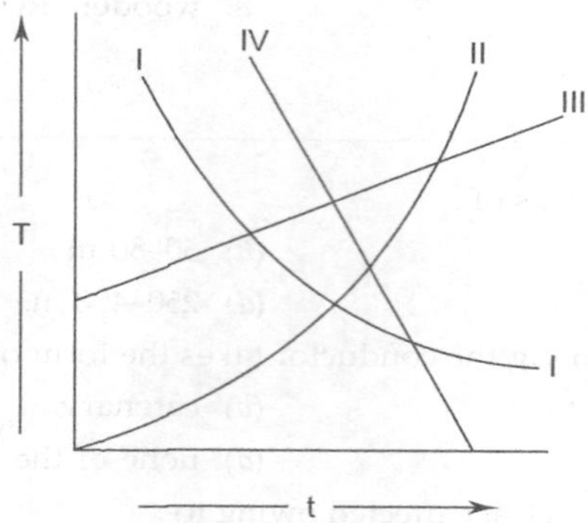

13. Which of the following statements is correct ?
   (*a*) Ice on conductors reduces sag
   (*b*) Ice on conductors increases skin effect
   (*c*) Wind pressure is taken to act at perpendicular to that for ice
   (*d*) Wind pressure reduces corona effect.

14. Steel poles for transmission lines need protection against :
   (*a*) termites (*b*) borer
   (*c*) corrosion (*d*) all of these.

15. Steel poles need galvanizing for longer life, galvanizing is the process of applying a layer of :
(*a*) point (*b*) zinc
(*c*) tarcoal (*d*) varnish.

16. Increase in temperature in overhead transmission lines causes :
(*a*) decrease in stress but increase in length (*b*) decrease in stress and length
(*c*) increase in stress and length (*d*) none of above.

17. Hot template curves are plots of :
(*a*) conductor weight and sag (*b*) conductor sag and span lengths
(*c*) temperature and humidity (*d*) none of above.

18. In a transmission line sag depends upon :
(*a*) length of span (*b*) conductor material
(*c*) the tension in conductors (*d*) all of the above.

19. In order to protect the wooden poles from the action of termites etc. It is necessary to treat them with :
(*a*) creosite oil (*b*) lime water
(*c*) sodium hydroxide (*d*) bleaching powder.

20. The supports used for transmission lines should have the characteristic (s) of :
(*a*) cheap in initial as well in maintenance cost
(*b*) high mechanical strength and longer life
(*c*) good looking, light in weight and easily accessible for painting and erection of line conductors.
(*d*) all of the above.

Answers to Objective Questions

1. (*c*)	**2.** (*b*)	**3.** (*d*)	**4.** (*d*)
5. (*b*)	**6.** (*a*)	**7.** (*b*)	**8.** (*b*)
9. (*d*)	**10.** (*d*)	**11.** (*c*)	**12.** (*a*)
13. (*c*)	**14.** (*c*)	**15.** (*b*)	**16.** (*a*)
17. (*b*)	**18.** (*d*)	**19.** (*a*)	**20.** (*d*).

EXERCISE

1. What is a stringing chart ? How can it be prepared? What is its use?
2. Write a short note on conductor vibrations.
3. Discuss the considerations which govern the selection span and conductor configuration of a high voltage line.
4. Show how to allow for the presence of ice on the conductor for calculation of sag in vertical direction.
5. What are the sources of vibrations in a transmission line? Explain the methods used to damp out these vibrations.
6. How sag is affected by wind pressure?
7. Prove that a transmission line conductor between two supports at equal heights takes the form of a catenary?
8. Why steel towers are used? Sketch and explain any one such tower.
9. How does the formations of an ice coating on a line conductor affect the sag?
10. Show how the sag of an overhead line can be calculated in case of supports at different levels?

11. An overhead line has a span of 220 meters, the line conductor weighs 684 kg per 1000 meters. Calculate the maximum sag in the line, if the maximum allowable tension in the line is 1450 kg. [**Ans:** 2.85 m]

12. Calculate the sag under the following conditions: Length of span = 100 m.
Cross-sectional area of the conductor = 1.2 sq.cm
Breaking strength = 4220 kg/sq.cm.
Factor of safety = 5
Weight of conductor = 1.10 kg/metres
Maximum wind pressure = 100 kg / sq.cm. [**Ans:** 3.18 m]

13. In the case of transmission line having the following data, calculate the safety factor allowed for:
Span across level supports = 200 m
Cross-sectional area of conductor = 1.2 sq.cm
Weight of conductor = 1 kg/ metre
Vertical sag = 3.95 metre
Wind pressure = 82.5 kg/sq.cm
Breaking stress = 4220 kg/sq.cm. [**Ans** = 4]

14. An overhead transmission line has a span of 220 metres, the conductor weighing 604 kg/km calculate the maximum sag if the ultimate tensile strength of conductor is 5758 kg. Assume a factor of safety = 2. [**Ans:** 1.27 metres]

15. Estimate the vertical sag for an overhead line with the following data: ACSR conductor, diameter 2.226 cm, weight 1097 kg/m, permissible tension 3.5×10^4 N, wind pressure 380 N/m^2 of projected area, ice coating 0.96 cm thick, weight of ice 8900 N/m^3, span 325 m. [**Ans.** 7.29 m]

16. An OH line has a span of 200 metres between level supports. The conductor diameter is 1 cm and weight 0.65 kg/metre length. The allowable tension 550 kg. Calculate the maximum sag [**Ans.** 5.91 m]

17. A transmission line conductor having a diameter of 19 mm weighs 0.82 kg/m. The span is 250 m. The wind pressure is 39 kg/m^2 of projected area with ice coating of 12 mm. The ultimate strength of the conductor is 8000 kg. Calculate the maximum sag if the factor of safety is 2.5 and ice weighs 900 kg/m^3. [**Ans:** 6.135 m]

18. An overhead transmission line spanning over a river is supported at two ends by towers at 45 m and 75 m. Find the clearance between the conductor and water at a point midway between the towers. The tension is to be limited to 2500 kg. The weight of the conductor is 0.9 kg/m and the distance between the towers is 300 metres. [**Ans:** 55.95 m]

11

Underground Cables

11.1 INTRODUCTION

In big cities and densely populated areas, overhead lines become impractical owing to safety regulations. In such places insulated conductors are usually laid underground and are called cables. All electric cables consist of 3-essential elements.

(*a*) the conductor for transmitting electrical power,

(*b*) the insulation needed to insulate the conductor from direct contact with earth or other objects and

(*c*) external protection against mechanical damage, chemical attack, fire etc.

11.2 UNDERGROUND CABLES

An underground cable essentially consists of one or more conductors covered with suitable insulation and surrounded by a protecting cover.

Many types of cables are used depending upon the mechanical properties required, the transmission voltage range and the type of insulation used. In general, a cable must fulfil the following requirements :

1. The conductor size should be such that the cable carries the desired load current without overheating and causes voltage drop within permissible limits.
2. The cable must have proper thickness of insulation in order to give high degree of safety and reliability at the voltage for which it is designed.
3. The insulating material used for cables should have good insulation resistance and high dielectric strength.
4. The conductor used in cables must be stranded because it helps in carrying out more current.
5. The insulating material used for cables should be non-hygroscopic and should not react with acids and alkalies.
6. The cable must be provided with suitable mechanical protection so that it may withstand the rough use and any chemical attack.

11.3 INSULATING MATERIALS FOR CABLES

The conductors of a cable have to be covered by an insulation to isolate the conductors from each-other and from their surroundings. The satisfactory operation of a cable depends to a great extent upon the characteristics of insulation used.

The main requirements of the insulating materials used for cable are :

1. High dielectric strength in order to avoid electrical breakdown of the cable.
2. Good mechanical properties *i.e.*, tenacity and elasticity.
3. High insulation resistance to avoid leakage current.
4. It should not be affected by chemicals around it.
5. It should be non-hygroscopic *i.e.*, should not absorb the moisture from air or soil, because the dielectric strength of any material goes down with moisture content.
6. It should be able to withstand temperatures from about – 30°C to over 100°C.
7. It should be non-inflammable.

The principal insulating materials used in cables are given below :

(*a*) **Rubber :** Rubber may be obtained from milky sap of tropical trees or it may be produced from oil products. Although pure rubber has reasonably high insulating properties but it suffers from some major drawbacks viz., readily absorbs moisture, maximum safe temperature is low and liable to damage due to rough handling.

(*b*) **Vulcanized India Rubber (VIR) :** When the pure rubber is mixed with sulphur with other gradients like zinc oxide, red lead etc., and is subjected to a particular temperature, it changes into vulcanized rubber which does not absorb moisture and has better insulating properties than the pure rubber. VIR has greater mechanical strength, durability and wear resistant property. The insulation is limited to wiring cables for lightning and minor power installations.

(*c*) **Impregnated Paper :** A suitable layer of paper is lapped on the conductor depending upon the operating voltage. It is then dried by combined application of heat and vacuum. Paper is most important insulating material used in manufacture of power cables upto 500 KV rating. Paper insulation has advantages of low cost, low capacitance, high dielectric strength, ability to withstand high temperature and high insulation resistance. Thus, the paper insulation has superseded the rubber insulation. The main disadvantage of this insulation is that paper is hygroscopic and if it is impregnated with suitable compound, it absorbs moisture and thus lowers the insulation resistance of the cable. For this reason, paper insulated cables are always provided with some protective covering and are never left unsealed.

(*d*) **PolyVinyl Cloride (PVC) :** It is a synthetic material obtained from the polymerisation of acetylene and can be produced in different grades depending upon the polymerization process. For use in cable industry, the polymer must be compounded with a plasticisers whose function is to form a gel and make the material plastic over the desired range of temperature.

PVC has high insulation resistance, good dielectric strength and mechanical toughness over a wide range of temperature. It is inert to oxygen and almost inert to oils, many acids and alkalies. Therefore, it is preferable to rubber in many environmental conditions.

(*e*) **Polythene :** It is a straight chain polymer derived from ethylene. It has very low dielectric constant and high resistivity. Its main use is in the manufacture of high frequency cables. The maximum operating temperature of this cable under short-circuits is 100°C.

(*f*) **XLPE Cables (Cross Linked Polythene Cables) :** Low density polythene when vulcanized under controlled conditions results in cross linking of carbon atoms and produces cross-linked polythene, which does not melt but carbonizes at 250°C to 300°C.

XLPE cables have light weight, small overall cable dimensions, low dielectric constant and good mechanical strength. These cables permit conductor temperatures of 90°C and 250°C under normal and short-circuit conditions and can be buried directly in soil. These cables are very suitable for all voltages upto 11 KV, 33 KV, 66 KV, 132 KV, 220 KV and 400 KV.

11.4 PROTECTIVE COVERINGS

1. **Sheath :** A metal sheath is provided over the cable insulation so that moisture may not be able to affect the insulation and may also protect the insulation from mechanical damage. Many years ago lead was used as a suitable material for sheathing. The main advantages of lead sheaths are their flexibility and high corrosion resistance. Some of disadvantages of lead are low mechanical strength and its large mass. Now a days, Aluminium is being increasingly used as a sheathing material. Aluminium sheaths have low weight, greater mechanical strength and costs about 3.5 times less than the lead sheaths.
2. **Armour :** Cables are protected against mechanical damage by armouring the cables with steel tapes or galvanized steel wires. A bedding of compounded fibrous material under the armour is used to provide a cushion between the sheath and tapes or wires.

   Steel tape is the cheapest material for armouring a cable and is useful for damage against direct blows. This is used normally for cables with conductor diameter more than half an inch. Galvanized steel wires are used for longer length of cables and is therefore more suitable for installation where longitudinal stresses are involved.

11.5 CLASSIFICATION OF CABLES

Underground cables may be classified in two ways according to :

(*a*) the type of insulating material used in their manufacture.

(*b*) the voltage for which they are manufactured.

However, the latter method of classification is generally preferred, according to which cables can be divided into the following groups :

1. Low-Tension (LT) cables – up to 1000 V
2. High-Tension (HT) cables – up to 11000 V
3. Super-Tension (ST) cables – from 22 KV to 33 KV
4. Extra high-Tension (E.H.T.) cables – from 33 KV to 66 KV
5. Extra super voltage cables – beyond 132 KV.

A cable may be single-core, two-core, three-core etc., depending upon the type of service for which it is intended.

Figure 11.1 shows single-core cable. A solid single-core cable has a central conductor with an insulation of impregnated paper. The insulation is surrounded by a lead sheath which prevents the entry of moisture into the inner parts. In order to protect the lead sheath from corrosion, plastic oversheath is provided. Single-core cables are not usually armoured in order to avoid excessive sheath losses.

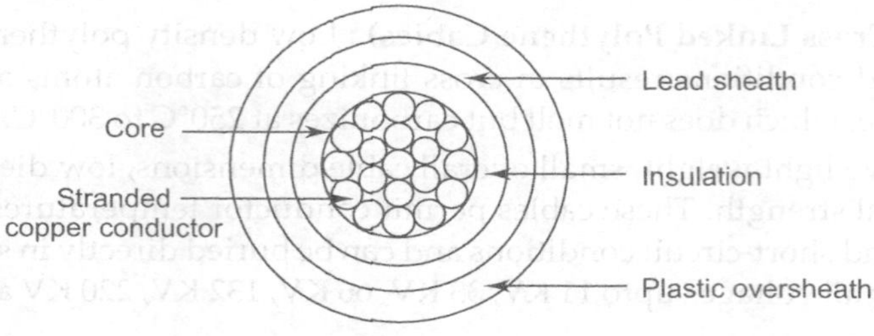

Fig. 11.1

Paper insulated metal sheathed power cables can be classified as solid types and pressurized cables.

(*a*) **Solid Type Cables :** In these cables the pressure within the cable does not go above atmospheric pressure and may even fall below it in case of voids. The cables for voltages up to 33 KV belong to this category.

(*b*) **Pressurized Cables :** In these cables, pressure is maintained above atmospheric either by oil in oil filled cables or by gas in gas filled cables. Gas filled cables are used up to 275 KV while oil filled cables are used up to 500 KV.

11.5.1 Solid Type Cables

It may be a belted type cable or H-type cable.

1. **Belted cables :** A belted cable consists of three separately insulated conductors with an overall insulating tape enclosing all the three conductors and finally the metallic sheath is applied. The cores are generally stranded and may be of non-circular shape to make the better use of available space. Figure 11.2 shows the constructional details of a 3-core belted cable.

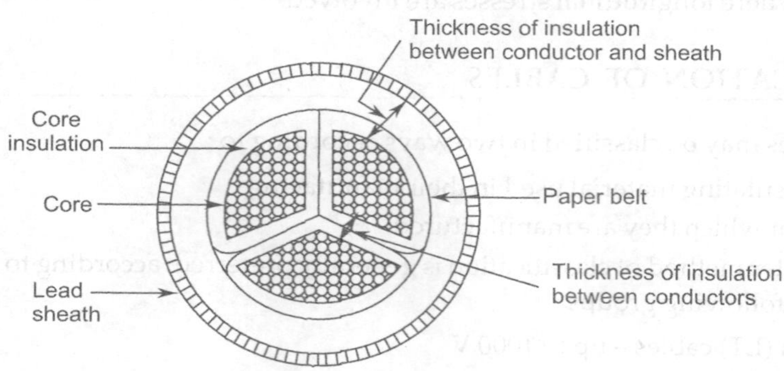

Fig. 11.2

Belted cables are used upto 11 KV. The major disadvantage of belted cables is that the electric stress is not purely radial. The existence of tangential stresses forces a leakage current to flow along the layers of paper. The leakage current causes local heating resulting in the risk of breakdown of insulation at any moment. In order to overcome this difficulty, screened or H-type cables are used where leakage currents are conducted to earth through metallic screens.

2. **H-type cables :** H-types cables also known as screened cables were introduced by Hochstadter in 1914. Figure 11.3 shows the constructional details of a typical 3-core, H-type cable. In these cables each core has an insulation layer but no belt insulation is used. The insulation on each core is covered with a metallic screen which usually consists of a perforated aluminium foil. The cores are laid in such a way that metallic screens are in good contact with each other. An

additional conducting belt (copper tape) is wrapped around the three cores. Then the lead sheath and armour are applied. As all the four screens (3-core screens and one conducting belt) and the lead sheath are at earth potential, the electric stresses are purely radial and dielectric losses are reduced. This reduces the possibility of breakdown.

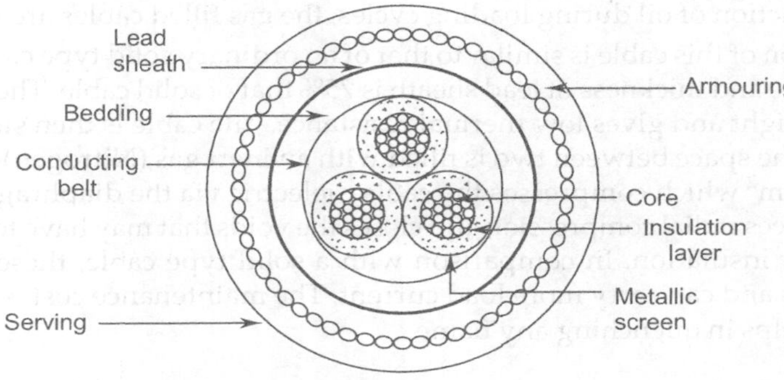

Fig. 11.3

H-type cables have been used up to 66 KV but are mostly used up to 33 KV because pressurized cables have better performance than H-types cables for voltages above 33 KV. H-type cable may be designated as SL (Separated Lead) type cable if each core has a separate lead sheath and also an overall lead sheath. It may be designated as SA cable if each core has a separate aluminium sheath and also an overall aluminium sheath.

11.5.2 Pressurized Cables

In these cables voids are eliminated by maintaining a higher pressure than atmosphere. Pressurized cables may be oil filled or gas filled cables :

1. **Oil-filled cables :** In these cables, channels of ducts are provided in the cable for oil circulation and the ducts being connected to oil reservoirs. The expansions and contractions merely produce changes in oil level in the reservoir and therefore do not result in voids. Due to elimination of voids, oil-filled cables can be used for higher voltages up to 500 KV.

   Figure 11.4 shows the constructional details of a single-core conductor channel, oil filled cable. The oil channel is formed at the centre by stranding the conductor wire around a hollow cylinderical steel spiral tape. The oil under pressure is supplied to the channel by means of external reservoir. The system is so designed that when the oil gets expanded due to increase in cable temperature, the extra oil collects in reservoir. However, when the cable temperature falls during light load conditions, the oil from the reservoir flows to the channel.

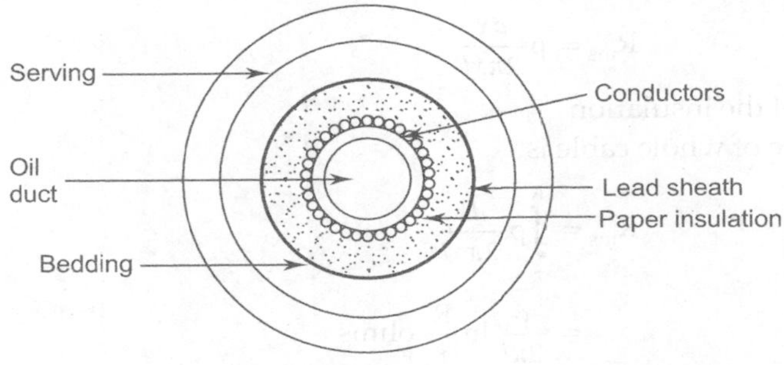

Fig. 11.4

Leakage of oil in these cables is a very serious problem. Therefore, automatic signalling is installed to indicate the fall in oil pressure in any of the phases. The main disadvantages of these cables are high initial cost and complicated system of laying.

2. **Gas filled cables :** In order to obviate the disadvantages of oil filled cables in terms of expansion and contraction of oil during loading cycles, the gas filled cables are used.

   The construction of this cable is similar to that of an ordinary solid type cable except that it of triangular shape and thickness of lead sheath is 75% that of solid cable. The triangular section reduces the weight and gives low thermal resistance. The cable is then surrounded with an envelope and the space between two is filled with an inert gas (Nitrogen) at a nominal pressure of 14 kg/cm^2 which compresses the cable dielectric via the diaphragm sheath. The gas pressure produces radial compression and closes the voids that may have formed between the layers of paper insulation. In comparison with a solid type cable, these cables operate at higher voltages and can carry more load current. The maintenance cost is also small and the nitrogen gas helps in quenching any flame.

11.6 INSULATION RESISTANCE OF A SINGLE CORE CABLE

The opposition offered by insulation to leakage current is known as insulation resistance of the cable. For satisfactory operation, the insulation resistance of the cable should be very high. Figure 11.5 shows a single core cable of conductor radius r and internal sheath radius R.

Consider a very small layer of insulation of thickness dx at a radius x. The length through which leakage current tends to flow is dx and the area of cross-section offered to this flow is $2\pi xl$.

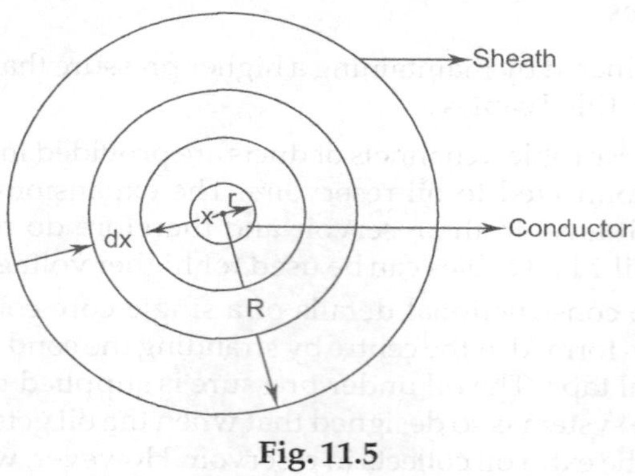

Fig. 11.5

Insulation resistance of layer is

$$R_{ins} = \rho \frac{dx}{2\pi xl}$$

where ρ is resistivity of the insulation

Insulation resistance of whole cable is

$$R_{ins} = \int_r^R \rho \frac{dx}{2\pi xl}$$

$$= \frac{\rho}{2\pi l} \ln \frac{R}{r} \text{ ohms.}$$

It is to be noted that insulation resistance of a cable is inversely proportional to its length.

Example 11.1. *A single-core cable has a conductor diameter of 1 cm and insulation thickness of 0.4 cm. If specific resistance of insulation is 5×10^{12} Ωm. Calculate the insulation resistance for a 3 km length of the cable.*

Solution.

$$r = \frac{1}{2} = 0.5 \text{ cm}$$

$$l = 3000 \text{ m}$$

$$\rho = 5 \times 10^{12} \ \Omega\text{-m}$$

$$R = 0.5 + 0.4 = 0.9 \text{ cm}$$

$$R_{ins} = \frac{\rho}{2\pi l} \ln \frac{R}{r}$$

$$= \frac{5 \times 10^{12}}{2 \times 3.142 \times 3000} \ln \frac{0.9}{0.5}$$

$$= 0.156 \times 10^9 \ \Omega = 156 \text{ M}\Omega.$$

11.7 CAPACITANCE OF A SINGLE CORE CABLE

A single core cable is in effect an electrostatic capacitor because it has two electrodes, the core of the cable and sheath separated by a dielectric material. Let the charge on the surface of the conductor be q coulomb per metre length of the cable. From Fig. 11.5, the electric flux density at radius x is :

$$D_x = \frac{q}{2\pi x} \text{ C/m}^2.$$

The electric field intensity at radius x is :

$$E_x = \frac{D_x}{\varepsilon_o \varepsilon_r}$$

$$= \frac{q}{2\pi \varepsilon_o \varepsilon_r x} \text{ V/m}$$

Where ε_r is the relative permittivity of the insulation and ε_o is the permittivity of free space and is equal to 8.85×10^{-12} F/m.

Potential difference between the core and sheath is

$$V = \int_r^R E_x \, dx$$

$$= \frac{q}{2\pi \varepsilon_o \varepsilon_r} \ln \frac{R}{r} \text{ volts}$$

Capacitance of the cable is

$$C = \frac{q}{V}$$

$$= \frac{2\pi \varepsilon_o \varepsilon_r}{\ln \frac{R}{r}} \text{ F/m}$$

If the cable has a length of l metres, the capacitance of the cable is

$$C = \frac{2\pi\varepsilon_o \varepsilon_r l}{\ln \frac{R}{r}} \text{ F.}$$

If the conductor is not circular, the radius r may be taken as the $\frac{1}{2\pi} \times$ actual periphery.

Example 11.2. *A 33 KV, 3-phase underground feeder 4.0 km long, uses three single core cables. Each cable has a conductor diameter of 2.5 cm and the radial thickness of insulation is 0.6 cm. The relative permittivity of dielectric is 3.1. Find (a) capacitance of the cable per phase and (b) charging current per phase.*

Solution.

(*a*) Given :

$$r = 1.25 \text{ cm}$$

$$R = 1.25 + 0.6 = 1.85 \text{ cm}$$

$$\varepsilon_r = 3.1$$

$$C = \frac{2\pi\varepsilon_o \varepsilon_r}{\ln \frac{R}{r}}$$

$$= \frac{2\pi \times 8.85 \times 10^{-12} \times 3.1}{\ln \frac{1.85}{1.25}} = 439.7 \times 10^{-12} \text{ F/m.}$$

For 4.0 km length,

$$C = 439.7 \times 10^{-12} \times 4 \times 10^3$$

$$= 1.758 \times 10^{-6} \text{ F/phase.}$$

(*b*) Voltage between core and sheath

$$V_{ph} = \frac{33 \times 10^3}{\sqrt{3}}$$

Charging current

$$= \frac{V_{ph}}{X_c}$$

$$= 2\pi f C V_{ph}$$

$$= 2\pi \times 50 \times 1.758 \times 10^{-6} \times \frac{33 \times 10^3}{\sqrt{3}}$$

$$= 10.3 \text{ A.}$$

11.8 DIELECTRIC STRESS IN A SINGLE CORE CABLE

Under operating conditions, the insulation of a cable is subjected to electrostatic forces. This is known as dielectric stress. The dielectric stress at any point in a cable is in fact the potential gradient at that point. From the equations :

$$E_x = \frac{q}{2\pi\varepsilon_o \varepsilon_r x}$$

and

$$V = \frac{q}{2\pi\varepsilon_o \varepsilon_r} \ln \frac{R}{r}$$

The potential gradient at a radius x is :

$$E_x = \frac{V}{x \ln \frac{R}{r}} \text{ V/m.}$$

It is clear that potential gradient varies inversely as the distance x. Therefore, potential gradient will be maximum when x is minimum *i.e.*, when $x = r$ or at the surface of the conductor. On the other hand, potential gradient will be minimum at $x = R$ or at the sheath surface.

Maximum potential gradient is

$$E_{max} = \frac{V}{r \ln \frac{R}{r}}$$

Minimum potential gradient is

$$E_{min} = \frac{V}{R \ln \frac{R}{r}}$$

The ratio of maximum and minimum potential gradient is

$$\frac{E_{max}}{E_{min}} = \frac{R}{r}$$

The variation of stress in dielectric is shown in Fig. 11.6. It is clear that the dielectric stress is maximum at the conductor surface and its value goes on decreasing as we move away from the conductor. The above analysis assumes that the conductor is perfectly circular. Stranding of the conductor increases E_{max} by about 20%.

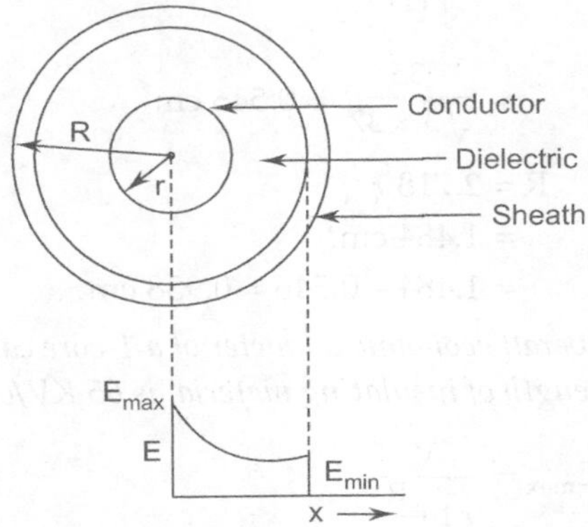

Fig. 11.6

In order to keep a fixed overall size of the cable (R) for a particular operating voltage V, there is a particular value of the radius of conductor which gives minimum gradient at the surface of conductor. Thus, E_{max} is minimum when $r \ln \left(\frac{R}{r}\right)$ is maximum. This happens when

$$\frac{d}{dr}\left[r \ln \frac{R}{r}\right] = 0$$

or $$-r\frac{r}{R}\frac{R}{r^2}+\ln\frac{R}{r}=0$$

or $$\ln\frac{R}{r}=1$$

or $$\frac{R}{r}=e$$

or $$R=2.718r$$

Most economical conductor diameter

$$r=\frac{R}{2.718}.$$

Example 11.3. *Find the most economical conductor diameter for single core cables to be used on 35 KV, 3-phase system, if the rms value of maximum stress is not to exceed 37 KV/cm. Also find the radial thickness of insulation.*

Solution. $$E_{max}=\frac{V}{r\ln\frac{R}{r}}$$

For most economical condition,

$$\ln\left(\frac{R}{r}\right)=1$$

$$37=\frac{35/\sqrt{3}}{r(1)}$$

or $$r=\frac{35}{\sqrt{3}\times 37}=0.546\text{ cm}$$

$$R=2.718\,r$$
$$=1.484\text{ cm.}$$

Thickness of insulation $= 1.484 - 0.546 = 0.938$ cm.

Example 11.4. *Determine the overall economic diameter of a 1-core cable metal sheathed for a working voltage of 85 KV if the dielectric strength of insulating material is 65 KV/cm.*

Solution. $$E_{max}=\frac{V}{r\ln\frac{R}{r}}$$

or $$65=\frac{85}{r(1)}\qquad\left[\because\ \ln\frac{R}{r}=1\right]$$

or $$r=\frac{85}{65}=1.3\text{ cm.}$$

Diameter of the conductor $= 2\times 1.3 = 2.6$ cm

Diameter of the sheath $= 2.718\times 2.6$

$= 7.06$ cm.

11.9 DIELECTRIC LOSS

A cable has a capacitance between the core and the sheath. When a voltage is applied to an unloaded cable, a capacitive current flows and power loss occurs. Thus, a cable behaves as an imperfect capacitor and the total current under unloaded conditions leads the voltage not by 90° but by an angle (90 – δ) as shown in Fig. 11.7. The angle δ is termed as loss angle of dielectric. The dielectric loss is :

$$\begin{aligned} P_d &= V I \cos \phi \\ &= V I \sin \delta \qquad [\because \ \delta + \phi = 90°] \\ &= \omega C V^2 \sin \delta \end{aligned}$$

where C is the capacitance of cable
and V is the line to neutral voltage

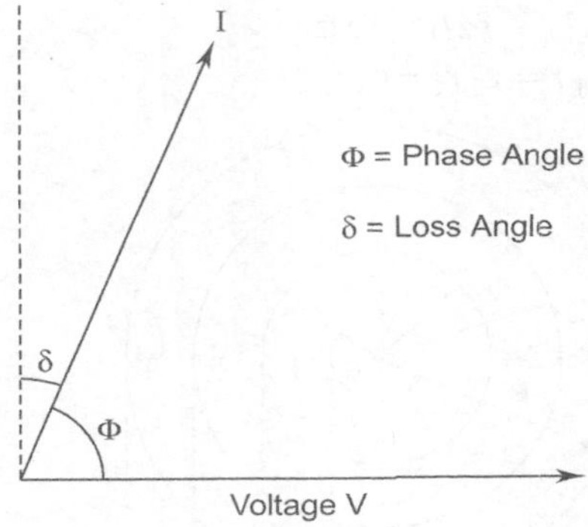

Fig. 11.7

11.10 GRADING OF CABLES

The process of achieving uniform electrostatic stress in the dielectric of the cables is known as grading of cables. By grading of a cable means the distribution of dielectric material such that the difference between the maximum and minimum gradient is reduced, thereby a cable of the same size could be operated at higher voltages or for the same operating voltage, a cable of relatively smaller size could be used.

There are two methods used for grading of cables :

(*a*) Capacitance Grading
(*b*) Intersheath Grading.

11.10.1 Capacitance Grading

The process of achieving uniformity in the dielectric stress by using layers of different dielectrics is known as capacitance grading. This method involves the use of two or more layers of dielectrics having different permittivities, those with higher permittivities being nearer to the conductor.

The capacitance grading can be explained by Fig. 11.8. There are three dielectrics of outer diameter r_1, r_2 and R and of relative permittivity ε_1, ε_2 and ε_3.

As $$E_x = \frac{q}{2\pi\varepsilon_o \varepsilon_r x}$$

At $x = r$, $$E = \frac{q}{2\pi\varepsilon_o \varepsilon_1 r}$$

At $x = r_1$, $$E = \frac{q}{2\pi\varepsilon_o \varepsilon_2 r_1}$$

At $x = r_2$, $$E = \frac{q}{2\pi\varepsilon_o \varepsilon_3 r_2}$$

If the permittivities are such that $\varepsilon_1 > \varepsilon_2 > \varepsilon_3$ and the three dielectrics are worked at the same maximum stress, then,

$$\frac{1}{\varepsilon_1 r} = \frac{1}{\varepsilon_2 r_1} = \frac{1}{\varepsilon_3 r_2}$$

or $$\varepsilon_1 r = \varepsilon_2 r_1 = \varepsilon_3 r_2.$$

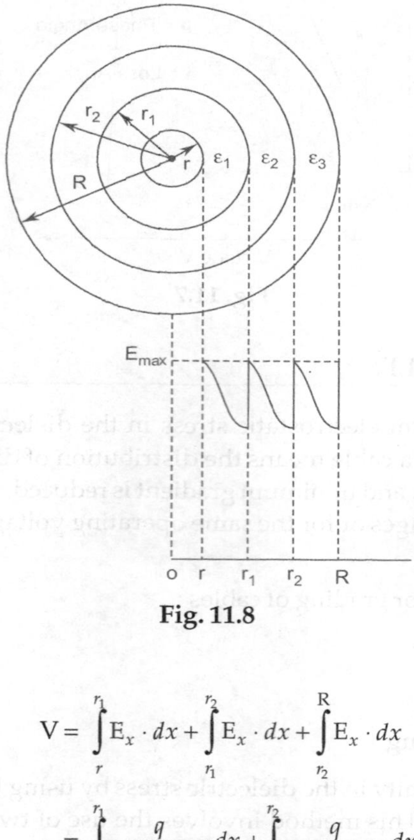

Fig. 11.8

The operating voltage is

$$V = \int_r^{r_1} E_x \cdot dx + \int_{r_1}^{r_2} E_x \cdot dx + \int_{r_2}^{R} E_x \cdot dx$$

$$= \int_r^{r_1} \frac{q}{2\pi\varepsilon_o \varepsilon_1 x} dx + \int_{r_1}^{r_2} \frac{q}{2\pi\varepsilon_o \varepsilon_2 x} dx + \int_{r_2}^{R} \frac{q}{2\pi\varepsilon_o \varepsilon_3 x} dx$$

$$= \frac{q}{2\pi\varepsilon_o \varepsilon_1} \ln\frac{r_1}{r} + \frac{q}{2\pi\varepsilon_o \varepsilon_2} \ln\frac{r_2}{r_1} + \frac{q}{2\pi\varepsilon_o \varepsilon_3} \ln\frac{R}{r_2}$$

or $$V = \frac{q}{2\pi\varepsilon_o \varepsilon_1 r} r \ln \frac{r_1}{r} + \frac{q}{2\pi\varepsilon_o \varepsilon_2 r_1} r_1 \ln \frac{r_2}{r_1} + \frac{q}{2\pi\varepsilon_o \varepsilon_3 r_2} r_2 \ln \frac{R}{r_2}$$

$$= E_{max} \left[r \ln \frac{r_1}{r} + r_1 \ln \frac{r_2}{r_1} + r_2 \ln \frac{R}{r_2} \right].$$

Example 11.5. *A single core lead sheathed cable is graded by using three dielectrics of relative permittivity 5, 4 and 3. The conductor diameter is 2 cm and overall diameter is 8 cm. If the three dielectrics are worked at the same maximum stress of 40 KV/cm, find the safe working voltage of the cable.*

Solution. Given $\varepsilon_1 = 5, \varepsilon_2 = 4, \varepsilon_3 = 3$

As $$\varepsilon_1 \frac{d}{2} = \varepsilon_2 \frac{d_1}{2} = \varepsilon_3 \frac{d_2}{2}$$

$$\frac{5 \times 2}{2} = \frac{4 \times r_1}{2} = \frac{3 \times r_2}{2}$$

∴ $$r_1 = 2.5 \text{ cm}$$

$$r_2 = 3.34 \text{ cm}$$

$$V = E_{max} \left[r \ln \frac{r_1}{r} + r_1 \ln \frac{r_2}{r_1} + r_2 \ln \frac{R}{r_2} \right]$$

$$= \frac{E_{max}}{2} \left[d \ln \frac{r_1}{r} + d_1 \ln \frac{r_2}{r_1} + d_2 \frac{R}{r_2} \right]$$

$$= \frac{40}{2} \left[2 \ln \frac{2.5}{2} + 2.5 \ln \frac{3.34}{2.5} + 3.34 \ln \frac{8}{3.34} \right]$$

$$= 81.808 \text{ KV}.$$

Safe working voltage (rms) for cable

$$= \frac{81.808}{\sqrt{2}} = 57.84 \text{ KV}.$$

Example 11.6. *Find the maximum working voltage of a single core cable having two insulating materials A and B with conductor radius 0.5 cm, inside sheath radius 2.5 cm, maximum working potential gradient of A 60 KV/cm, maximum working potential gradient of B 50 KV/cm, relative permittivities of A and B, 4 and 2.5.*

Solution. Given

$$r = 0.5 \text{ cm}$$

$$R = 2.5 \text{ cm}$$

$$E_A = \frac{q}{2\pi\varepsilon_o \varepsilon_A r} = \frac{q}{2\pi\varepsilon_o 4 \times 0.5}$$

or $$\frac{q}{2\pi\varepsilon_o} = 2 \times 60 = 120$$

$$E_B = \frac{q}{2\pi\varepsilon_o \varepsilon_B r_1}$$ [Dielectric B is between r_1 and R]

or $$50 = \frac{q}{2\pi\varepsilon_o} \frac{1}{\varepsilon_B} \cdot \frac{1}{r_1}$$

$$= 120 \times \frac{1}{2.5} \times \frac{1}{r_1}$$

or $$r_1 = 0.96 \text{ cm}$$

$$V = \frac{q}{2\pi\varepsilon_o \varepsilon_A} \ln\frac{r_1}{r} + \frac{q}{2\pi\varepsilon_o \varepsilon_B} \ln\frac{R}{r_1}$$

$$= \frac{120}{4} \ln\frac{0.96}{0.5} + \frac{120}{2.5} \ln\frac{2.5}{0.96} = 65.50 \text{ KV}.$$

11.10.2 Intersheath Grading

In this method, only one dielectric is used but it is divided into various layers by placing metallic intersheaths between the core and lead sheath. The intersheaths are held at suitable potentials by connecting them to tappings on the windings of the transformer feeding the cable. This arrangement improves voltage distribution in the dielectric of the cable and consequently more uniform potential gradient is obtained.

Figure 11.9 shows a single core cable with two intersheaths of radius r_1 and r_2. Let V, V_1, V_2 be voltage between core and intersheath 1, between intersheath 1 and 2, and between intersheath 2 and outer sheath. Thus, potential difference between conductor and first intersheath is $V - V_1$, between intersheath 1 and 2 is $V_1 - V_2$ and between intersheath 2 and outer sheath is V_2. Thus, the maximum voltage gradients in the three sections are :

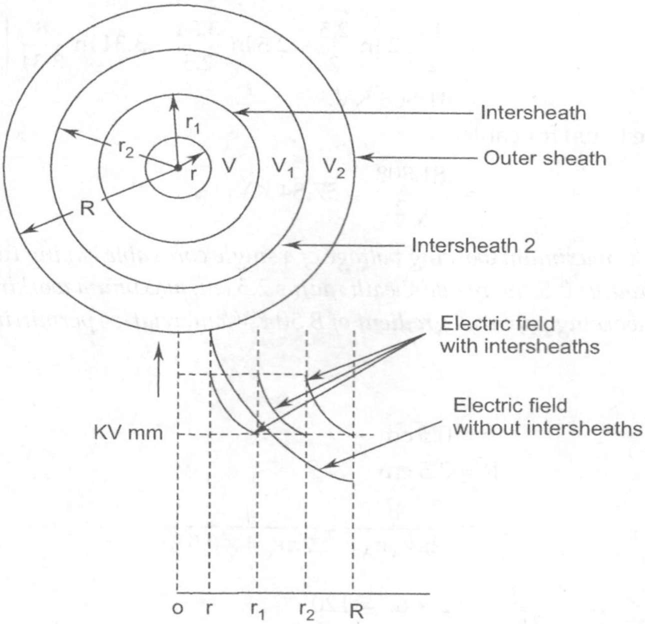

Fig. 11.9

$$E_{max_1} = \frac{V - V_1}{r \ln\left(\frac{r_1}{r}\right)}$$

$$E_{max_2} = \frac{V_1 - V_2}{r_1 \ln\left(\frac{r_2}{r_1}\right)}$$

$$E_{max_3} = \frac{V_2}{r_2 \ln\left(\frac{R}{r_2}\right)}.$$

Let the thickness of the materials be such that

$$\frac{r_1}{r} = \frac{r_2}{r_1} = \frac{R}{r_2} = \alpha$$

Then,
$$\frac{V - V_1}{r \ln \alpha} = \frac{V_1 - V_2}{r_1 \ln \alpha} = \frac{V_2}{r_2 \ln \alpha}$$

or
$$\frac{V - V_1}{r} = \frac{V_1 - V_2}{r_1} = \frac{V_2}{r_2}.$$

Intersheath grading has three principal disadvantages :

(*a*) There are complications in fixing the sheath potentials.

(*b*) There is possibility of damage of intersheath during laying operation and transportation.

(*c*) Since charging current flows through the intersheath which may result in overheating.

For these reasons, intersheath grading is rarely used.

Example 11.7. *A 66 KV concentric cable with two intersheaths has a core diameter 1.8 cm. Dielectric material 3.5 mm thick constitutes the three zones of insulation. Determine the maximum stress in each of the three layers if 20 KV is maintained across each of the inner two.*

Solution. From Fig. 11.10,

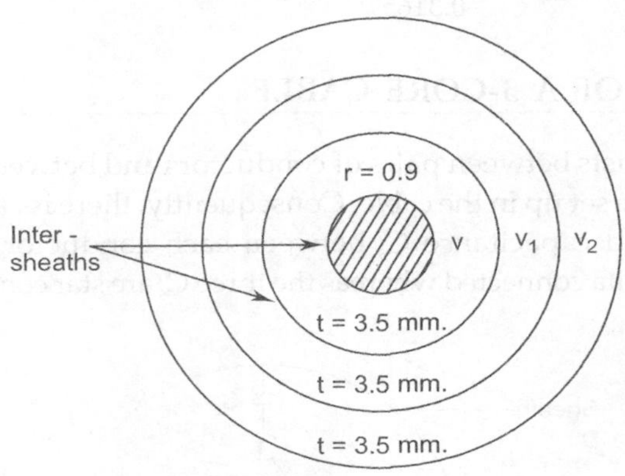

Fig. 11.10

Overall radius of cable

$$0.9 + 3 \times 0.35 = 1.95 \text{ cm}.$$

$$E_{\phi\, max_1} = \frac{V - V_1}{r \ln\left(\frac{r_1}{r}\right)}$$

or $$V - V_1 = E_{max_1} r \left(\ln \frac{r_1}{r} \right)$$

or $$20 = E_{max_1} 0.9 \left(\ln \frac{1.25}{0.9} \right)$$

or $$E_{max_1} = \frac{20}{0.2956} = 67.60 \text{ KV/cm.}$$

$$E_{max_2} = \frac{V_1 - V_2}{r_1 \ln \left(\frac{r_2}{r_1} \right)}$$

or $$V_1 - V_2 = E_{max_2} r_1 \left(\ln \frac{r_2}{r_1} \right)$$

or $$20 = E_{max_2} \times 0.3085$$

or $$E_{max_2} = \frac{20}{0.3085} = 64.83 \text{ KV/cm}$$

$$E_{max_3} = \frac{V_2}{r_2 \ln \left(\frac{R}{r_2} \right)}$$

or $$V_2 = E_{max_3} r_2 \ln \left(\frac{R}{r_2} \right)$$

$$(66 - 40) = E_{max_3} \times 0.3165$$

or $$E_{max_3} = \frac{26}{0.3165} = 82 \text{ KV/cm.}$$

11.11 CAPACITANCE OF A 3-CORE CABLE

Since potential difference exists between pairs of conductors and between each conductor and the sheath, electrostatic fields are set up in the cable. Consequently, there is capacitance C_c between any two pairs of conductors and capacitance C_s between each conductor and sheath as shown in Fig. 11.11. The three C_c are delta connected whereas the three C_s are star connected, the sheath forming the star point.

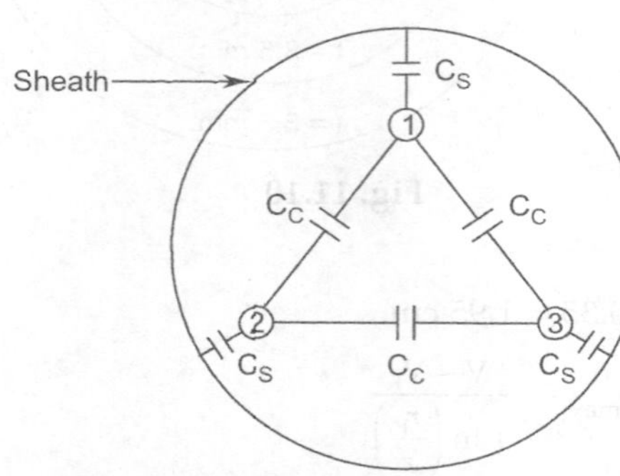

Fig. 11.11

The three delta connected capacitances C_c can be converted into equivalent star connected capacitances as shown in Fig. 11.12. The capacitance to sheath can be assumed to be in parallel with the star connected capacitance of each core to earthed neutral. Therefore, the whole cable is equivalent to three star-connected capacitors each of capacitances shown in Fig. 11.13.

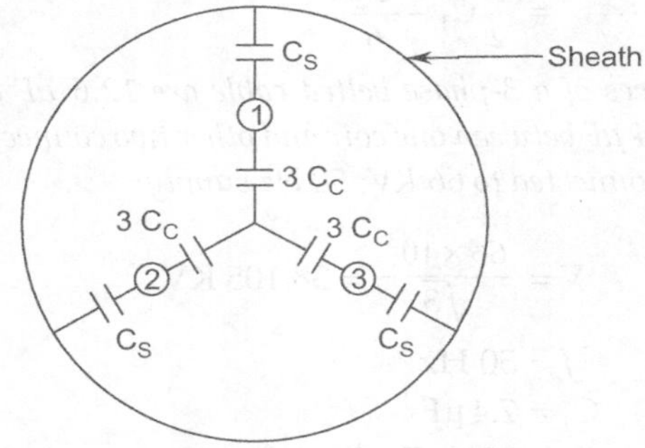

Fig. 11.12

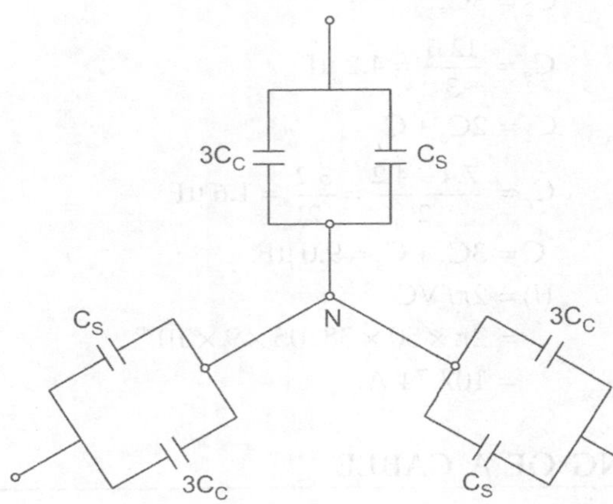

Fig. 11.13

Hence, $$C = 3C_c + C_s$$

The capacitances C_c and C_s are obtained by following measurements :

(*a*) Any two conductors are connected to the sheath and the capacitance C_1 between this combination and third conductor is measured. Then,

$$C_1 = 2C_c + C_s \quad \text{...(1)}$$

(*b*) All the three conductors are joined together and the capacitance C_2 between this combination and sheath is measured. Then,

$$C_2 = 3C_s \quad \text{...(2)}$$

From equations (1) and (2)

$$C_s = \frac{C_2}{3} \text{ and } C_c = \frac{C_1}{2} - \frac{C_s}{2}$$

$$\therefore \quad C_c = \frac{C_1}{2} - \frac{C_2}{6}.$$

The effective capacitance between each core and earthed neutral is :

$$C = 3C_c + C_s$$
$$= 3\left[\frac{C_1}{2} - \frac{C_2}{6}\right] + \frac{C_2}{3}$$
$$= \frac{3}{2}C_1 - \frac{C_2}{6}.$$

Example 11.8. *The capacitances of a 3-phase belted cable are 12.6 μF between the three cores joined together and the lead sheath and 7.4 μF between one core and other two connected to sheath. Find the charging curernt drawn by the cable when connected to 66 KV, 50 Hz supply.*

Solution. Given $V = \frac{66 \times 10^3}{\sqrt{3}} = 38.105 \text{ KV}$

$f = 50 \text{ Hz}$
$C_1 = 7.4 \, \mu F$
$C_2 = 12.6 \, \mu F$

As $C_2 = 3C_s$

∴ $C_s = \frac{12.6}{3} = 4.2 \, \mu F$

and $C_1 = 2C_c + C_s$

or $C_c = \frac{7.4 - 4.2}{2} = \frac{3.2}{2} = 1.6 \, \mu F$

Hence, $C = 3C_c + C_s = 9.0 \, \mu F.$

Charging Current $(I) = 2\pi f VC$
$$= 2\pi \times 50 \times 38105 \times 9 \times 10^{-6}$$
$$= 107.74 \text{ A}.$$

11.12 CURRENT RATING OF A CABLE

It is of great importance to determine the optimum current carrying capacity of the cables as the capital investment on underground system is too much. The following factors decide the safe continuous current in a cable :

(*a*) The maximum permissible temperature at which the insulation of the conductor can be operated.

(*b*) The installation and ambient conditions.

(*c*) The method of heat dissipation through the cable.

11.13 OVERHEAD LINES VERSUS UNDERGROUND CABLES

1. The insulation cost is more in case of cables than the overhead lines as overhead lines use bare conductors supported on steel towers insulated from the towers through the porcelain insulators.
2. A cable has high capacitance which results in charging current and reactive power. The flow of charging current causes heating of cable. In the case of overhead lines, inductance is more predominant.

3. The size of conductor for the same power transmission is smaller in case of overhead lines than the cables because of the better heat dissipation in overhead lines.
4. Underground cables give greater safety to the public than the overhead system.
5. The erection cost of an overhead line is much less than the underground cable.
6. Accessibility of fault in cable is poor and repairing of damaged cable is costly, difficult and time consuming.

11.14 HVDC CABLES

Main merits of HDVC cables are :

1. HVDC cables have no practical limit of rating.
2. Losses are limited to I^2 R losses only.
3. HVDC cables do not take continuous charging currents.
4. For monopolar HVDC, only one cable is used. For bipolar HVDC, only two cables are used. For A.C., minimum three cables are used.
5. HVDC cables have lesser insulation thickness and are more compact.
6. D.C. cables can be designed for withstanding high voltage stress, almost five times the voltage stress used in A.C. cables.

D.C. cables are always single core and HVDC cables designed is similar to A.C. power cables. Four different types of designs are adopted :

(*a*) Paper-insulated solid type
(*b*) Oil-filled
(*c*) Gas pressure type
(*d*) Plastic insulated.

11.15 XLPE CABLES

Cross Linked Poly Ethylene (XLPE) is a synthetic insulating material preferred for high voltage cable insulatin. XLPE is technically superior or economical over oil-filled cables, low pressure oil cables, and high pressure oil cables.

Technical merits of XLPE are :

(*a*) Higher temperature withstand during normal loads and short time overloads.
(*b*) Better thermal properties.
(*c*) Smaller thickness due to higher dielectric strength. Thus, more compact for same voltage.
(*d*) More homogeneous, dense and free from voids, moisture and defects.
(*e*) Excellent moisture resistance.
(*f*) High impact strength and abrasion resistance.
(*g*) Resistance to deformation at higher temperatures.
(*h*) Low dielectric loss, high dielectric strength and low p.f.

SELF TEST

Fill in the blanks by inserting appropriate words / figures :

1. In capacitance grading of cables, we use a __________ dielectric.
2. Under operating conditions, the maximum stress in a cable is at __________.
3. A metallic sheath is provided over the insulation to protect the cable from __________.
4. Single core oil filled cables can be employed upto __________ KV.
5. If the length of a cable increases, its insulation resistance __________.
6. If the length of a cable is doubled, its capacitance is __________.
7. The process of achieving uniformity in dielectric stress is known as __________ of cables.
8. The multicore cables for use upto 11000 V are of __________ type.
9. The underground system is __________ costly than the equivalent overhead line system.
10. For operating voltages beyond 66 KV, __________ cables are used.
11. Belted cables are generally used upto __________ KV.
12. There are two methods of grading viz intersheath grading and __________ grading.
13. The most commonly used insulation in high voltage cables is __________.
14. The minimum dielectric stress in a cable is at __________.
15. __________ gas is often used in external pressure cables.
16. In single core cables, armouring is not in order to avoid __________.

Answers to Self Test

1. composite
2. conductor surface
3. moisture
4. 132
5. decreases
6. doubled
7. grading
8. belt
9. more
10. oil filled
11. 11
12. capacitance
13. impregnated paper
14. lead sheath
15. nitrogen
16. excessive sheath losses.

OBJECTIVE QUESTIONS

1. The surge impedance of cables is around :
   (*a*) 20 Ω (*b*) 50 Ω
   (*c*) 100 Ω (*d*) 250 Ω.
2. The surge impedance of a 50 miles long underground cable is 50 Ω. For a 25 miles length it will be :
   (*a*) 50 Ω (*b*) 75 Ω
   (*c*) 25 Ω (*d*) 100 Ω.
3. The charging current drawn by the cable :
   (*a*) lags behind the voltage by 90° (*b*) leads the voltage by 90°
   (*c*) leads the voltage by 180° (*d*) None of these.
4. In capacitance grading of cables, the dielectric used is :
   (*a*) Homogeneous (*b*) Porous
   (*c*) Composite (*d*) Hygroscopic.

5. The material commonly used for sheaths of underground cable is :
(a) Copper (b) Lead
(c) Rubber (d) Steel.

6. Sheaths are used in cables to
(a) Provide mechanical strength (b) Prevent ingress of moisture
(c) Provide paper insulation (d) None of the above.

7. The size of conductor used in power cables depends on :
(a) Type of insulation used (b) Power factor
(c) Opeating voltage (d) Current to be carried.

8. The maximum safe temperature of paper insulated cables is about :
(a) 60°C (b) 95°C
(c) 135°C (d) 165°C.

9. Which of following protects the underground cables against mechanical injury :
(a) Armouring (b) Bedding
(c) Sheath (d) None of these.

10. The cables used for 132 KV are :
(a) High tension (b) Super tension
(c) Extra super voltage (d) Extra high tension.

11. The insulation resistance of the cable decreases with :
(a) decrease in length of cable (b) increase in length of cable
(c) do not change with length of cable (d) None of these.

12. With the rise in temperature, the insulation resistivity :
(a) Increases linearly (b) Decreases linearly
(c) Remains unchanged (d) Reduces exponentially.

13. Cables used for 220 KV lines are invariably :
(a) Paper insulated (b) Mica insulated
(c) Compressed oil or gas insulated (d) None of the above.

14. Single core cables are usually not provided with :
(a) steel armouring (b) paper insulation
(c) rubber insulation (d) lead sheath.

15. Oil filled cables have the advantage (s) of :
(a) smaller overall size (b) most perfect impregnation
(c) easy detection of fault (d) all of the above.

16. Single core cables are usually not provided with armouring in order to :
(a) make the cable more flexible (b) make the cable non-hygroscopic
(c) avoid excess loss in the armour (d) none of the above.

17. The insulation used in a cable designed for use on 1000 KV is usually :
(a) compressed SF_6 gas (b) PVC
(c) impregnated paper (d) any of the above.

18. SF_6 - cable is insulated by :
(a) High pressure oil (b) PVC
(c) impregnated paper (d) compressed gas.

19. Metallic shielding is provided on underground cables to :
(a) reduce corona effect (b) control the electrostatic voltage stress
(c) reduce thermal resistance (d) all of the above.

20. Dielectric strength of rubber is about
(a) 10 KV/mm (b) 20 KV/mm
(c) 30 KV/mm (d) 60 KV/mm.

21. Copper used as conductor in a cable is :
(a) annealed (b) hard drawn
(c) hardened and tempered (d) alloyed with chromium.

22. Normally expected life of a cable is around :
(a) 5 years (b) 15 years
(c) 25 years (d) 50 years.

23. The thickness of insulation layer povided on the conductor in cables depends upon :
(a) operating voltage (b) current to be carried
(c) power factor (d) both (a) and (b).

24. Paper as an insulating material has the main drawback that it :
(a) has poor dielectric strength (b) has high capacitance
(c) is hygroscopic (d) has low insulation resistivity.

25. The insulation resistance of a cable of length 10 km is 1 MΩ, its resistance for 50 km length will be:
(a) 5 M Ω (b) 0.2 MΩ
(c) 1 MΩ (d) None of the above.

26. Underground cables are laid at sufficient depth so as to :
(a) minimise effects of shocks and vibratons owing to passing vehicles.
(b) minimise temperature stresses
(c) avoid being unearthed easily owing to removal of soil
(d) all of the above.

27. The leakage resistance of a 50 km long cable is 1 MΩ. For a 100 km long cable it will be :
(a) 2 MΩ (b) 0.66 MΩ
(c) 1 MΩ (d) None of the above.

28. The dielectric stress in a cable is maximum at :
(a) armour (b) bedding
(c) conductor surface (d) lead sheath.

29. The capacitance of a cable increases :
(a) linearly with the increase in cable length
(b) linearly with the decrease in cable length
(c) exponentially with the increase in cable length
(d) none of the above.

30. To obtain the maximum value of stress in cable the ratio $\frac{R}{r}$ should be :
(a) 1.1 (b) 1.96
(c) 2.718 (d) 2.13.

31. The insulating material most commonly used for power cables is :

(*a*) PVC (*b*) paper

(*c*) rubber (*d*) VIR.

32. In a cable the sheath radius is R and conductor radius is r. As r changes from 0.5 R, to 0.25 R, the maximum voltage gradient in the dielectric :

(*a*) increase by about 6% (*b*) decrease by about 6%

(*c*) increase by about 15% (*d*) decrease by about 15%.

33. The source (*s*) of heat geneation in cables is / are :

(*a*) dielectric losses in cable insulation

(*b*) losses in metallic sheathings and armourings

(*c*) copper loss in conductor

(*d*) all of the above.

34. If *r* is the radius of the conductor and R the radius of the sheath of the cable, the cable operates stably from the view point of dielectric strength if :

(*a*) $\frac{r}{R} < 1.0$ (*b*) $\frac{r}{R} > 1.0$

(*c*) $\frac{r}{R} < 0.368$ (*d*) $\frac{r}{R} < 0.632$.

35. The dielectric strength of impregnated. Paper is about :

(*a*) 5 KV/mm (*b*) 15 KV/mm

(*c*) 20 KV/mm (*d*) 30 KV/mm.

36. The capacitance of a cable depends upon the

(*a*) ratio of sheath diameter and core diameter

(*b*) length of the cable

(*c*) relative permittivity of dielectric used in cable

(*d*) all of the above.

37. The capacitance of a 3-core cable between any two conductors with sheath earthed is 2 μF. The capacitance per phase will be :

(*a*) 1 μF (*b*) 4 μF

(*c*) 1.414 μF (*d*) 0.667 μF.

38. Capacitance grading of cable means :

(*a*) introduction of capacitances at various lengths of cable to counter the effect of inductance.

(*b*) grading according to capacitance per km length of the cable.

(*c*) use of dielectrics of different permittivities.

(*d*) use of dielectrics in different concentrations.

39. If δ is the loss angle of the cable, its power factor is :

(*a*) sin δ

(*b*) cos δ

(*c*) power factor is independent of δ

(*d*) power factor depends upon δ but it is not as per (*a*) and (*b*).

40. Cables generally used beyond 66 KV are :

(*a*) belted (*b*) oil filled

(*c*) SL type (*d*) H-type.

41. In order to eliminate sheath losses, a successful method is :
   (a) cross bonding the cables is enough
   (b) transpose the cables only
   (c) to transpose the cable along with cross bonding
   (d) none of above is effective.
42. In a cable having homogeneous insulating material, the maximum stress is of 20 KV/mm. The dielectric strength of the insulating material should be at least :
   (a) 10 KV/mm (b) 20 KV/mm
   (c) 40 KV/mm (d) 60 KV/mm.
43. Three insulating materials with same maximum working stress and permittivities 2.0, 3.0, 4.0 are used in a single core cable. The location of the materials with respect to the core of the cable will be :
   (a) 2.0, 3.0, 4.0 (b) 3.0, 2.0, 4.0
   (c) 4.0, 3.0, 2.0 (d) 4.0, 2.0, 3.0.
44. Sheaths are used in cables to :
   (a) prevent ingress of moisture (b) provide mechanical strength
   (c) provide proper insulation (d) none of above.

Answers to Objective Questions

1. (b)	2. (a)	3. (b)	4. (c)
5. (b)	6. (b)	7. (d)	8. (b)
9. (a)	10. (c)	11. (b)	12. (d)
13. (c)	14. (a)	15. (d)	16. (c)
17. (a)	18. (d)	19. (d)	20. (c)
21. (a)	22. (b)	23. (a)	24. (c)
25. (b)	26. (a)	27. (d)	28. (d)
29. (a)	30. (c)	31. (b)	32. (a)
33. (d)	34. (c)	35. (d)	36. (d)
37. (b)	38. (c)	39. (a)	40. (b)
41. (c)	42. (b)	43. (c)	44. (b).

EXERCISE

1. What materials are used for the insulation of cables? Discuss the advantages and disadvantages of each.
2. Derive a formula for the electric stress in a single core cable. Where is the stress maximum? Where is it minimum?
3. What is meant by dielectric loss and power factor in cables?
4. On what factors does the insulation resistance of a cable depend?
5. What is the purpose of using intersheath(s) in a cables?
6. What are oil filled cables? Describe the construction of an oil filled cable.
7. For what purpose metallic sheath is provided in underground cables?
8. What is meant by capacitance grading of a cable? Derive expressions for capacitance of and maximum potential gradients in two (or more) dielectrics of a graded cable in terms of dielectric constants and radius of core and overall radius etc.

9. Show that in a 1-core cable the ratio of maximum to minimum potential gradients $g_{max} | g_{min} \frac{R}{r}$, where r and R are the core radius and overall sheath radius respectively.

   Also show that $g = \frac{V}{r \ln R/r}$ .Where V is working voltage of the cable.

10. State the classification of cables and discuss their general construction.
11. For a single-core lead sheathed cable derive the expressions for (*i*) maximum dielectric stress and (*ii*) capacitance.
12. Derive an expression for the insulation resistance of a single core metal sheathed cable.
13. A concentric cable has a conductor diameter of 1 cm and an insulation thickness of 1.5 cm. Find the maximum filed strength when the cable is subjected to a test pressure of 33 KV. **[Ans:** 47,600 volts]
14. Find the most economical diameter of a single core cable to be used on a 132 KV, 3-phase system. Find also the overall diameter of the insulation if the peak permissible stress is not to exceed 60 KV per cm. **[Ans:** d = 3.6 cm, D = 9.786 cm]
15. A single core cable, 2.1 km long, has a conductor radius of 13 mm and an insulation thickness of 5.8 mm. The dielectric has a relative permittivity of 2.8. Find the capacitance per metre length of cable. **[Ans:** 0.886 μF]
16. A 2.2 km long 11 KV, 3-phase, 3-core belted cable gave the following results in a test for capacitance:
    (*i*) Capacitance between two conductors joined to sheath and the third conductor 1.5 μF.
    (*ii*) Capacitance between all the three conductors joined and sheath 1.8 μF. Find
    (*a*) effective capacitance of each core to neutral
    (*b*) Capacitance between any two cores. **[Ans:** 1.95 μF, 0.975 μF]
17. Three single core cables are used for a 132 KV 3-phase underground line. The core diameter of the cable is 2.5 cm and the insulation thickness is 4 cm. Calculate the maximum and minimum electrical stress in the dielectric. Derive formula used if any. **[Ans:** 4.248×10^6 V/m; 1.01×10^6 V/m]
18. Find the overall diameter of single core cable and its most economical core diameter when working on 220 KV, 3-phase system. The maximum permissible stress in the dielectric is not to exceed 250 KV cm. **[Ans:** 3.9 cm, 1.44 cm]
19. A 3-phase, 3-core metal sheathed cable gave the following results for measurement of capacitance:
    (*a*) Capacitance between all the three bunched conductors and sheath 0.8 μF.
    (*b*) Capacitance between any one conductor and sheath with the other two conductors insulated is 0.5 μF. Calculate for the sheathed cable:
    (*i*) The capacitance between any two bunched conductors and the third conductor.
    (*ii*) between any two conductors.

12

HIGH VOLTAGE DC TRANSMISSION

12.1 INTRODUCTION

The initial application of electricity started with the use of direct current. The first central electric station was installed by Edison in New York in 1882 which operated at 110 V DC The invention of transformer and induction motor initiated the use of AC. The polyphase induction motors which serve the majority of industrial and residential purposes are simpler and rugged in construction and cheaper as compared to DC motors of the same ratings. The advantages of 3-phase AC almost eliminated the use of DC systems except for some specific applications in electrolytic processes and adjustable speed motor drives. The main technical problems of long distance power transmission using AC are :- voltage regulation associated with reactive power balance, steady state, transient state and dynamic stability of the system under different load conditions and also under outage conditions. In view of these problems with AC, the DC transmission has staged a come back in the form of high voltage DC transmission.

The first commercially used HVDC link (20 MW, 100 KV), in the world was built in 1954 between the mainland of Sweden and the island of Gotland. In 1961, an under water DC link was set up between England and France. Since then the technique of power transmission by HVDC has been continuously developed. In 1970, thyristor valves replaced the valves based on the mercury-arc technique. To date the biggest HVDC transmission is ITAIPU in Brazil (two bipoles, 6300 MW and ± 300 KV). In India, the first HVDC line is Rihand-Delhi (± 500 KV, 800 MW).

The use of an HVDC link in an AC system requires converter stations at each end of the line. The main equipments in a converter station are transformers and thyistor valves. At the sending end the thyristor valves act as rectifier to convert AC into DC which is transmitted over the line. At the receiving end the thyristor valves act as inverter to convert DC into AC which is utilized at the receiving end. Each converter can function as rectifier or inverter and thus power can be transmitted in either direction.

12.2 PRINCIPLES OF AC/DC CONVERSION

HVDC transmission consists of two converter stations which are connected to each-other by a DC cable/line. Main components of an HVDC transmission is shown in Fig. 12.1.

Two series connected 6-pulse converters consisting of valves and converter transformers are used. A smoothing reactor in the DC circuit reduces the harmonic currents in the DC line and possible transient over currents.

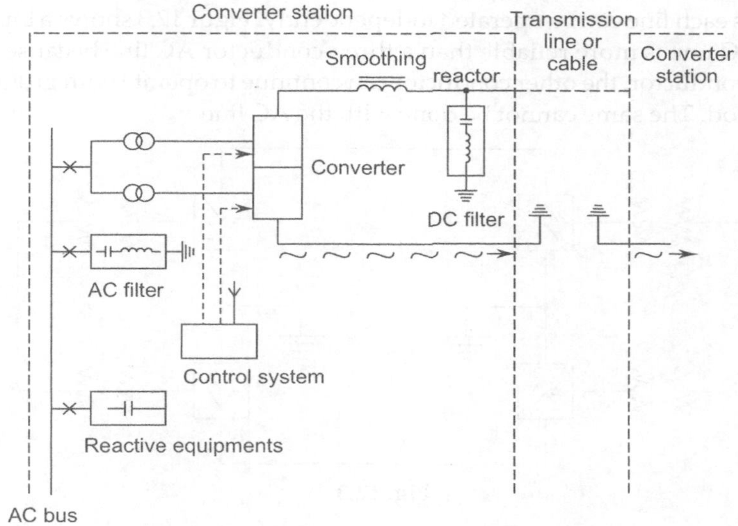

Fig. 12.1

Filters are used to take care of harmonics generated at the conversion. Thus we see that in an HVDC transmission, power is taken from one point in an AC network where it is converted to DC in a converter station (rectifier), transmitted to another converter station (inverter) via line or a cable and injected into an AC system. By varying the firing angle α, the DC output voltage can be controlled between two limits, + ve and – ve. When α is varied, we get :

Maximum DC voltage when $\alpha = 0°$

Rectifier operation when $0 < \alpha < 90°$

Inverter operation when $90° < \alpha < 180°$.

12.3 TYPES OF HVDC TRANSMISSION SYSTEMS

1. **Monopolar System :** As the name suggests, in this system the line has one conductor only and the earth is used as the return conductor. The conductor is usually of negative polarity and the negative polarity is preferred on overhead lines due to lesser radio interference. Figure 12.2 shows a monopolar link.

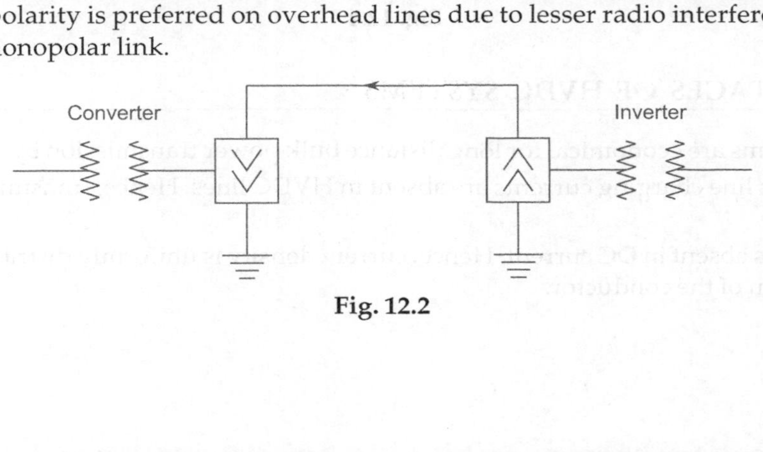

Fig. 12.2

2. **Bipolar System :** The bipolar lines have two conductors—one operating with positive polarity and the other negative polarity. At each terminal, two converters of equal rated voltages are connected in series on the DC side. The rating of the bipolar line is expressed as ± 650 KV. The junction of the converters may be grounded at one end or at both the ends. If it is grounded at both the ends each line can be operated independently. Figur 12.3 shows a bipolar link. A two-conductor DC line is more reliable than a three-conductor AC line because in the event of a fault on one conductor, the other conductor can continue to operate with ground return during the fault period. The same cannot be done with the AC line.

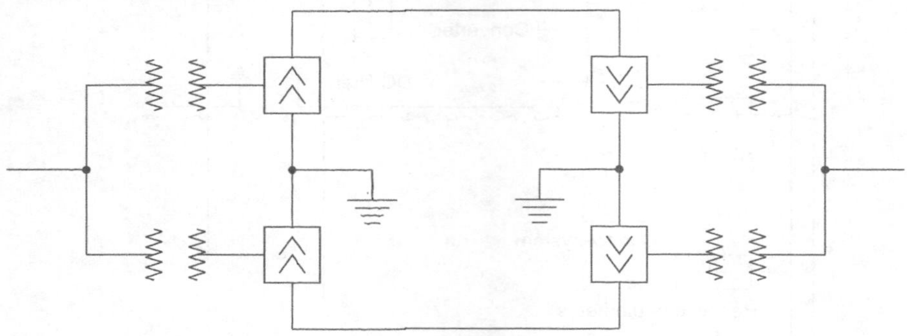

Fig. 12.3

3. **Homopolar System :** The homopolar lines have two or more conductors having the same polarity usually negative for the reason of reduced corona and radio interference and always operate with ground as the return. If a fault develops on one conductor, the converter equipment can be reconnected so that the healthy conductor can supply more than half the rated power. Figure 12.4 shows a homopolar link.

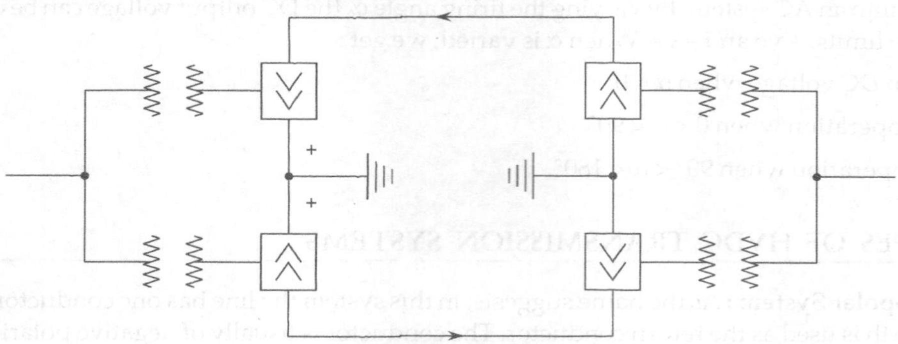

Fig. 12.4

12.4 ADVANTAGES OF HVDC SYSTEMS

1. These systems are economical for long distance bulk power transmission by overhead lines.
2. Continuous line charging currents are absent in HVDC lines. Hence, transmission losses are low.
3. Skin effect is absent in DC current. Hence, current density is uniformly distributed across the cross-section of the conductor.

4. HVDC transmission can utilize earth return and therefore does not need a double circuit.
5. There is easy reversibility and controllability of power flow through a DC link.
6. For the same power transfer and same distance, the corona losses and radio interference of DC systems is less than that of AC systems.
7. No reactive compensation of DC lines is required.
8. The phase-to-phase clearance, phase-to-ground clearances and tower size is smaller for DC transmission as compared to equivalent AC transmission for same power and distance.
9. The voltage regulation problem is much less serious for DC since only the IR drop is involved (IX = 0).
10. The line losses in a DC line are lower than the losses in an AC line of the same power transmission capability.
11. There is greater power per conductor and simpler line construction.
12. Smaller amount of right of way is required. The right of way requirement of a DC line is about 20–40% lesser than that of an AC line of the same power transmission capability.
13. A two-conductor bipolar DC line is more reliable than a 3-conductor, 3-phase AC line because in the event of a fault on one conductor, the other conductor of the DC line can continue to operate with ground return.
14. DC cables can be worked at higher voltage gradient.
15. HVDC line does not need intermediate compensating substation while EHV-AC transmission needs intermediate substations at an interval of 300 km for compensation.
16. For a single DC line between two converter stations, Circuit breakers are unnecessary since control of the converters can be used to block current flow during faulted conditions.
17. HVDC gives asynchronous tie and transient stability does not pose any limit. Line can be loaded up to thermal limit of the line or valves.
18. There is considerable insulation economy. The peak voltage of 400 KV AC line is $\sqrt{2} \times 400 = 564$ KV. So, AC line requires more insulation between the tower and conductors as compared to corresponding 400 KV DC line.
19. An HVDC transmission line can be built in stages. Initially the line can be built as a monopolar line with ground return. Later on the line can be converted into a bipolar line. As a result, the investment for future extension is not required to be made initially.
20. Line losses are smaller.
21. If two AC systems are interconnected by a DC line, a fast change in the direction of energy flow is possible. This allows a more flexible coordination of system control at the two ends.

12.5 DISADVANTAGES OF HVDC SYSTEMS

1. Cost of HVDC terminal substations is very high. There are several additional equipments and auxillaries.
2. HVDC substation require additional harmonic filters and shunt capacitors.
3. HVDC is generally used only for point-to-point transmission. The power transmitted can be used at lower voltage only. Voltage transformation is not easier in case of DC and hence it has to be done on the AC side of the system.

4. Converters require considerable reactive power.
5. HVDC substation has additional losses in converter transformers and valves. These losses are continuous. Elaborate cooling system is necessary to dissipate the heat.
6. Converters do not have overload capability.
7. Operation of HVDC transmission requires continuous firing of thyristor valves.
8. Circuit breaking for multi-terminal lines is difficult.
9. Contamination of insulators is polluted in some areas or along the sea coast. Pollution affects DC more than AC. More frequent cleaning of insulators is required.

12.6 ECONOMIC CONSIDERATIONS

Consider an AC line and a DC line having the same number of conductors and insulators. Now, we will compare the power per conductor on the two lines. Assuming that the current is limited by temperature rise, the direct current equals the rms alternating current. Assuming that the insulators withstand the same peak voltage to ground in each case, the direct voltage is $\sqrt{2}$ times the rms AC voltage.

The power per conductor in case of DC is

$$P_d = V_d I_d$$

Power per conductor in AC is

$$P_a = V_a I_a \cos\phi$$

where I_a and I_d are the currents per conductor and V_a and V_d are line-to-ground voltages and $\cos\phi$ the power factor.

Since,
$$V_d = \sqrt{2}\, V_a$$

and
$$I_d = I_a$$

$$\therefore \quad \frac{P_d}{P_a} = \frac{V_d I_d}{V_a I_a \cos\phi}$$

$$= \frac{\sqrt{2}}{\cos\phi}$$

Since $\cos\phi \le 1.0$, the power per conductor in case of DC is more as compared to AC.

Now compare the power transmission capabilities of a 3-phase single circuit line and a bipolar line. The power capabilities of the respective circuits are :

$$P_d = 2\, p_d$$

and
$$P_a = 3\, p_a.$$

Where p_d and p_a are the power transmitted per conductor of DC and AC lines.

$$\frac{P_d}{P_a} = \frac{2 p_d}{3 p_a}$$

$$\frac{P_d}{P_a} = \frac{2 V_d I_d}{3 V_a I_a \cos\phi}$$

$$= \frac{2\sqrt{2}}{3\cos\phi} = \frac{2.828}{3\cos\phi} \approx 1 \qquad [\cos\phi \le 1]$$

Therefore, the power transmission capability of the bipolar line is same as that of the 3-phase single circuit line. However, the DC line is simpler and cheaper having two conductors instead of three. The percentage loss of the DC line is only two-thirds that of AC line.

12.7 ECONOMIC DISTANCES FOR DC TRANSMISSION

The cost per unit length of a DC line is less as compared to AC line for the same power capacity and comparable reliability whereas the cot of terminal equipment is much more in case of DC (converters and inverters) than in case of AC (Transformers). Figure 12.5 shows a comparison of the total cost for DC and AC transmission. The vertical intercept of each curve is the cost of the terminal equipment and the slope gives cost per unit length of the line and other accessories.

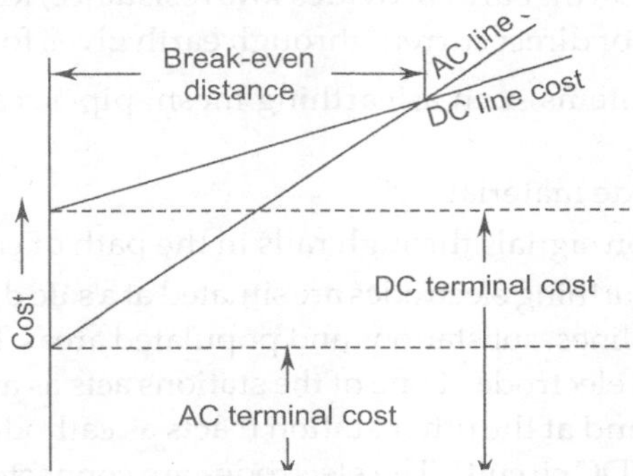

Fig. 12.5

The curve for AC transmission intersects that of DC at an abscissa called a breakeven point. For distances above the breakeven distance, the total cost of DC transmission is lesser than the total cost of AC transmission and thus it is preferable to use DC, otherwise AC should be used. The breakeven distance is generally different for different projects. For overhead lines the breakeven distance generally lies in the range of 600–900 km. For submarine cables the breakeven distance lies in the range of 25–50 km and for underground cables it lies in the range of 50–100 km.

In some cases, it is necessry to adopt DC transmission even for distances less than the breakeven distance because of technical advantages. To give an approximate idea of economic distance of transmission using DC is that for 400 km of distance, the power to be transmitted should be atleast 100 MW.

12.8 HVDC APPLICATIONS

In following particular applications, HVDC is a strong alternative to EHV-AC transmission :

(*a*) Long distance high power transmission by overhead lines.

(*b*) Interconnection between power systems: For smooth interchange of power between neighbouring grids irrespective of voltage and frequency fluctuations, such links ensure retention of the tie under most stringent conditions of the constituent grids.

(c) Multiterminal HVDC system for interconnecting three or more 3-phase AC systems.
(d) Medium and long high power submarine or underground cables.
(e) System interconnection by means of overhead lines or underground cables or back to back HVDC coupling stations.
(f) HVDC links are used to evacuate power from the remote super power stations to the load centres situated several hundred kilometers away.
(g) Frequency conversion (60 Hz – 50 Hz ; 50 Hz – 25 Hz).
(h) Incoming lines in megacities.

12.9 EARTH ELECTRODE AND EARTH RETURN

Earth return is used universally for Monopolar and Bipolar HVDC transmission systems. In monopolar system, full return current flows through the earth. In bipolar systems, only small differential current flows through return earth. Earth provides low resistance, low loss, reliable inexpensive return path. However, the flow of direct current through earth gives following problems:

1. Corrosion of buried metallic items such as earthing mesh, pipes, cable sheaths etc., due to electrolytic action.
2. Rapid consumption of electrode material.
3. Disturbance in communication signals through rails in the path of earth current.

To avoid the above problems, the earthing electrodes are situated at a safe distance (about 5 to 30 km) from the terminal station, major pipelines, substations and populated areas. It is installed in a marshy soil having low resistivity. The earth electrode at one of the stations acts as anode (where DC current from terminal station enters earth) and at the other station it acts as cathode (from which the direct current returns from the earth to HVDC circuit). The electrodes are connected to the neutral point of convertor via electrode line. The electrode line is lightly insulated from the ground. The electrodes are designed such that at the location, the touch potential and step potential are within specified limits.

12.10 LOCATION AND DESIGN OF EARTH ELECTRODE

The location and design of the earth electrode is based on the following design objectives :

1. Low current density at the surface of the earth electrode.
2. Temperature rise of the electrode and surroundings should be with in permissible limits.
3. Step voltage on the ground above the earth electrode should be with in safe limits.
4. Minimum risk of electrolytic corrosion of buried metallic objects such as cables, pipelines etc.
5. Earth resistance should be low.

12.11 CONFIGURATION OF EARTH ELECTRODES

An earth electrode may be a straight electrode, a ring electrode or a radial star electrode buried horizontally in earth at about 1 m depth. Figure 12.6 (a, b, c) shows these configurations :

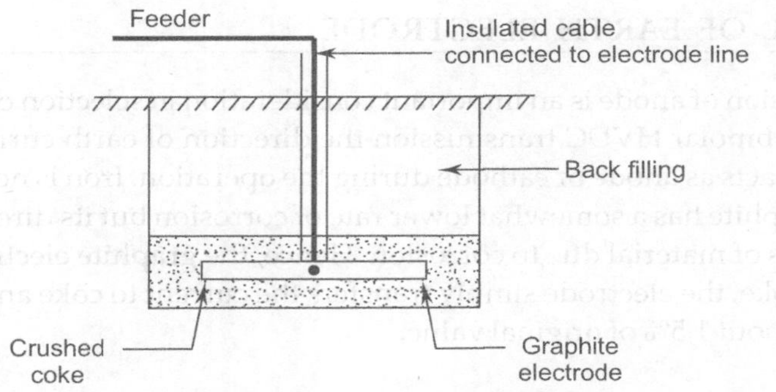

Fig. 12.6(*a*) *Straight*

The back filling is made by original grade of soil. The back filling is also compacted. Graphite electrode elements are used.

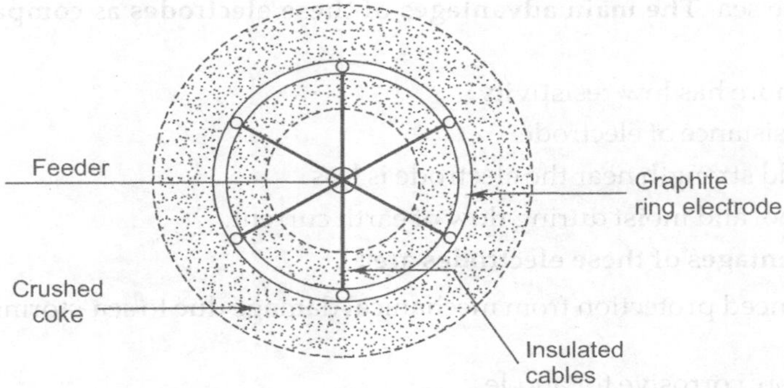

Fig. 12.6(*b*) *Ring*

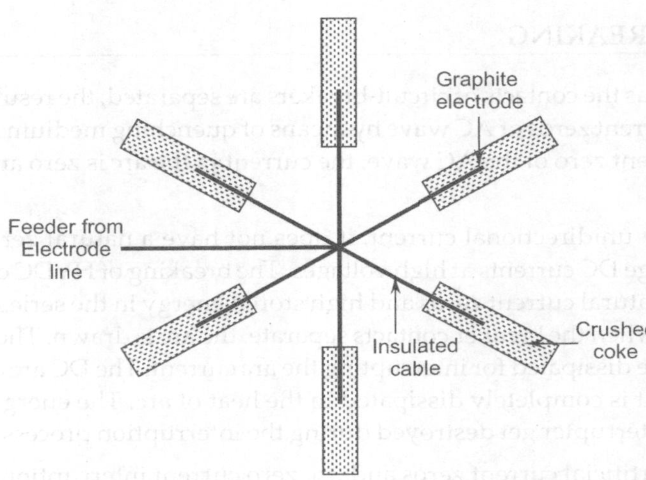

Fig. 12.6(*c*) *Radial Star*

The radial star arrangement uses land area more effectively and is very commonly used.

12.12 MATERIAL OF EARTH ELECTRODE

The electrolytic corrosion of anode is an important consideration in selection of material and design of earth electrode. In bipolar HVDC transmission the direction of earth current is unpredictable. Hence each electrode acts as anode or cathode during the operation. Iron is not used due to its high rate of corrosion. Graphite has a somewhat lower rate of corrosion but its direct burial in earth also causes significant loss of material due to corrosion. Hence, the graphite electrode is buried in a pit filled with crushed coke, the electrode simply transfers the current to coke and the loss of graphite anode is reduced to about 1.5% of original value.

12.13 SEA ELECTRODE AND SHORE ELECTRODE

For submarine cable HVDC transmission, the earth electrode are in the form of sea electrode or shore electrode. The sea electrode is held in the sea water near the shore. The shore electrode is buried on the sea shore near the sea. **The main advantages of these electrodes as compared with the land electrode are :**

(*a*) Soil near sea shore has low resistivity.

(*b*) Lower earth resistance of electrode.

(*c*) The electric field strength near the electrode is less.

(*d*) Soil remains cool and moist during flow of earth current.

The main disadvantages of these electrodes are :

(*a*) Sea electrodes need protection from mechanical damage due to sea-storms and ice formation in winter.

(*b*) Sea water is more corrosive for anode.

(*c*) The sea electrode needs protective covering so that fish do not come in contact.

12.14 CIRCUIT BREAKING

In AC circuit-breakers, as the contacts of circuit-breakers are separated, the resulting arc is quenched at one of the natural current zeros of AC wave by means of quenching medium like vacuum, oil, SF_6 gas etc. At natural current zero of the AC wave, the current in the arc is zero and arc is not allowed to reappear again.

Since DC is a steady unidirectional current, it does not have a natural zero and therefore it is difficult to interrupt large DC currents at high voltages. The breaking of HVDC circuit is difficult due to non-availability of natural current zeros and high stored energy in the series smoothing reactors in HVDC pole circuit. When the breaker contacts separate, the arc is drawn. The entire energy in the series reactors should be dissipated for interrupting the arc current. The DC arc continues to burn till the energy in the circuit is completely dissipated in the heat of arc. The energy is so large and the breaker contacts and interrupter get destroyed during the interruption process.

Solution is to have artificial current zeros and use zero current interruption as in the case of AC circuit Breakers. Some switches have been suggested wherein an artificial zero of current is created through the contacts of the switch by the oscillatory discharge of a capacitor. The crest value of the oscillatory currents should be greater than the direct current to be interrupted. A schematic diagram of such a switch is shown in Fig. 12.7.

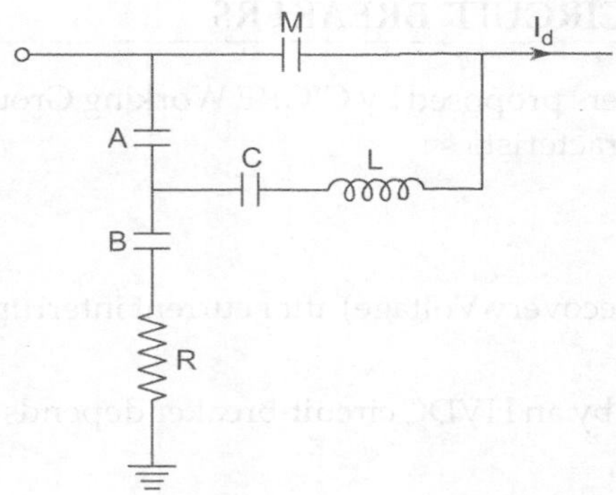

Fig. 12.7

A is a normally open contact whereas M and B are normally closed contacts. As a result, the capacitor C is charged to line voltage through the high resistance R. When it is desired to interrupt the current I_d, the operating mechanism opens contact B and closes A, thus initiating the oscillations in the circuit consisting of M, A, C and L and immediately afterwards the contact M opens which interrupt the current at a current zero such as P as shown in Fig. 12.8. After this, contact A is opened and B closed.

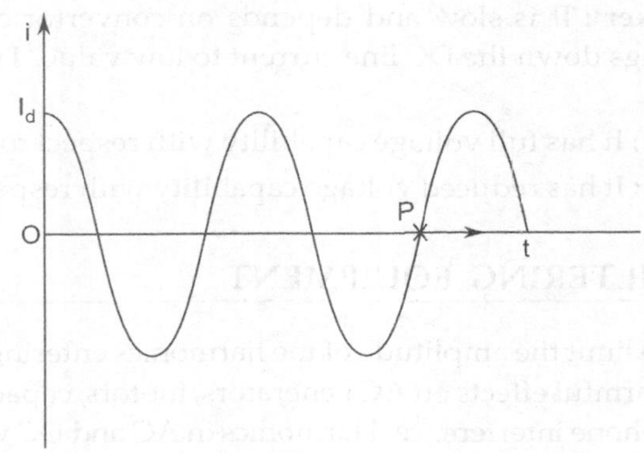

Fig. 12.8

Another switch is shown in Fig. 12.9. Here M is the main contact which is normally closed and C is a capacitor which is normally uncharged. When it is desired to interrupt current I_d, the contacts M open, thereby a part of the current is diverted to the capacitor and as a result the current to be interrupted by the contact M is small. The rate of rise of recovery voltage across M is $\frac{dV_c}{dt} = \frac{I_d}{C}$. The non-linear resistor R absorbs energy without greatly adding to the voltage across M.

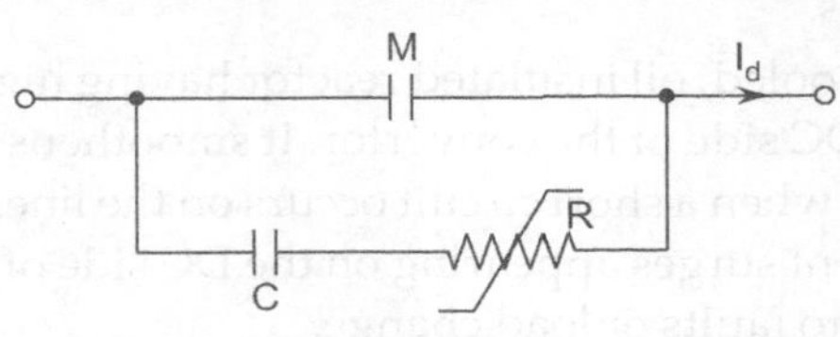

Fig. 12.9

12.15 TYPES OF HVDC CIRCUIT BREAKERS

The type of HVDC circuit-breakers proposed by CIGRE Working Group (1986) are identified on the basis of following principle characteristics :

1. Switching time
2. Current to be interrupted
3. Voltage (TRV-Transient Recovery Voltage) after current interruption.
4. Switching energy.

The current to be interrupted by an HVDC circuit-breaker depends upon :

(a) Switching time

(b) Action of convertor control

(c) Short-circuit ratio of HVDC system w.r.t. AC network

CIGRE working group on HVDC circuit-breakers has given following nomenclature for the types of proposed HVDC breaker.

1. **Type A HVDC Breaker :** It is very fast and does not wait for convertor action to reduce DC voltage. It interrupts line current within 15 ms and before the fault current rises to prospective peak.
2. **Type B HVDC Breaker :** It is slow and depends on convertor control and waits till the convertor control brings down the DC line current to low value. Type B is further classified as :

   (a) **Type B1 Breaker :** It has full voltage capability with respect to TRV withstand.

   (b) **Type B2 Breaker :** It has reduced voltage capability with respect to TRV withstand.

12.16 HARMONICS FILTERING EQUIPMENT

It is necessary to use filters to limit the amplitude of the harmonics entering the AC network and the DC line as harmonics have harmful effects on AC generators, motors, capacitors, supply circuits etc., and cause unacceptable telephone interference. Harmonics in AC and DC waveforms are minimised by following means :

(a) Use of 3-phase convertor bridges of higher pulse number. However, pulse number 12 is most economical and is used in present HVDC systems. In 12-pulse valve group operation, the orders of main harmonics are 12th and 24th on the DC side and 11th, 13th, 23rd and 25th on the AC side.

(b) Use of DC smoothing reactor.

(c) Use of AC harmonic filters.

(d) Use of DC harmonic filters.

DC smoothing reactor is oil cooled, oil insulated reactor having high inductance (0.35 H to 1 H). It is connected in series on the DC side of the convertor. It smoothens the ripple in DC current and limits the current in the rectifier when a short circuit occurs on the line. The DC reactor also helps in reducing the rate of rise of current surges appearing on the DC side of the convertor due to sudden changes in DC power flow due to faults or load change.

DC harmonic filters are shunt filters and are installed in DC yard. They are connected between pole bus and earth bus. They are tuned to certain characteristic DC harmonic frequencies so that these harmonics are diverted to earth and do not pass on to the DC line. They do not contribute to DC shunt compensation as DCline does not need shunt compensation.

AC harmonic filters are installed on AC yard side. These are connected in shunt branches to AC busbars. The AC filters reduce the harmonic voltages and currents in the AC power network to acceptable levels and also provide whole or part of the reactive power needed by the convertor. AC harmonic filters are subdivided in complete harmonic filter banks. The size of each filters bank is specified in terms of reactive power at rated fundamental frequency and rated voltage. The AC filters which have been used in HVDC systems are :

1. Tuned filters for suppressing lower order characteristic harmonics.
2. High pass filters to suppress harmonics of higher order say 24th and above.

The tuned filters and high pass filters are suitably grouped to permit the switching of the groups in accordance with MVAr requirements and harmonic filteration requirements.

12.17 REACTIVE POWER COMPENSATION

Supply of leading Q (Reactive Power) is called Reactive Power Compensation. Reactive Power Compensation has significant influence on the cost of HVDC Transmission Terminal Substation in Bipolar systems as well as Back to back coupling stations. No reactive power is transmitted over a DC line. However, the converters at the two ends draw reactive power from the AC systems. The reactive power balance must be maintained at the two ends to ensure that AC voltage is held within specified limits. Hence, additional compensation is provided on AC side of convertor by means of AC filter capacitors, shunt capacitors, synchronous condensers or Static VAr Sources (SVS).

Reactive powe compensation consumption of convertors vary mainly with the following :

(*a*) Active power ($Q \propto P$)

(*b*) $\sin \propto$ (in case of rectifier) where $\propto$ is delay angle of rectifier and $\sin \gamma$ (in case of inverter) where γ is extinction angle of inverter.

Various methods of compensation of reactive power required by AC side of HVDC convertor are shown in Fig. 12.10.

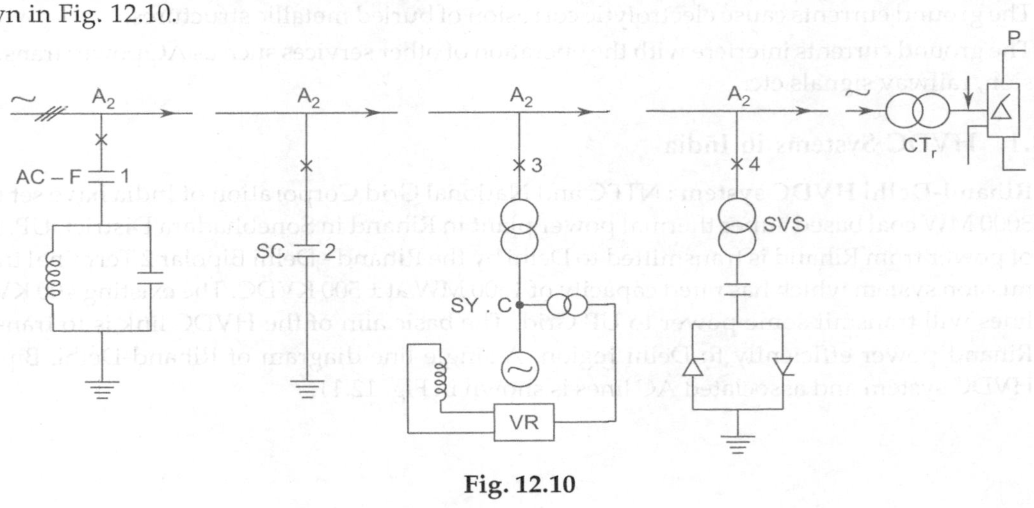

Fig. 12.10

1. AC Filter Capacitors (AC-F)
2. Shunt Capacitors (SC)
3. Synchronous Condender (SY-C) with Voltage-Regulator (VR)
4. Static VAr Source (SVS).

12.18 GROUND RETURN

HVDC transmission lines use ground or sea water as the return conductor continuously (monopolar) or for short times of emergency (bipolar). These return paths are called ground return even if sea water is used as a return path. For the same length of transmission the resistance offered by the ground in case of DC is much less as compared to AC transmission because the DC spreads over a very large cross-sectional area in both depth and width as compared to AC. The resistance of the ground return path is independent of the length of the lines and is essentially the sum of the grounding resistances of the grounding electrodes at the two ends of the line. Since the resistance in case of DC is low as compared to AC, there is low power loss in comparison with a metallic line conductor.

A line with ground return (monopolar) is more economical than a bipolar line because the ground return saves most of the cost of one metallic conductor and the losses in it. In a bipolar link, the main advantage of the ground return is that the line can be built in stages. Initially the power capability requirement is generally less and it will operate as a monopolar line with ground as return and later on in the second stage when the load requirement increases, it can be built as a bipolar line. Thus, a considerable part of the total investment can be deferred until the second stage. The another advantage is the reliability of the system *i.e.*, in the event of one of the lines being under repair or maintenance, the line can be operated as a monopolar pink with ground return and about half the rated load of the bipolar link can be supplied even under such a condition.

The ground return lines have the following disadvantages :

1. It is difficult to design grounding electrodes for low cost of installation and maintenance.
2. Ground currents cause dangerous step and touch voltages.
3. The ground currents cause electrolytic corrosion of buried metallic structures.
4. The ground currents interfere with the operation of other services such as AC power transmission, railway signals etc.

12.18.1 HVDC Systems in India

1. **Rihand-Delhi HVDC system :** NTPC and National Grid Corporation of India have set up a 3000 MW coal based super thermal power plant in Rihand in Sonebhadara District, UP. Part of power from Rihand is transmitted to Delhi by the Rihand - Delhi Bipolar 2 Terminal transmission system which has rated capacity of 1500 MW at ± 500 KVDC. The existing 400 KVAC lines will transmit some power to UP Grid. The basic aim of the HVDC link is to transmit Rihand power efficiently to Delhi region. A single line diagram of Rihand-Delhi. Bipolar HVDC system and associated AC lines is shown in Fig. 12.11.

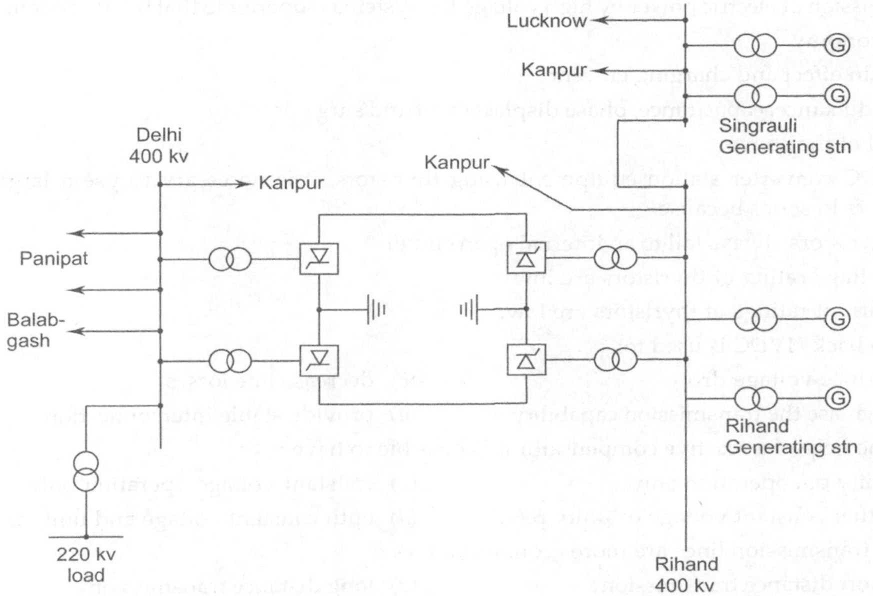

Fig. 12.11

2. **Vindhyachal HVDC back to back system :** This is the first HVDC system in India and this link is for exchange of power between Northern and Western regions. It has two 250 MW converter 12-*p* blocks. Each block consists of rectifier and inverter installed in the same valve hall. Each block of 250 MW is capable of operating independently in either direction and can transfer power in the range of 25 MW to 250 MW depending on system conditions.

SELF TEST

Fill in the blanks by inserting appropriate words/figures :

1. __________ voltage is used for power transmission as a matter of economy.
2. The highest transmission voltage in the world is __________ KV.
3. The higher the transmission voltage, the __________ is the conductor material required.
4. One of the advantage of high voltage DC transmission is that there is no __________ effect.
5. DC transmission is __________ to AC transmission.

Answers to Self Test

1. high
2. 750
3. lesser
4. skin
5. superior.

OBJECTIVE QUESTIONS

1. The main reason for using high voltage for long distance power transmission is:
   (*a*) increase in system reliability
   (*b*) reduction in transmission losses
   (*c*) reduction in time of transmission
   (*d*) none of the above.

2. Transmission of electric power by high voltage DC system is superior to that by AC system in respect of:
   (*a*) economy
   (*b*) skin effect and charging current
   (*c*) inductance, capacitance, phase displacement and surge problem
   (*d*) all of the above.
3. In HVDC converter station equipment using thyristors, it is necessary to use a large number of thyristors in series because :
   (*a*) thyristors always fail to an internal open circuit
   (*b*) voltage rating of thyristors are low
   (*c*) current ratings of thyristors are low.
4. Back to Back HVDC is used to:
   (*a*) reduce voltage drop
   (*b*) decrease line losses
   (*c*) increase the transmission capability
   (*d*) provide stable interconnection.
5. With the help of a reactive compensator it is possible to have :
   (*a*) unity p.f. operation only
   (*b*) constant voltage operation only
   (*c*) either constant voltage or unity p.f.
   (*d*) both constant voltage and unity p.f.
6. HVDC transmission lines are more economical for :
   (*a*) short distance transmission
   (*b*) long distance transmission
   (*c*) inter connected system.
7. The main consideration for higher and higher operating voltage of transmission is to :
   (*a*) reduce power losses
   (*b*) increase power transfer capability
   (*c*) increase efficiency of transmission
   (*d*) both (*a*) and (*c*).
8. In 12 pulse valve group operation, the most trouble some harmonics on AC side are :
   (*a*) 11th and 13th
   (*b*) 5th and 7th
   (*c*) 3rd
   (*d*) 23rd and 25th.
9. Which one of the following is not true regarding HVDC transmission ?
   (*a*) There is no distance limitation for HVDC transmission by underground cable.
   (*b*) The power transmission capability of bipolar line is almost the same as that of single-circuit AC line.
   (*c*) HVDC link can operate between two AC system whose frequencies need not be equal.
   (*d*) Corona loss is much more in HVDC transmission.
10. In a HVDC transmission, the equivalent circuit representing the operation of a bridge rectifier is as shown in the given figure :
    The symbols in the figure have the usual meanings. In this circuit, the voltage drop $\frac{3XI_d}{\pi}$ represents a drop due to

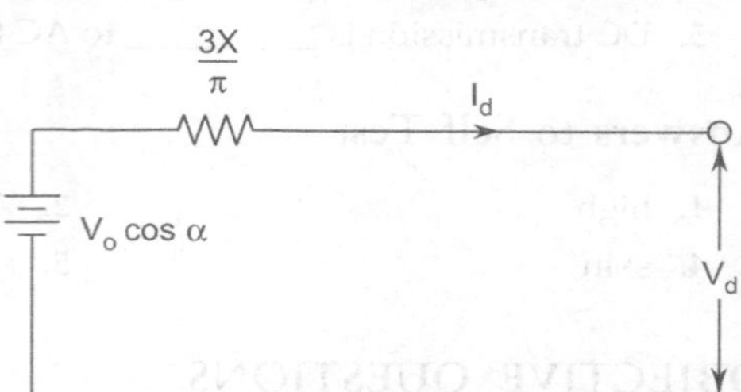

   (*a*) ohmic resistance of the thyristors
   (*b*) commutation in the rectifier
   (*c*) leakage reactance of the transformer
   (*d*) ohmic resistance of the rectifier transformer.
11. In HVDC transmission system, rectifier firing angle alpha is kept near :
    (*a*) 90°
    (*b*) 30°
    (*c*) 15°
    (*d*) 0°.

12. For transmission of power over a distance of 500 km, the transmission voltage should be in the range of :

(*a*) 11-22 KV (*b*) 33-66 KV

(*c*) 66-110 KV (*d*) 132-220 KV.

13. A DC reactor is connected in series with each pole of a converter station in order to :

(*a*) prevent commutation failures in the inverter

(*b*) improve system stability

(*c*) supply reactive power to the converter

(*d*) increase the power transfer capability.

14. In India high voltage transmission is mostly by :

(*a*) underground system (*b*) overhead system

(*c*) both (*a*) and (*b*) (*d*) none of above.

15. In a high voltage DC transmission scheme, reactive power is needed both for the rectifier at the sending-end and for the inverter at the receiving-end. During the operation of such a DC link the rectifier receives:

(*a*) lagging reactive power and the inverter supplies lagging reactive power

(*b*) lagging reactive power and the inverter supplies leading reactive power

(*c*) leading reactive power and the inverter supplies lagging reactive power

(*d*) leading reactive power and the inverter supplies leading reactive power.

16. The power transmission capability of bipolar lines is approximately :

(*a*) half that of 3-phase single circuit line

(*b*) twice that of 3-phase single circuit line

(*c*) thrice that of 3-phase single circuit line

(*d*) the same as that of 3-phase single circuit line.

17. If a power of 30 MW is to be transmitted over a distance of 160 km, the desirable transmission voltage will be :

(*a*) 33 KV (*b*) 66 KV

(*c*) 110 KV (*d*) 220 KV.

18. Which one of the following V_{dc} vs I_{dc} characteristics of a HVDC link represents the one that would provide stable steady-state operation ?

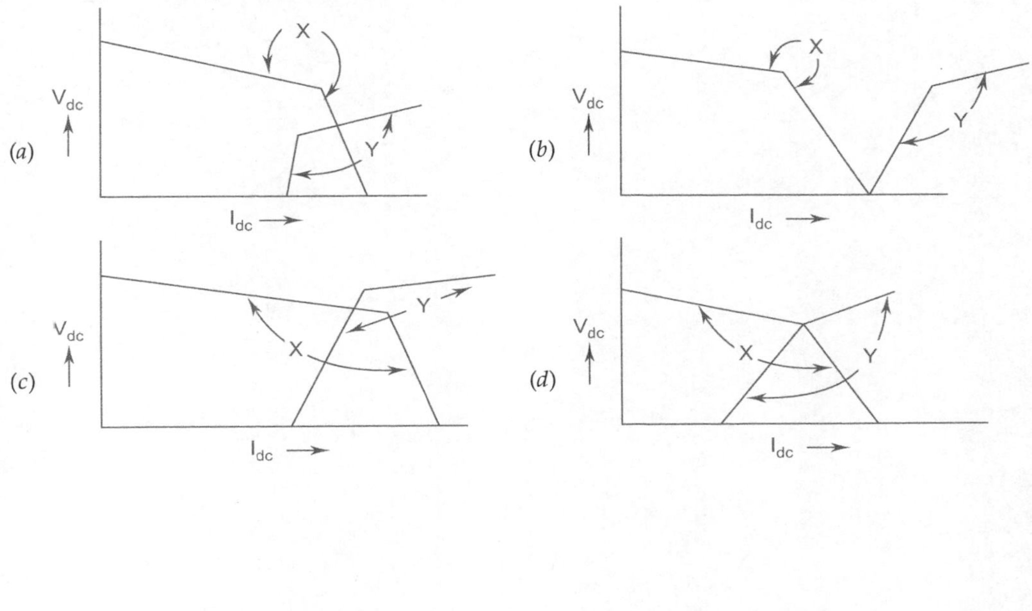

Answers to Objective Questions

1. (*b*)	2. (*d*)	3. (*b*)	4. (*d*)
5. (*c*)	6. (*b*)	7. (*b*)	8. (*a*)
9. (*d*)	10. (*b*)	11. (*c*)	12. (*d*)
13. (*a*)	14. (*b*)	15. (*b*)	16. (*d*)
17. (*c*)	18. (*a*).		

EXERCISE

1. What are the advantages and disadvantages of HVDC transmission?
2. Why is voltage regulation better in case of DC transmission ?
3. Give the applications of HVDC transmission.
4. What type of DC links are used in present day power systems?
5. Briefly state the principle of HVDC system operation.
6. What is a DC link? Show by a single line diagram of a 2-pole HVDC transmission line including the features of the line.
7. Why HVDC line do not require any reactive power compensation?
8. Why DC transmission is superior to AC transmission for very long distances?
9. What is ground return and why is it employed? Give the problems associated with the use of ground as the return conductor.

13

SUBSTATIONS

13.1 STRUCTURE OF A POWER SYSTEM

Electricity is generated at 11 KV by electrical generators which utilize the energy from thermal, hydro nuclear and renewable energy resources. To transmit electricity over long distances, the supply voltage is stepped up to 132/220/400/800 KV as required. Electricity is carried through a transmission network of high voltage lines usually, these lines run into hundreds of kilometers and deliver the power into a common power pool called the grid. The grid is connected to load centres through a sub-transmission network of usually 33 KV (or sometimes 66 KV) lines. These lines terminate into a 33 KV (or 66 KV) substation where the voltage is stepped-down to 11 KV for power distribution to load points through a distribution network of lines at 11 KV and lower.

The power network of concern to the end-user is the distribution network of 11 KV feeder which emanates from the 33 KV substation branches further into several subsidiary 11 KV to carry power close to the load points (localities, industrial areas, villages etc.). At these load points, a transformer further reduces the voltage from 11 KV to 415 V to provide the last-mile connection through 415 V feeders (also called Low-Tension, LT feeders) to individual consumers either at 240 V (as single-phase supply) or at 415 V (as three-phase supply). The utility voltage of 415 V, 3-phase is used for running the motors for industry and agricultural pump sets and 240 V, single phase is used for lighting in houses, schools, hospitals and for running industries, commercial establishments etc.

A feeder could be either an overhead line or an underground cable. In urban areas, owing to the density of consumers, the length of an 11 KV feeder is generally upto 3 km. On the other hand, in rural areas, the feeder length is much larger upto 20 km. A 415 V feeder should normally be restricted to about 0.5 – 1.0 km. Unduly long feeders lead to low voltage at the consumer end. The power supply system including the distribution network is shown in Fig. 13.1.

The main components of the power distribution system and their brief descriptions are given in below table:

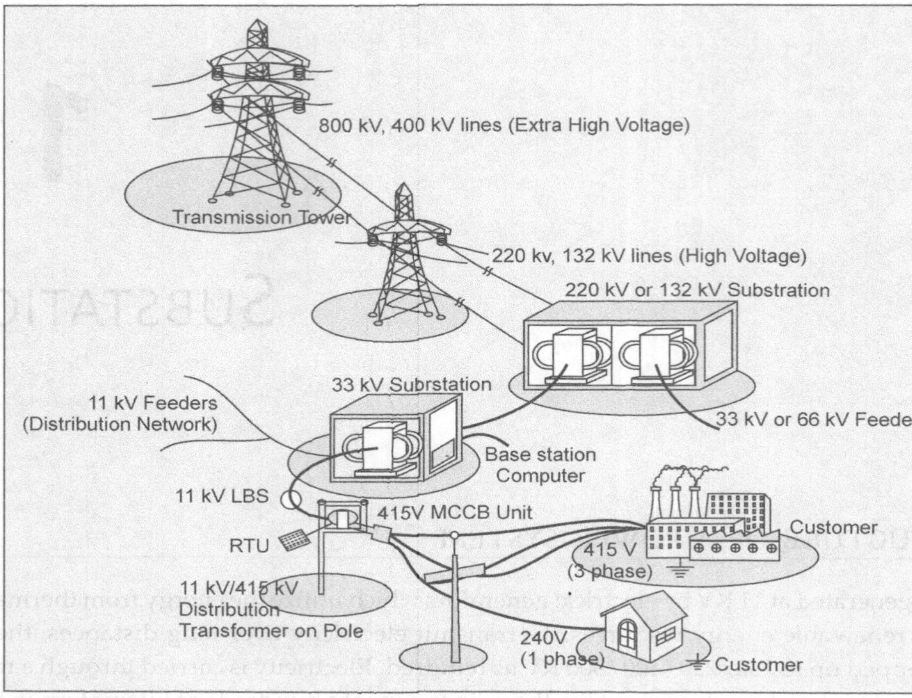

Fig. 13.1

Component	*Description*
Grid substation (GSS)	Power from transmission network is delivered to sub transmission network after stepping down the voltage to 66 KV or 33 KV though 220/132/66/33 KV Grid substations
Sub-transmission Network	Power is carried at 66 or 33 kV by overhead lines or underground cables.
Power sub-transmission (PSS)	Power is stepped-down by 66–33/11 KV to 11 KV for distribution.
Primary Distribution Feeders	Power is delivered from PSS through primary feeders at 11 or 6.6 KV to various distribution transformers.
Distribution substation (DSS)	Power is further stepped down by 11/0.4 KV transformers to utististation voltage of 415V.
Secondary distribution Network	It carries power from DSS at 415 V (240 V single phase) to various consumers through service lines and cables.

13.1.1 Voltage Levels

The voltage range varies widely in various parts of the power supply system. Voltages at different segments in the Power Distribution System are given in below table:

Power System Segment	*Voltage*
Generation voltages	415 V, 6.6 KV, 10.5 KV, 11 KV, 13.8 KV, 15.75 KV, 21 KV and 33 KV.
Transmission voltages	3.3 KV, 6.6 KV, 132 KV, 220 KV and 400 KV,
High voltage primary distribution or sub-transmission	3.3 KV, 6.6 KV, 11 KV, 22 KV, 33 KV, 66 KV
Low voltage distribution phase	415 V (3-phase) and 240 V (1 – phase)

Higher voltages are used for 3-phase, 3- wire supply to large consumers. Low voltage distribution of generally 415 V, 3-phase 4-wire system and 240 V, single phase 2-wire, phase to neutral system is used for small and medium consumers. The size and voltage of supply to a consumer is decided by the load of the consumes.

13.1.2 Conductors

The 11 kV feeders carry comparatively bulk power from secondary substation (33/11 kV) to distribution substation transformers (DTRs). Distribution (or secondary network) carry power from DTRs through service lines (or LT feeders) which deliver power from the supplier's nearest support to the consumer's premises up to the energy meter through a weather - proof service line/wire.

All lines have inherent resistances, inductances and capacitances resulting in a voltage drop in the line. Thus, to minimize voltage drop in a line, the values of these parameters should be carefully selected. For LT supply, the declared voltages at the consumer premises are 415/240 V. All appliances and motors give good performance for long duration if this voltage is maintained.

The following factors should be considered for the proper selection of conductor size:

(*a*) Current carrying capacity &

(*b*) Tensile strength of the conductor.

The size of conductor for a distributor is determined in the following manner:

(*a*) The current that the distributor has to carry is calculated on the basis of the load incident on the conductor.

(*b*) The conductor size capable of carrying this current at the ambient temperature of the area is selected from standard tables.

(*c*) The voltage drop is calculated taking products of loads and their distances.

Mainly the following types of conductors are available:

- All Aluminium Conductor (AAC)
- Aluminium Conductor steel Reinforced (ACSR Conductors)
- All Aluminium Alloy Conductors (AAAC).

ACSR and AAAC conductors are used for secondary distribution systems. ACSR conductors are preferred to AAC conductors for long spans owing to their greater tensile strength. The current carrying capacity of ACSR conductors is as follows:

Squirrel	(7/2.11)	–	115 A
Weasel	(7/2.59)	–	150 A
Rabbit	(7/3.35)	–	208 A

The numbers in bracket indicates the number of strands/diameter in mm.

A simple schematic diagram depicting power system structure is shown is Fig. 13.2.

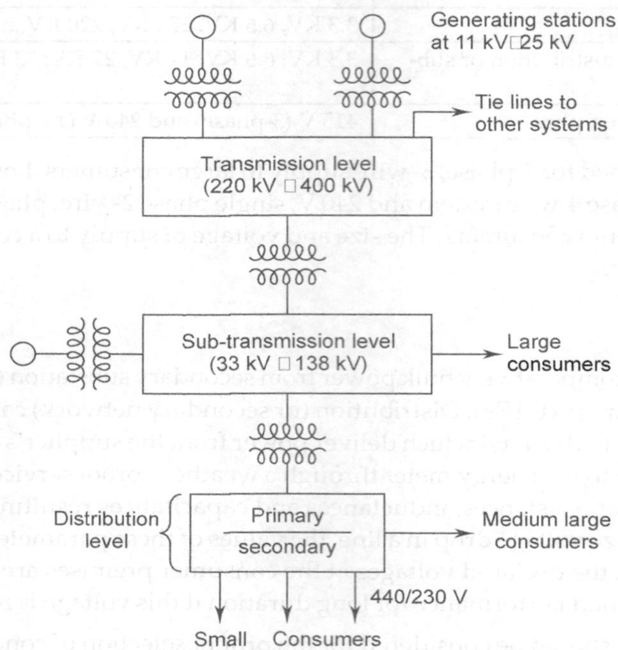

Fig. 13.2

13.2 SUBSTATION

A substation is the meeting point between the transmission grid and the distribution feeder system. This is where a fundamental change takes place within most T & D systems.

The substations may also be defined "as assembly of apparatus which transforms the characteristics of electrical energy from one form to another, say for example, from a.c to d.c or from one voltage to another."

Substations are important part of power system. The continuity of supply depends to a considerable extent upon the successful operation of substations. It is therefore essential to exercise to exercise utmost care while designing a substation. The following important points must be kept in view while laying out a sub-station:

(*a*) It should be located at a proper site

(*b*) It should be easily operated and maintained

(*c*) It should provide safe and reliable arrangement

(*d*) It should involve minimum capital cost.

Typically, a substation consists of high and low voltage racks and buses for power flow, circuit breakers at the transmission and distribution level, metering equipment and the control house where the relaying, measurement and control equipment is located. The most important piece of equipment that gives the substation its capacity rating is the substation transformer. It converts the incoming power from transmission voltage levels to the primary voltage for distribution. Very often, a substation has more than one transformer. Apart from the transformer, a substation has other equipment such as lightning arrestors, isolators etc.

Power distribution substations are diagrammatically shown in Fig. 13.3

Fig. 13.3

As the cost of material used in a distribution circuit per KVA of distributed power varies as the square of the linear dimensions of the supply area. If the numbers of substations is less (*i.e.*, supply area per substation is large), the cost of distribution circuit will be high. It the number of substations is large, the cost of distribution network will be less but the substation cost will be high. The below mentioned table shows the size of substations usually employed:

Type of Consumer	*Size of Substation*	
	Distribution 11000/400 V	Secondary 33/11 KV
Domestic commercial (small and medium) and small industrial	25–1000 KVA	–
Commercial (large) and industrial (medium)	300–3000 KVA	2–10 MVA
Industrial (large) HV supply	–	2–30 MVA

13.3 INDOOR AND OUTDOOR SUBSTATIONS

(*a*) **Indoor Substations:** In such substations, the equipments are installed indoor or with in the substation building. Such substations are usually for a voltage upto 11 kV. However, when the atmosphere is contaminated with impurities, these substations can be erected for voltages upto 66 kV.

According to construction, indoor distribution transformer substations and high voltage switch boards are further sub divided into the following categories:

- *Substations of the integrally built type:* In which the apparatus is installed on site.
- *Substations of the composite built-up type:* In which the assemblies and parts are factory or workshop prefabricated but are assembled on site with in a substation switchgear room.

- *Unit type factory fabricated substations:* In which the assemblies and parts are built in electrical workshops and are shipped to the site of installation fully pre-assembled. After installation of substations and switchboards, only connection to the incoming and outgoing power circuits are required to be mode

  The auxiliaries and equipments of indoor type of substations are discussed later on in this unit

(*b*) **Outdoor Substations:** For voltages beyond 66 KV, equipment is invariably installed outdoor. It is because for such voltages, the clearances between conductors and the space required for switches, circuit breakers and other equipment becomes so great that it is not economical to install the equipment indoor. These substations are further subdivided into:

- *Pole mounted substations:* A pole mounted substation consists of transformer and the isolating switch mounted on an H-pole or an iron platform formed by four poles. The H-pole mounting is limited to transformers upto 160 KVA capacity. Such substations are cheapest, simple and smallest of substations and used for voltages not exceeding 11 KV (or 33 KV in some cases). Electric power is almost distributed in localities through such substations. Figure 13.4 shows the layout of a pole-mounted substation and Fig. 13.5 shows the schematic connections.

  The 11 KV line is connected to the transformer (11 KV/400 V) through gang isolator and fuses. HT fuse unit is installed for protection of HT-side. Lightning arresters are installed on the HT-side to protect the transformer from surges. Substation is earthed at two or more places.

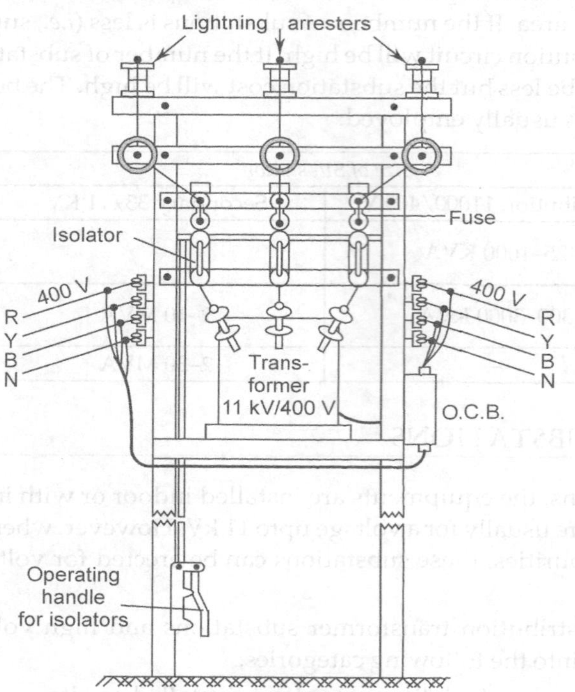

Fig. 13.4

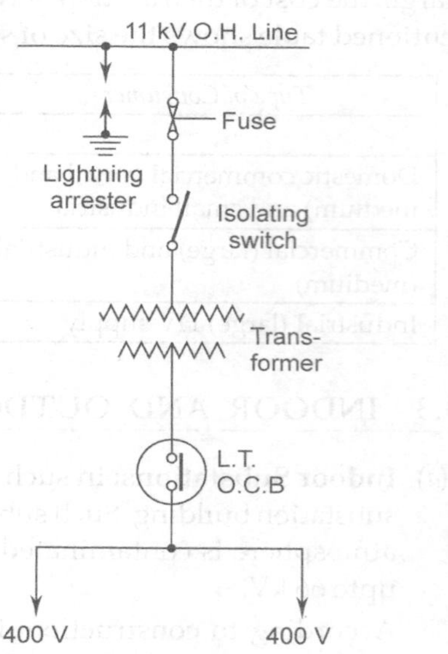

Fig. 13.5

The Oil Circuit Breaker (OCB) is installed on the L.T. side and it automatically isolates the transformer form the consumers in the event of any fault. Generally, transformers of capacity upto 125 KVA are mounted on double pole structure and for capacity upto 125 KVA but not exceeding 250 KVA, 4 - pole structure with suitable platform is used. This type of pole mounted substation is erected in very thickly populated areas. The maintenance cost of such substations is low.

- *Floor/foundation mounted substations:* For transformers of capacity above 250 KVA, the transformers are too heavy for pole mounting. Such substations are usually for voltages of 33 kV and above.

These substations are built entirely in open and in such substations all the equipment is assembled into one unit usually enclosed by a fence from the point of view of safety. Substations for primary and secondary transmission and for secondary distribution are foundation mounted outdoor type.

13.4 ADVANTAGES AND DISADVANTAGES OF OUTDOOR SUBSTATION AS COMPARED TO INDOOR SUBSTATION

Advantages:

(*a*) The constructional work needed is much smaller as compared to indoor substations.
(*b*) Less quantity of building material is needed.
(*c*) Cost of the switchgear installation is low.
(*d*) Fraction of such substations can be completed in much lesser time.
(*e*) Whole of the equipment can be viewed which makes fault location easier.
(*f*) The extension of the installation is easier if required.
(*g*) Adequate space between two adjoining equipments can be provided which will reduce the possibility of faults.

Disadvantages:

(*a*) A much more space is needed for arranging the equipment.
(*b*) Protection devices are required to be installed for protection against lightning surges.
(*c*) The dirt and dust deposits over the contact switches etc, thus the maintenance cost increases.
(*d*) The length of control cables required is more.
(*e*) The switching operation becomes difficult during rainy season as the chances for leakage increase.

13.5 COMPARISON BETWEEN OUTDOOR AND INDOOR SUBSTATIONS

It is shown in tabular form as below:

S.NO	*Particulars*	*Outdoor s/str*	*Indoor s/str*
1.	Space required	More	Less
2.	Fault location	Easier	Difficult
3.	Time required for erection	Less	More
4.	Capital Cost	Low	High
5.	Operation	Difficult	Easier
6.	Future extension	Easy	Difficult

13.6 SYMBOLS FOR EQUIPMENT IN SUBSTATIONS

Symbols of important equipment used in substation are given below:

S.No	*Circuit Element*	*Symbol*
1.	Earthing	
2.	Bus-bar	
3.	Current transformer	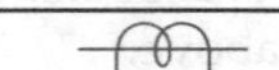
4.	Potential transformer	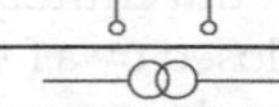
5.	Single break isolating switch	
6.	Double-break isolating switch	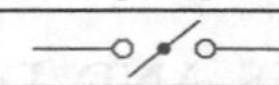
7.	Oil circuit breaker	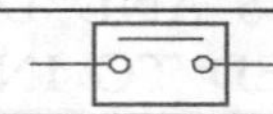
8.	Air circuit breaker	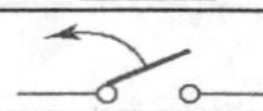
9.	Air-blast circuit breaker	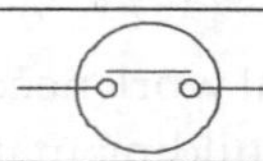
10.	Fuse	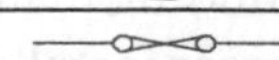
11.	Lightning arrester (active gap)	
12.	Lightning arrester (valve type)	
13.	Arcing horn	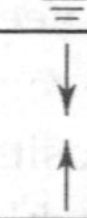
14.	Earth fault relay	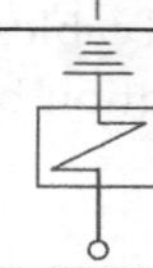
15.	Over current relay	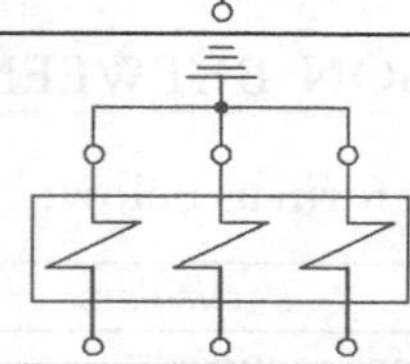
16.	Capacitive voltage transformer	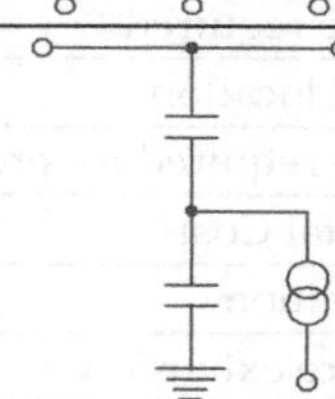

13.7 SUBSTATION EQUIPMENTS

13.7.1 66 – 33/11 KV Substation Equipment

Main equipments used in 66 – 33/11 KV substation are:

1. **Structures:** The main structures required for 33/11 KV substations are:
    - Incoming and outgoing gantries
    - Support structures for breakers, isolators, insulators, CTs and PTs
    - Bus-bars.

   Structures are required to provide entry from the overhead line to the substation and to extend out required number of feeders. The numbers of structures should be kept to a minimum as large number of structures may be uneconomical and also create obstructions in extending bus-bar, lines etc.
2. **Power Transformers:** Power Transformers are used for stepping-up the voltage for transmission at generating stations and stepping down voltages for further distribution at main step down transformer substations. Power transformers are most of the time loaded to levels just below the rated power and accordingly they are designed to operate at maximum possible flux density level with maximum efficiency at near full load. For ratings upto 10 MVA, naturally cooled, oil immersed transformers are used. For ratings higher than 10 MVA, the transformers are generally air-blast cooled.

   The important specifications for a transformer are:
    - KVA rating
    - Rated voltage
    - Rated frequency
    - Efficiency
    - Output type (3-wire or 4-wire)
    - Connection (Δ or Y)
    - Type of core
    - Toppings if any
    - Temperature rise above ambient in °C.
3. **Bus-bars:** A bus-bar in electrical power distribution refers to the thick strips of copper or aluminum that conduct electricity within the substation [Fig. 13.6]. Bus-bars are merely convenient means of connecting switches and other equipment into various arrangements. The usual arrangement of connections in most of the substations permits working on almost any piece of equipment without interruption to incoming or outgoing feeders.

   The size of the bus-bar is important in determining the maximum amount of current that can be safely carried. The bus-bar should be able to carry the expected maximum load current without exceeding the temperature limit. The capacity of bus should also be checked for maximum temperature under short circuit conditions.

   Bus-bars maybe of copper, aluminum or steel. Copper has a comparatively low resistively and also the advantage of relatively high mechanical strength, this makes it economical to use copper bus-bars in installations of very large capacity where the currents are particularly heavy. Now a days, aluminum is being increasingly used for various switchgear installations due to its numerous advantages over copper.

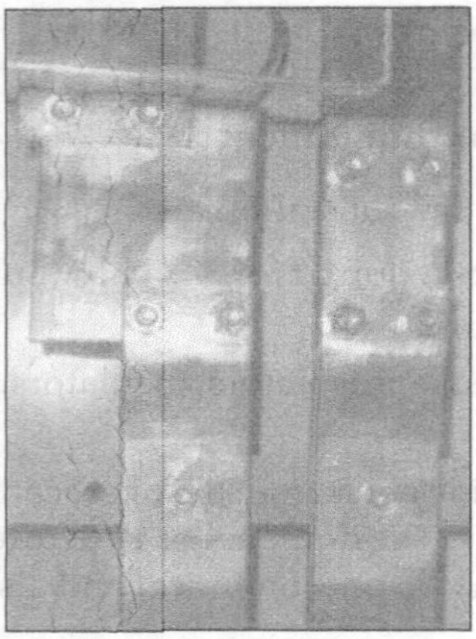

Fig. 13.6

Such as lower cost for equal current carrying capacity and excellent corrosion resistance steel bus bars are used in low capacity installations where the load currents donot exceed 200 – 300 A. In steel bus-bars, the losses due to hysteresis and eddy currents when carrying AC are considerable.

Different types of bus-bars namely single bus bar, double bus-bar, single bus-bar with bus sectionalizer, double bus-bar with double breaker scheme and mesh scheme are used in a substation in accordance with the safety and reliability considerations.

Figure 13.7 represents most simple connections for primary substation having two step-down transformers which are fed through two sections of single bus-bar. Similarly, there are two sections of low voltage bus-bar. The two sections (both on high and low voltage side) can be connected together though sectionalizing circuit breakers SCB_1 and SCB_2 in which case both the transformers can be operated in parallel or independently is that in case of a fault, the fault current will be smaller.

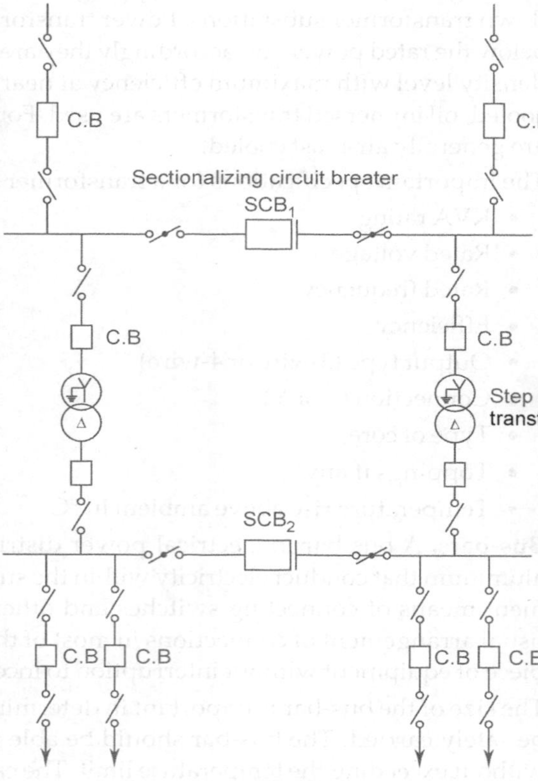

Fig. 13.7

4. **Circuit Breakers:** A circuit breaker is an equipment which can open or close a circuit under normal as well as fault conditions. It is so designed that it can be operated manually under normal conditions and automatically under fault conditions.

A circuit breaker contains both fixed and moving contacts. The circuit breakers are found at the arrivals and departures of all the lines incident on a substation. When the circuit breaker is closed, these contacts are held together. The mode of action of all circuit breakers consists in the breaking of the fault current by the separation of the moving contacts away from the fixed ones. An are is immediately established on separation of the contacts Interruption of the current occurs after the arc at these contacts is extinguished and current becomes zero.

The types of breakers used in a distribution system are:

- Air break type
- Oil break type
- Vaccum type
- SF6 gas breaker.

The rated voltage of circuit breakers for 33 KV level is 36 KV and for 11 KV, it is 12 KV. The short circuit current rating is 25 KA.

Oil break type and SF6 breakers are shown in Fig. 13.8 and Fig. 13.9

Fig. 13.8

Fig. 13.9

5. **Insulators (Isolating Switches):** Isolators are mechanical switching devices capable of opening or closing a circuit
   - When a negligible current is broken or made or
   - Only a small charging current is to be interrupted.

Isolators operate under no-load conditions. In other words. Isolator switches are operated only when the lines in which they are connected carry no current. They are not equipped with arc-quenching devices They donot have any specified current breaking capacity or current making capacity

Isolators arc employed in addition to circuit breakers and are provided on each side of every circuit breaker to provide isolation. While opening a circuit, the circuit breaker is opening first and then isolator. It an isolator is opened carelessly, when carrying a heavy current, the resulting arc could easily cause a flash-over to ground and may shatter the supporting insulators and even cause a fatal accident to the operator. While closing is circuit, the isolator is closed first and then circuit breaker. Consider Fig. 13.10 which shows the use of isolators in a typical substation.

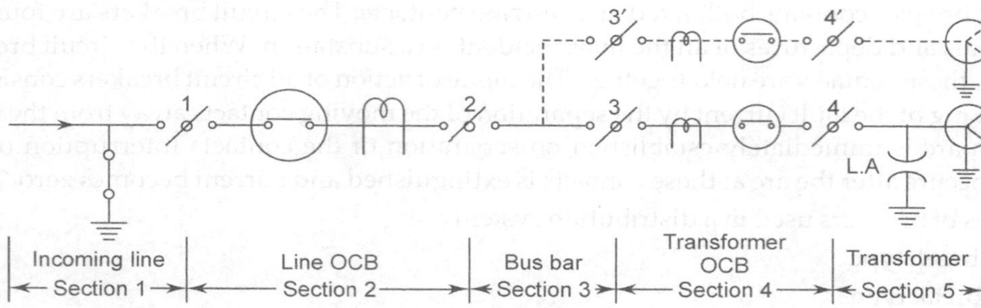

Fig.13.10

The entire substation is divided in to 5 sections. Each section can be disconnected with the help of isolators for repair and maintenance. For instance, if it is desired to repair section No. 2, the procedure of disconnecting this section will be:

First of all open the circuit breaker in this section and then open the isolators 1 and 2. This will disconnect section 2 for repairs.

After the repair, close the isolators 1 and 2 first and then the circuit breaker.

6. **Earthing Switches:** Earthing switch is connected between the line conductor and earth. Normally it is open and it is closed to discharge the voltage trapped on the isolated or disconnected line. When the line is disconnected line. When the line is disconnected from the supply end, there is some voltage on the line to which the capacitance between the line and earth is charged. This voltage is significant in case of HV systems. Before commencement of maintenance work, it is necessary that these voltages are discharged to earth by closing the earthing switch. Normally, the earthing switches are mounted on the frame of the isolator.
7. **Insulators:** An electrical insulator resists the flow of electricity. Application of voltage difference across a good insulator results in negligible electrical current. Adequate insulation is of prime importance for reliability of supply and safety of personal as well as equipment also.

   The insulators in use at substations are post insulators of pedestal type. The station design should be such that the number of insulators is kept at a minimum at the same time ensuring security of supply. In the areas where the problem of insulator pollution is expected such as near the sea, thermal station, railway station, industrial area etc., special insulators with higher leakage resistance should be used.
8. **Protective Relays:** Whenever a fault occurs, the protective relay operates and sends a trip signal to the circuit breaker. The relays are housed in panels in the control room.

   Time graded over current protection is used for feeders. Distance protection and carrier current protection are used for transmission lines. Differential protection is used for transformer protection.
9. **Indicating and Metering Instruments:** Ammeters, voltmeters, wattmeters, kwh meters, KVARh meters, power factor meters, reactive- volt- ampere meters are installed in substations to control and maintain a watch over the currents flowing through the circuits.
10. **Surge (Lightning) Arresters:** Large over voltages that develop suddenly on electric transmission and distribution system are referred as surges or transients. These are caused by lightning strikes or by circuit switching operations. Surge arresters protect the substation equipment from lightning and switching surge. Present day power systems invariably use valve type arresters.

    The rated voltage of arresters for 33 KV should be 30 KV for use on 33 KV systems and with nominal discharge current rating of 10 KA. The rated voltage of lightning arresters for effectively earthed 11 KV system should be 9 KV (r.m.s.) with nominal discharge current rating of 5 KA.

Typical surge arresters used in substation are shown in Fig. 13.11

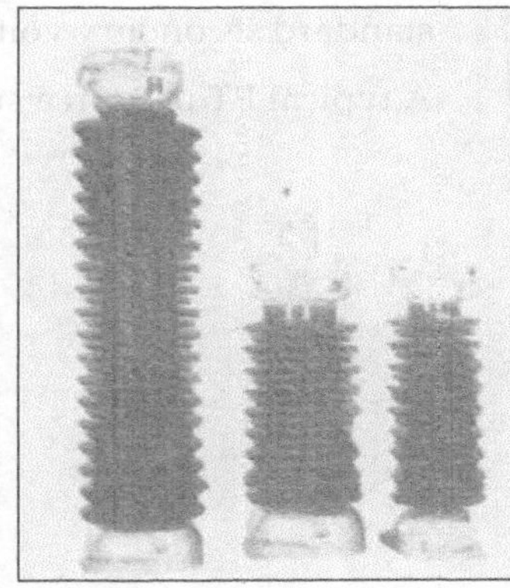

Fig. 13.11

11. **Instrument Transformers:** The lines in substations operate at high voltages and carry current of thousands of amperes. The measuring instruments and protective devices are designed for low voltages, generally 110 V and currents, about 5 A. Therefore, they will not work satisfactorily if mounted directly on the power lines. This difficulty is overcome by installing instrument transformers on the power lines. The function of these instrument transformers is to transfer voltages or currents in the power lines to values which are convenient for the operation of measuring instruments and relays.

    There are 2 types of instrument transformers:

    (*a*) *Current Transformer (CT):* A current transformer is essentially an instrument transformer which steps down the current in a known ratio for the purpose of metering and relaying. The primary winding of this transformer is connected in series with the load and carries the load current to be measured. The secondary winding is connected to the measuring instrument or relay which together with the winding impedance of the transformer and lead resistance constitute the burden of the transformer.

    Fig. 13.12

    CTs are used in power installations for supplying the current circuits of indicating instruments (ammeter, wattmeter etc), energy meters and protective relays. These transformers are designed to provide a standard secondary current output of 1 or 5A when rated current flows through the primary. A fundamental characteristic of CT is its transformation ratio expressed as the ratio of the rated primary to rated secondary current. The secondary current ratings are of the order of 1 A or 5 A. Primary current ratings vary from 10 A to 3000 A or more.

    CT-units used in a substation are shown in Fig. 13.12

    Suppose a CT rated at 100/5 A is connected in the line to measure current. If the current in the line is 100 A, then current in the secondary will be 5 A. If the current in the line is 50 A, then secondary of C.T. will have a current of 2.5 A.

    (*b*) *Voltage (Potential) Transformer (PT):* It is an instruments transformer which steps down the voltage in a known ratio for metering and relaying. The primary winding of the PT is connected directly to the power circuits either between two phases or between a phase and ground depending on the rating of the transformer. To the secondary windings various indicating and metering instruments and relays are connected.

PTs are used for supplying the voltage circuit of indicating instruments, integrating meters, other measuring apparatus and protective relays or trip coils. These may be of single phase or three phase design and of the dry or oil immersed types. A PT is rated by the secondary output it will deliver without exceeding specified temperature rise. All PTs are designed for a standard secondary voltage of 110 V or $110/\sqrt{3}$ V.

A typical PT used in a substation is shown in Fig. 13.13

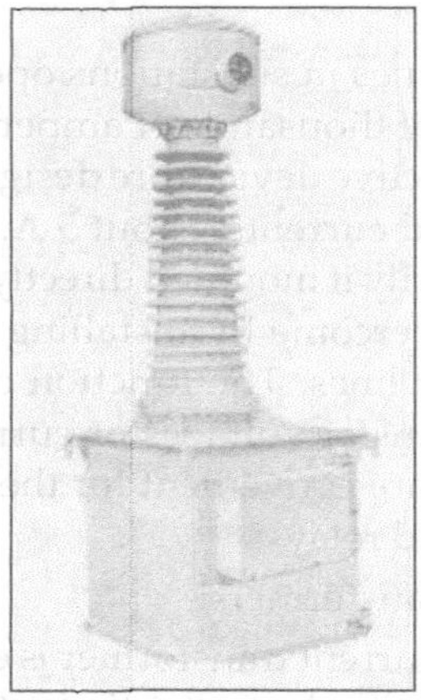

Fig. 13.13

Suppose a PT rated at 66 KV/110 V is connected to a power line. If the line voltage is 66 KV, then voltage across the secondary will be 110 V.

12. **Earthing Arrangements:** The function of station earthing system is to provide a low resistance path for flow of earth fault currents, for safety of equipment and personnel and for improved reliability of power supply.

    Earth mat is provided within the substation area. The earth rods are connected to the station earth mat. The earthing must be designed so as to keep the earth resistance as low as possible.

13. **Shunt Capacitors:** They supply the reactive power to the loads and thus help in improving the power factor. They are switched on during peak load conditions and switched off during opp-peak conditions. Shunt capacitors are connected on the secondary side (11 KV) of the 33/11 KV power transformers as shown in Fig. 13.14. The capacitors are generally of automatic switched type. The automatic system of the capacitor bank has the task of switching in the necessary capacitance according to the load requirements at each given moment.

Fig. 13.14

14. **Station Battery/DC Supply Arrangement:** Station batteries supply energy to operate protection equipment such as breakers, alarm and other control equipments. Battery bank is shown is Fig. 13.15. The station batteries are a source for operating DC control system equipment during system disturbances and outages. During normal conditions, the rectifier provides the required DC supply. However to take care of rectifier failure, a storage battery of adequate capacity is provided to meet the DC requirements. Normally, in a 33/11 KV substation, the DC system is of 30 cells consisting of 15 lead acid storage batteries or Nickel-cadmium batteries. The batteries are equipped with charging equipment 110 V DC supply is used for medium size substations and 240 V supply is used for large substations.

Fig. 13.15

15. **Fire Fighting System:** In view of the presence of oil filled equipment in a substation it is important that proper attention is given to isolation and extinguishing of fire so as to avoid damage to costly equipment and also ensure the safety of personnel. The layout of the substation itself should be such that the fire should not spread to other equipment as far as possible. Fire extinguishes of the following types must be used:
    - Carbon dioxide extinguisher and
    - Dry chemical Powder extinguisher.

    For oil fire, foam type extinguishers are used.

13.7.2 11/0.4 kV Substation Equipment

The main equipments used in a 11/0.4 kV substation are:

1. **Distribution transformer:** Transformers can be mounted outdoors in one of the following ways:
    - *Plinth mounting:* The transformer is mounted on a plinth made of concrete as shown in Fig. 13.16. The plinth has to be higher than the surroundings. The method can be used for all sizes of transformers. Where the distribution substations are plinth mounted, they are efficiently protected by fencing so as to prevent access to the apparatus by unauthorized persons.

Fig. 13.16

- *H-pole mounting:* The transformer can be mounted on cross arms fixed between two poles which are rigidly fastened to the poles. The transformer has two base channels which rest on the transformer mounting structure as shown in Fig. 13.17.

Fig. 13.17

- *Direct mounting:* The transformer is clamped directly to the pole by suitable clamps and bolts as shown in Fig. 13.18. This method is used for transformers upto 25 KVA only.

Fig. 13.18

2 **Earthing:** Pipe or rod earthing is provided for distribution substation. Three electrodes forming an equilateral triangle are provided so that adequate earth buffer is available.

3. **LT distribution box:** For transformers of 100 k and above, sheet metal LT distribution box consisting of LT breaker and fuse cut-outs is provided from where distribution feeders are to be taken out. The size of box has to be suitable for accommodating MCCB, fuse cut-outs, cable-connectors, bus-bars etc.

4. **Protection system:** The HT side of all transformer is normally protected by drop out expulsion type fuse. Three 11 KV drop out fuse units comprising a set are installed on mounting cross-arm. Horn gap fuses are also used on distribution transformers on HT side. The fuse wire is fixed between arcing horns. The advantage is that ordinary fuse wire of rated capacity can be used for replacement while for dropout fuses, fuse elements are required to be stocked for replacement.

5. **Lightning arrestors:** 11 kv lightning arrestors of outdoor type are used for diverting the lightning surges to earth resistance of earth resistance of earth. The lightning arrestor should be installed on the HT side and its lead should be kept at a minimum.
6. **Reactive compensation:** The load incident on the distribution system is mostly inductive requiring large reactive power. The best method is to compensate the reactive power at the load end it self but it is difficult to implement in practice. Hence, providing compensation on the distribution system is essential. So wherever the power factor is low, reactive compensation may be provided on the distribution transformers.

The shunt capacitors supplies constant reactive power at its location independent of the load. Fixed or automatic switched type capacitors of adequate rating are to be provided on the LT- side of the distribution transformers. In the switched capacitor system, the capacitors are switched on and off along with the load to avoid over-voltage during low load operation.

13.8 KEY DIAGRAM OF 33 KV/11KV SUBSTATION

Figure 13.19 shows a key diagram of a typical primary substation with two 33 KV incoming lines and four 11 KV outgoing feeders. A circuit breaker is connected in each incoming and outgoing line. On each side of breaker is an isolator. The isolator should never be opened on load. The feeder protection is by time graded over current relays. The circuit breakers have auto reclosing facility. A lightning arrestor is connected on each incoming and outgoing circuit for lightning protection. Each incoming line is connected of HV bus (33 KV) through a circuit breaker and two isolators. Each outgoing feeder is connected to 11 KV bus through a circuit breaker and two isolators.

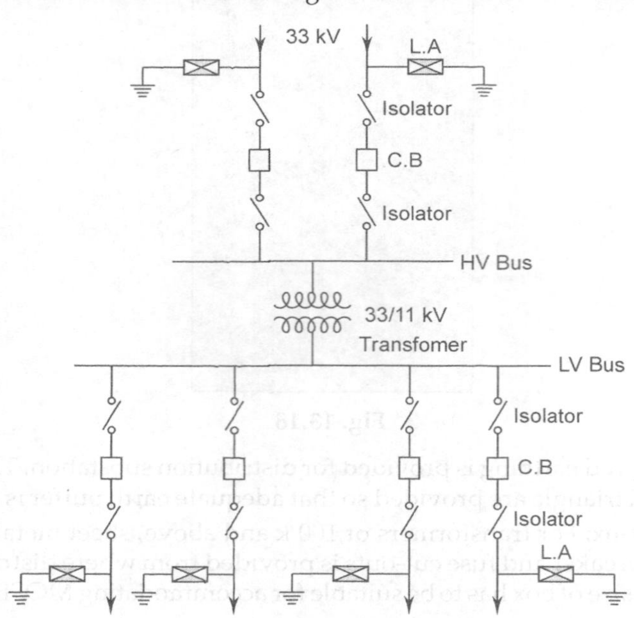

Fig. 13.19

13.9 KEY DIAGRAM OF 11 KV/400 V INDOOR SUBSTATION

Fig. 13.20 shows the key diagram of the same with details as under:

(*a*) 11 KV line is tapped and brought to the Gang Operated (G.O) switch installed near the substation. The G.O. switch consists of isolators connected in each phase of the 3-phase line

(*b*) From the G.O. switch, the 11 KV line is brought to the indoor substation as an underground cables. It is fed to the HT of transformer (11 KV/400 V) via the 11 KV O.C.B. The transformer steps down the voltage to 400 V, 3-phase, 4-wire.

(*c*) The secondary of transformer supplies to the bus-bars via the main O.C.B. From the bus-bars, 400 V, 3-phase, 4-wire supply is given to the various consumers via 400 V OCB.

(*d*) CTs are located at suitable places in the substation circuit and supply for the metering and indicating instruments and relay circuits.

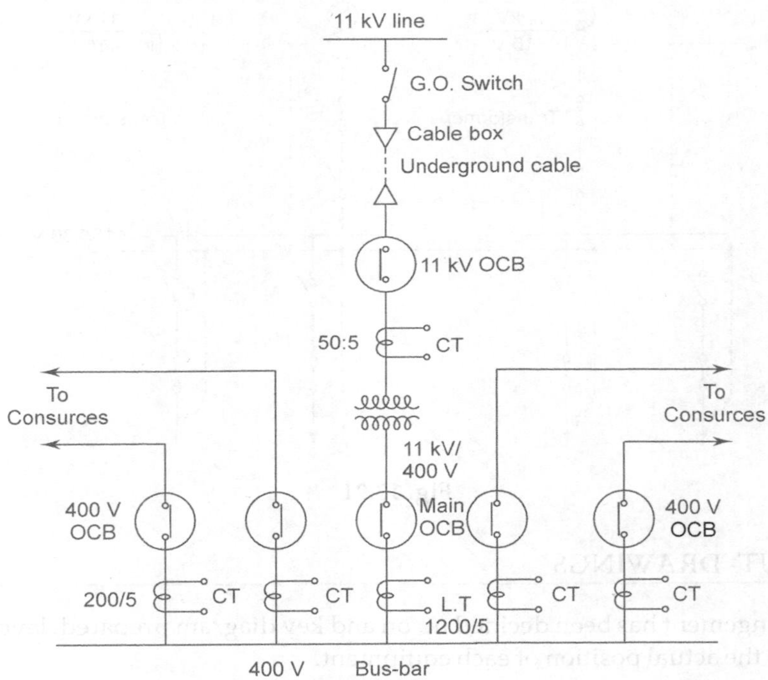

Fig. 13.20

13.10 SUBSTATION AUXILIARIES SUPPLY

In substations, the electric power is required for the auxiliaries like the lighting circuits, battery charging sets, air blast circuit breakers, water supply, oil servicing facilities, ventilating fans of the substation buildings etc. In substations incorporating synchronous condensers, the supply is also required for the operation of the auxiliary equipment of the synchronous condensers.

In large substations, it is a wide practice to connect two transformers to the 11 kv main busbars for supply of auxiliaries at 415/230 V as shown in Fig. 13.21

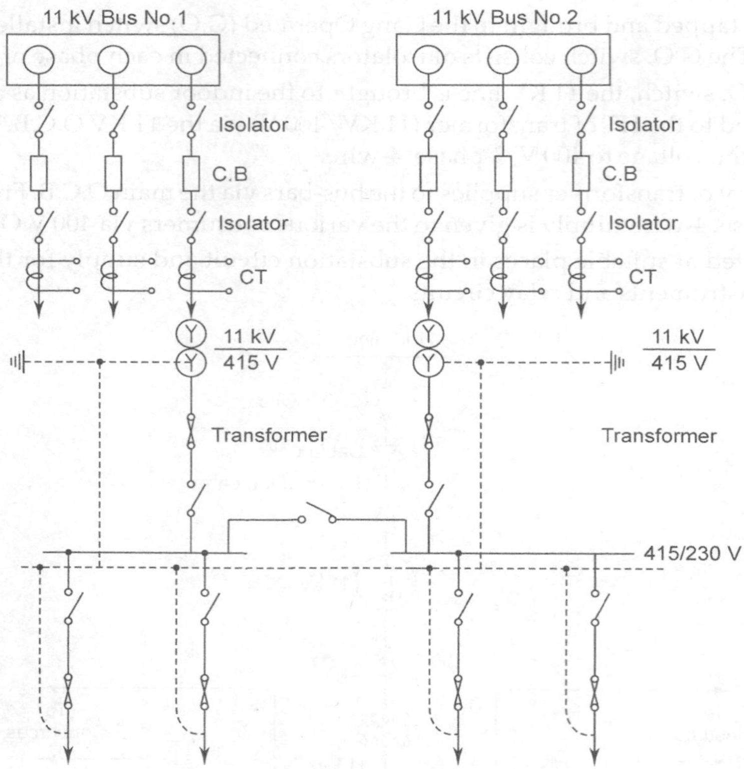

Fig. 13.21

13.11 LAYOUT DRAWINGS

After the bus arrangement has been decided up on and key diagram prepared, layout drawing are prepared to show the actual position of each equipment.

The land plan indicates the boundaries of the land. A general layout is first prepared on the land plan. Space is marked for civil works and electrical works and also the roads and paths for movement of men and material.

On the land marked for electrical works, detailed electrical layout which indicates the placing of various equipment in the switchyard is prepared. The number of columns and beams to support the bus bars and connections will depend on the system voltage and bus arrangement selected. On a copy of detailed electrical layout overhead ground wire, used for protection against direct lighting strokes is marked. This determines the position and number of columns which require peaks for binding the ground wire.

In addition to the main columns and beams, supporting structures are required for mounting individual items such as PTs, CTs, lighting arrestors, insulators, isolators, circuit-breakers etc. The design of such structures can be finalized only with the equipment drawings indicating dimensions and fixing details.

The supporting structures, columns and beams may be made of steel or reinforced cement concrete (RCC). The choice would depend on the cost as well as available time. As per cost, RCC structures are cheaper than the steel structures.

SELF TEST

Fill in the blanks by appropriate words/figures:

1. Out-door sub-station requires _______ space.
2. Underground sub-stations are generally located in _______.
3. Most of the sub-stations in the power system change _______ of electric supply.
4. Fault location is _______ in an outdoor sub-station than in indoor sub-station.
5. Power factor correction sub-station.
6. An indoor sub-station is _______ expensive than outdoor substation.
7. Majority of distribution sub-stations are of _______ type.
8. Pole-mounted sub-stations are used for _______ distribution.
9. A sub-station _______ some characteristic of electric supply.

Answers to Self Test

1. more
2. thickly populated areas.
3. voltage level
4. easier
5. receiving
6. more
7. pole-mounted
8. secondary
9. changes.

OBJECTIVE QUESTIONS

1. Feeder is designed mainly from the point of view of :
   (*a*) voltage drop in it   (*b*) operating voltage
   (*c*) its current carrying capacity   (*d*) operating frequency.
2. A device used for connecting a conductor to the earth for discharging the charges on the conductor to the earth is known as :
   (*a*) isolator   (*b*) limit switch
   (*c*) air-break switch   (*d*) earthing switch.
3. A circuit breaker is essentially :
   (*a*) a current interrupting device
   (*b*) an arc extinguisher.

(*c*) a power factor correcting device.
(*d*) a device for neutralising the effects of transients.

4. Circuit breakers usually operate under :
(*a*) sub-transient state of short-circuit current.
(*b*) steady-state short-circuit current.
(*c*) Transient state of short-circuit current.
(*d*) None of the above.

5. In a transmission system the feeder supplies power to :
(*a*) service mains (*b*) distributors
(*c*) transformer substations (step-up) (*d*) all of the above.

6. A 3-phase, 4-wire system is commonly used for :
(*a*) primary transmission (*b*) secondary transmission
(*c*) primary distribution (*d*) secondary distribution.

7. Which of the following devices automatically interrupts the supply in the event of surges?
(*a*) Isolator (*b*) Circuit breaker
(*c*) Lightning switch (*d*) Earthing switch.

8. The domestic consumers are supplied power at :
(*a*) 230 V (*b*) 400 V (*c*) 440 V (*d*) 600 V.

9. Earthing switch is usually installed on :
(*a*) main board (*b*) isolator frame
(*c*) circuit breaker frame (*d*) none of the above.

10. Distributors are designed from the point of view of :
(*a*) operating voltage (*b*) its current carrying capacity
(*c*) voltage drop in it (*d*) operating frequency.

11. The material used for bus-bars should have :
(*a*) low cost (*b*) low receptivity
(c) high softening temperature (*d*) all of the above.

12. Isolators are used for disconnecting a circuit when :
(*a*) line carries no current (*b*) line is energised
(*c*) line is on full load (*d*) any of these.

13. A bus-bar is rated by :
(*a*) voltage only
(*b*) current only
(*c*) current, voltage and frequency
(*d*) current, voltage, frequency and short-circuit current.

14. Bus-bars are usually made of :
(*a*) copper (*b*) zinc (*c*) aluminum (*d*) steel.

15. Bus-bars are made of aluminum mainly because of its :
(*a*) low density (*b*) low cost (*c*) high resistivity (*d*) low melting point.

16. Isolators used in transmission lines are capable of breaking :
(*a*) load current (*b*) fault current (*c*) no current (*d*) charging current.

17. An isolator is installed :
    (*a*) usually on both sides of a circuit breaker
    (*b*) to isolate one portion of the circuit from another.
    (*c*) as a substitute for a circuit breaker.
    (*d*) both (*a*) and (*b*).

Answers to Objective Questions

1. (*c*)	**2.** (*d*)	**3.** (*a*)	**4.** (*c*)	**5.** (*b*)
6. (*d*)	**7.** (*b*)	**8.** (*a*)	**9.** (*b*)	**10.** (*c*)
11. (*d*)	**12.** (*a*)	**13.** (*d*)	**14.** (*c*)	**15.** (*b*)
16. (*c*)	**17.** (*d*).			

EXERCISE

1. What is the necessity of a sub-station in a power system?
2. What are the factors governing the selection of site for substations?
3. Draw the layout and schematic connection of a pole-mounted sub-station.
4. Draw a neat layout of 132/33 KV substation having two 132 KV incoming lines and four 33 KV out going lines show the essential equipment in the diagram.
5. Compare outdoor and indoor substations.
6. What are bus-bars and what are their functions?
7. Describe briefly the equipment that must be available in a substation.
8. Draw the key diagram of a typical 11 KV/400 V indoor sub-station.

14

VOLTAGE CONTROL

14.1 INTRODUCTION

In a modern power system, electrical energy from the generating station is delivered to the ultimate consumers through a network of transmission and distribution. Practically all equipment used in power systems is rated for a certain voltage with a permissible band of voltage variations. For satisfactory operation of motors, lamps and other loads, it is desirable that consumers are supplied with substantially constant voltage. Too wide variations of voltage will affect the performance of equipments and the life of most of the equipments is also sacrificed. The picture on a television set starts rolling if the voltage is below a certain level. The fluorescent tube refuses to glow if the voltage is below a certain level. Thus the necessity of controlling the voltage on the system is very much strong. To safeguard the interest of the consumers, the government has enacted a law in this regard. The statutory limit of voltage variation is ± 6% of declared voltage at consumer's terminals.

The principal cause of voltage variation at consumer's premises is the change in load on the supply system. When the load on the system increases, the voltage at the consumer's terminals falls due to increased voltage drop in:

(*a*) Transmission line
(*b*) Transformer impedance
(*c*) Alternator synchronous impedance
(*d*) Feeder
(*e*) Distributors.

The higher the load with smaller power factor, the greater is the voltage variation. The voltage variation at a node is an indication of the unbalance between the reactive power generated and consumed by that node. To understand this, refer Fig. 14.1 where the two bus-bars are interconnected through a short line and node one is a generator node with reference voltage V_1 and node two is the load node with voltage V_2.

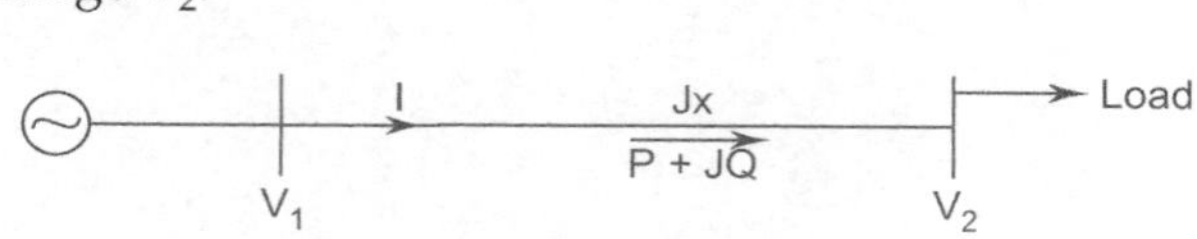

Fig. 14.1

Assuming the interconnector to be lossless (R = 0) and the voltage V_1 constant, the following relations hold good:

$$V_2 = V_1 - IZ$$

Assuming inductive load,

$$V_1^* I = P - JQ$$

or

$$I = \frac{P - JQ}{V_1^*}$$

As V_1 is the reference vector,

$$\therefore \quad V_1^* = V_1$$

Thus,

$$V_2 = V_1 - J\left(\frac{P - JQ}{V_1}\right)$$

$$= \left(V_1 - \frac{Q}{V_1} X\right) - J\frac{P}{V_1} X$$

The vector diagram is shown in Fig. 14.2

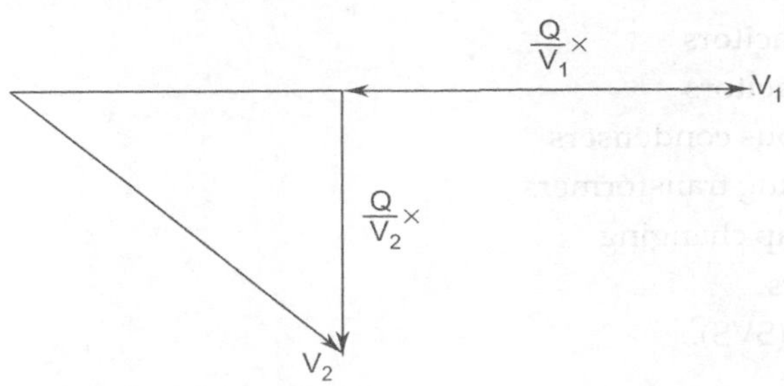

Fig. 14.2

Assuming the voltage drop due to real power negligible, the voltage drop is directly proportional to the reactive power Q, as shown in relationship:

$$V_2 = V_1 - \frac{Q}{V_1} X$$

For a particular sending end voltage V_1, the drop (Q/V_1) X must remain constant in order to keep the receiving end voltage V_2 fixed. In this only variable quantity is Q *i.e.*, reactive vars. In order to adjust Q, the shunt capacitors or synchronous capacitors or shunt inductors are used.

Again for fixed V_1, another possibility is that the product Q.X be kept constant in order to keep V_2 constant. This is achieved by using series capacitors which will reduce the net reactance of the system.

14.2 PERMISSIBLE STEADY-STATE VOLTAGE VARIATION

During heavy loads, the IX drop in transmission and distribution lines increases and the receiving-end voltage decreases.

During low loads, the IX drop in series reactance of lines is negligible. The shunt capacitance of transmission lines taking capacitive currents causes increase in receiving end voltage. Thus, the substation bus voltage experience:

Reduced Voltage	↔	High Load
Increased Voltage	↔	Low Load
Normal Voltage	↔	Normal Load

Low voltages cause higher current flow through supply line for same load causing higher line losses. Low voltages also cause increased current to deliver same power, hence increased heating of lines, motors and transformer. Below certain voltage (70 to 80% rated voltage), the motors get stalled and are tripped automatically by the over-current or under-voltage protection. Sustained high voltages cause failure of insulation of transformers and motors.

The power supplied to machines and passing through transformers, lines etc., may reduce due to fall in voltage. Rise in voltage is harmful to insulation of machines, transformers etc. It may lead to insulation failure or overstressing.

14.3 METHODS OF VOLTAGE CONTROL

There are various methods of voltage control used in power systems which are:

1. By excitation control
2. By using shunt capacitors
3. By using series capacitors
4. By using synchronous condensers
5. By using tap-changing transformers
6. Auto-transformer tap changing
7. Booster transformers
8. Static VAR Sources (SVS).

14.3.1 Excitation Control

When the load on the supply system changes, the terminal voltage of the alternator also varies due to the changed voltage drop in the synchronous reactance of the armature. The voltage of the alternator can be kept constant by changing the field current of the alternator in accordance with the load. This is known as excitation control method. There are also limits for the excitation. Excitation below a certain limit may result in unstability of the system and excitation above certain level will result in overheating of the rotor. The excitation of alternator can be controlled by the use of automatic or hand operated regulator acting in the field circuit of the alternator. There are two main types of automatic voltage regulators:

- Tirril regulator
- Brown-Boveri Regulator.

14.3.2 Shunt Capacitors

Shunt capacitor are located near load points and in receiving substations to supply reactive power to inductive loads and to improve power factor. Thereby the lines and transformers can supply more real power. Shunt capacitors are used in ratings from 15 KVAR to over 100 MVAR. Small banks of capacitors upto a few hundred KVAR rating are used on individual distribution circuits of consumers. Capacitor banks of 500–3000 KVAR are used in small distribution substations and those with still larger ratings are used in large substations. Shunt capacitors are switched in during heavy loads.

Most of the industrial loads like induction motors, welding sets, furnace transformers etc., draw inductive currents of poor power factor. The shunt capacitor provide leading voltampere reactive (VAR), thereby the total KVA loading of substation transformer and the current is reduced. Thereby IX_L drop of the line is reduced and the voltage regulation is improved.

Shunt capacitors are switched in when KVA demand on the distribution line goes up and voltage of the bus goes down. Switching in the shunt capacitor should improve the bus bar voltage. Switching may be done manually or automatically controlled either by time clocks or in response to voltage or reactive power requirements.

14.3.3 Series Capacitors

Series capacitors are used for 400 KV and 760 KV AC. transmission lines for increasing the power transfer ability P

$$P = \frac{|V_s||V_r|}{(X_L - X_C)} \sin\delta \text{ per phase}$$

where, P = Power transfer per phase

$|V_s|$ & $|V_r|$ = Sending-end and receiving-end voltages, phase to neutral.

X_L = Series inductive reactance of line, phase to neutral

X_C = Reactance of series capacitors, per phase.

X_C is usually 50% of X_L. These are installed in sending-end, receiving-end and intermediate substations of long EHV-AC and UHV-AC transmission lines. Power angle δ is the angle between V_s and V_r vectors which changes with power flow. The limit of δ is approximately 30°.

Figure 14.3 illustrates the difference between the principle of series capacitor installation and shunt capacitor installation. The series capacitor is connected in series with the power line. Hence the line current flows through the series capacitor. Shunt capacitor is connected between phase and earth. Hence phase to neutral voltage appears across shunt capacitor. Shunt capacitor improves the phase to neutral voltage. Series capacitor improves the power flow.

Vector equation of a series compensated line is given by:

$$V_r = V_s - I(X_L - X_C)$$

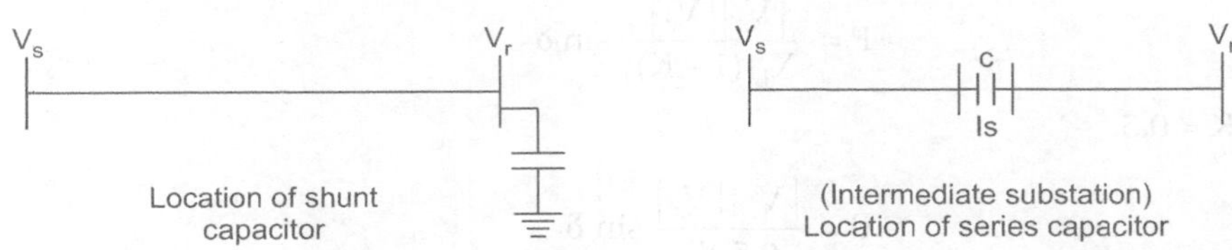

Fig. 14.3

where, V_r = Receiving-end voltage

V_s = Sending-end voltage

IX_L = Voltage drop in series inductive reactance of the line and $X_L = 2\pi f L$

IX_C = Voltage drop across series capacitor and $X_C = 1/2\pi f C$.

The vector diagram is shown in Fig. 14.4. Due to the effect of series capacitor, the receiving-end voltage will be V_r' instead of V_r. Angle δ between V_s and V_r is also reduced giving higher stability. Usually, the 40 to 60% of X_L is compensated by series capacitor.

With X_C in circuit, power flow P would be:

$$P = \frac{|V_s||V_r|}{X_L - X_C} \cdot \sin\delta$$

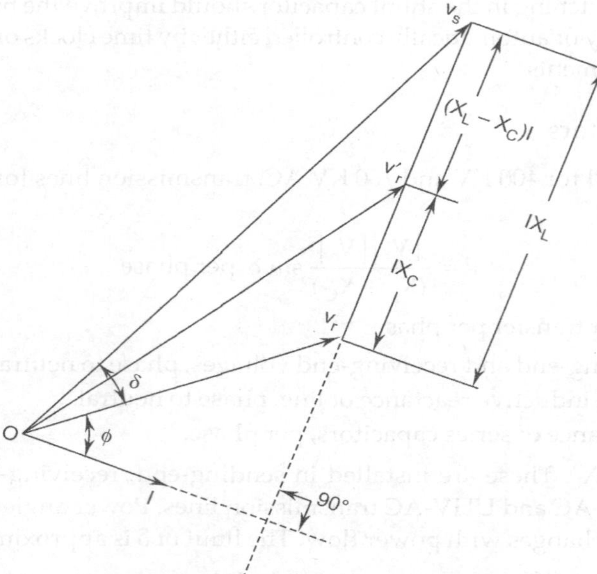

Fig. 14.4

Without X_C in circuit, power flow P would be:

$$P = \frac{|V_s||V_r|}{X_L} \cdot \sin\delta$$

Suppose $X_C = K X_L$

$$P = \frac{|V_s||V_r|}{X_L(1-K)} \cdot \sin\delta$$

For K = 0.5

$$P = \frac{|V_s||V_r|}{0.5\,X_L} \sin\delta$$

Thus, power flow is doubled.

Main drawback of series capacitors is the high over-voltage produced across the capacitor terminals under short circuit conditions. The drop across the capacitor is $I_f X_C$ where I_f is the fault current which is of the order of 20 times the full load current under certain circuit condition.

Comparison between series and shunt capacitors:

1. For the same voltage boost the reactive power capacity of a shunt capacitor is greater than that of a series capacitor.
2. The shunt capacitor improves the p.f. of the load whereas the series capacitor has little effect on p.f.
3. The change in voltage between the two ends of the series capacitor where it is connected, is sudden whereas the voltage boost due to a shunt capacitor is evenly distributed over the transmission line.

14.3.4 Synchronous Condensers

A synchronous motor at no load and taking a leading power factor is known as a synchronous condenser. Synchronous condensers (phase modifiers) are loadless synchronous motors connected to the line via a suitable transformer. The synchronous phase modifier has wide variation of excitation control. Under excited synchronous machine takes lagging p.f. currents. The synchronous condenser supplies wattless leading KVA to the line depending upon the excitation of the motor. The wattless leading KVA partly or fully cancels the wattless lagging KVA of the line, thus controlling the voltage drop in the line. Thus, by changing the excitation, the reactive power drawn/supplied by the synchronous condenser is varied. Synchronous phase modifiers are connected in receiving substations for voltage control. During low loads, they are operated with under excitation. During heavy loads, they are operated with over-excitation. The synchronous condenser is excellently suited for voltage regulation because it can supply lagging as well as leading vars. But high cost of synchronous condensers have made them nearly obsolete.

14.3.5 Tap-Changing Transformers

The voltage control of transmission and distribution systems is obtained basically by tap-changing. Transformers are provided with taps on windings to adjust the transformation ratio. Taps are usually on the high voltage winding since tappings on low voltage winding do not give sufficiently close control of voltage. Tap-changers are either on-load or off load tap-changers. By changing the turns ratio of transformer, the voltage ratio and the secondary voltage is changed and voltage control is obtained. Tap-changing is the most widely used method of controlling voltage at various levels.

The voltage control of the range ± 16% can be achieved by tap-changing transformers. Tap-changing requires 8 to 12 seconds per tap.

(*a*) ***Off-load tap-changing transformer:*** Figure 14.5 shows the arrangement where a number of tappings have been provided on the secondary. As the position of tap is changed, the effective number of secondary turns is varied and hence the output voltage of the secondary can be changed. In Fig. 14.5, when the movable arm makes contact with stud 1, the secondary voltage is minimum and when with stud 5, it is maximum. During the period of light load, the voltage across the primary is not much below the alternator voltage and the movable arm is placed on stud 1. Whenever a tapping is to be changed in this type of transformer, the load is kept off and hence the name off load tap-changing transformer.

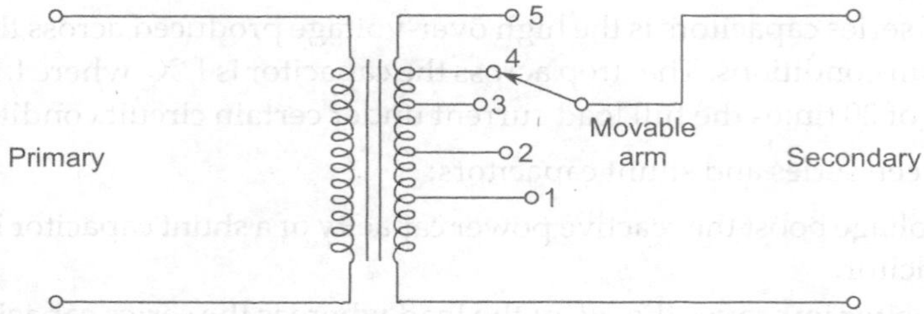

Fig. 14.5

The main disadvantage of this arrangement is that it cannot be used for tap-changing on load. Suppose that tapping is changed from position 1 to 2 when the transformer is supplying load. If contact with stud 1 is broken before contact with stud 2 is made, there is break in the circuit and arcing results.

(b) ***On-load tap-changing transformer:*** The daily voltage variation due to changing load and short period voltage variations are controlled by on-load tap changers automatically. On-load tap-changing is also used for industrial applications where variable voltage is required for the process *e.g.*, arc furnace duty. A tap changer is provided on a transformer for maintaining specified outgoing voltage where the incoming voltage is subjected to voltage variations. The tap-changer is mounted in/on the transformer tank. It comprises a motor driven mechanism and associated control circuit for starting and stopping the motor. The motor can be run in the direction for a 'raise' tap-change or in the reverse direction for a 'lower' tap-change.

Figure 14.6 shows one type of on-load tap changing transformer. The secondary consists of two equal parallel windings which have similar tappings 1*a* .... 5*a* and 1*b* .... 5*b*. In the normal working conditions, switches *a*, *b* and tappings with the same number remain closed and each secondary winding carries one-half of the total current. In Fig. 14.6, the secondary voltage will be maximum when switches *a*, *b* and 5*a*, 5*b* are closed. However, the secondary voltage will be minimum when switches *a*, *b* and 1*a*, 1*b* are closed.

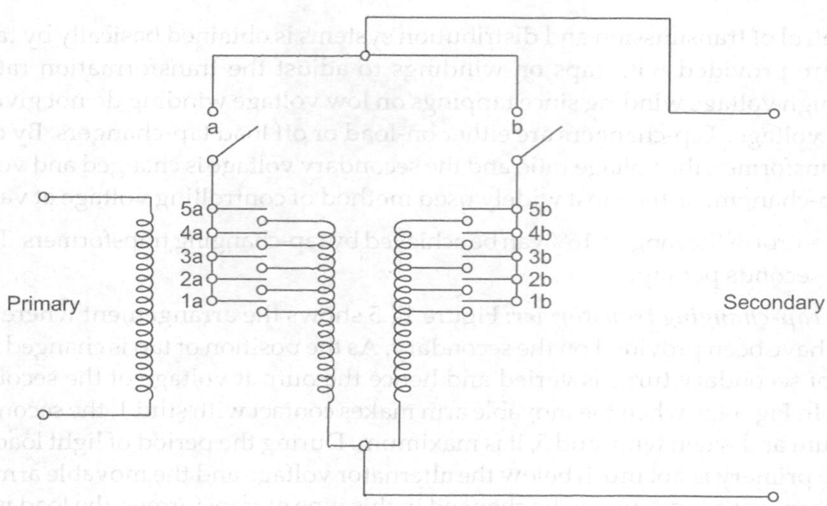

Fig. 14.6

Suppose that the transformer is working with tapping position at *4a*, *4b* and it is desired to alter its position to *5a*, *5b*. For this purpose, one of the switches out of *a* or *b* is opened. Suppose (*a*) is opened and this takes the secondary winding controlled by switch (*a*) out of the circuit. Now, the secondary winding controlled by switch *b* carries the total current which is twice its rated capacity. Then the tapping on the disconnected winding is changed to *5a* and switch *a* is closed. After this, switch *b* is opened to disconnect its winding tapping position on this winding is changed to *5b* and then switch *b* is closed. In this way, tapping position is changed without interrupting the supply.

Consider the operation of a radial transmission line with impedance R + JX having tap changing transformers at both ends as in Fig. 14.7. Let t_s and t_r be the fractions of the nominal transformation *i.e.*, tap ratio/nominal ratio, *e.g.*, a transformer with nominal ratio 11/33 KV when tapped to convert 11 KV to 36 KV has t_s = 36/33 = 1.09. V_1 and V_2 are the nominal voltages at the sending-end and receiving-end respectively. When taps are used, the actual sending-end and receiving-end voltages

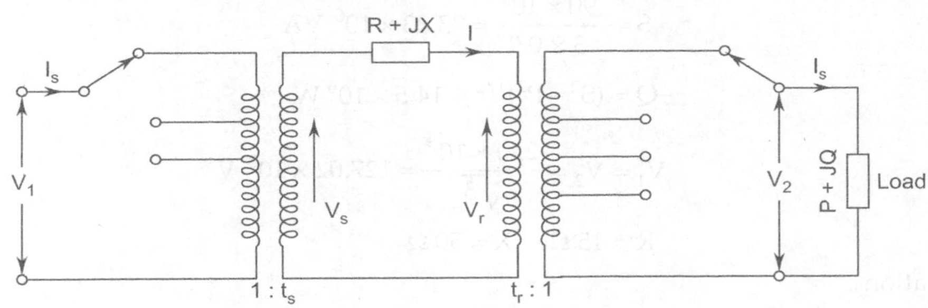

Fig. 14.7

be $t_s V_1$ and $t_r V_2$. We assume here that the product of t_s and t_r is unity as this ensures that the overall voltage level remains of the same order and minimum number of taps are used. The tap ratios t_s and t_r can be determined as under:

From Fig. 14.7,
$$t_s V_1 = t_r V_2 + IZ$$
$$= t_r V_2 + IR\cos\phi + IX\sin\phi$$

Since $I\cos\phi = \dfrac{P}{t_r V_2}$ and $I\sin\phi = \dfrac{Q}{t_r V_2}$

$$\therefore \quad t_s V_1 = t_r V_2 + \frac{RP}{t_r V_2} + \frac{XQ}{t_r V_2}$$

Substituting $t_r = 1/t_s$

$$t_s = \frac{1}{V_1}\left[\frac{V_2}{t_s} + \frac{t_s\,(RP + XQ)}{V_2}\right]$$

or
$$t_s^2 = \frac{V_2}{V_1} + \frac{t_s^2\,(RP + XQ)}{V_1 V_2}$$

or $$t_s^2\left[1-\frac{RP+XQ}{V_1\,V_2}\right]=\frac{V_2}{V_1}$$

Hence, for particular values of V_2 and V_1 and the load requirements P and Q, the value of t_s can be obtained.

Example 14.1. *A 220 KV line has tap changing transformers at both ends. The transformer at the sending end has a nominal ratio 11/220 KV and that the receiving end 220/11 KV. The line impedance is 15 + J 50 ohms and the load at the receiving end is 90 MW at 0.9 pf lagging. Assuming that the product of off-nominal tap settings is 1, find the tap settings to give 11 KV at load bus.*

Solution. $$P=\frac{90\times10^6}{3}=30\times10^6\ W$$

$$S=\frac{90\times10^6}{3\times0.9}=33.33\times10^6\ VA$$

$$Q=(S^2-P^2)^{1/2}=14.5\times10^6\ W$$

$$V_1=V_2=\frac{220\times10^3}{\sqrt{3}}=127.02\times10^3\ V$$

$$R=15\ \Omega,\quad X=50\ \Omega$$

Using equation

$$t_s^2\left[1-\frac{RP+XQ}{V_1\,V_2}\right]=\frac{V_2}{V_1}$$

or $$t_s^2\left[1-\frac{15\,(30\times10^6)+50\,(14.5\times10^6)}{(127.02\times10^3)^2}\right]=1$$

or $$t_s=1.0385$$

and $$t_r=\frac{1}{t_s}=\frac{1}{1.0385}=0.963$$

14.3.6 Auto-Transformer Tap-Changing

Figure 14.8 shows diagrammatically auto-transformer tap changing. Here, a mid tapped reactor or autotransformer is used. One of the lines is connected to its mid-tapping. One end, say *a* of this transformer is connected to a series of switches across the odd tappings and the other end *b* is connected to switches across even tappings. A short-circuiting switch S is connected across the auto-transformer and remains in the closed position under normal operation. In the normal operation, switch S remains closed so that half the total current flows through each half of the reactor. Since the currents in each half of the reactor are in opposition, no resultant flux is set up and consequently there is no inductive voltage drop across it. From Fig. 14.8, when switch 5 is closed, minimum secondary turns are in circuit and hence the output voltage will be the lowest. On the other hand, the output voltage will be maximum when switch 1 is closed.

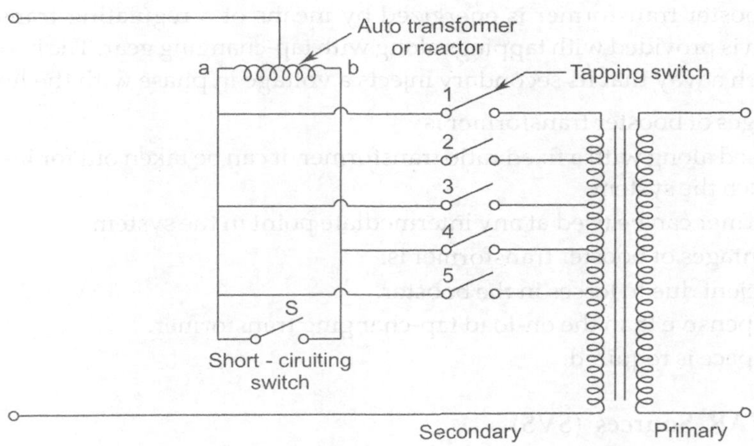

Fig. 14.8

Suppose now it is desired to alter the tapping point from position 5 to position 4 in order to raise the output voltage. For this purpose, short-circuiting switch S is opened, switch 4 is closed, then switch 5 is opened and finally short-circuiting switch is closed. In this way, tapping can be changed without interrupting the supply.

An autotransformer differs from the ordinary transformer in that the windings of the auto-transformer are electrically connected as well as being coupled by a mutual flux.

14.3.7 Booster Transformer

Sometimes it is desired to control the voltage of a transmission line at a point far away from the main transformer. This can be conveniently achieved by the use of a booster transformer as shown in Fig. 14.9. It can be installed at a substation as an additional equipment if voltage regulation is found to be necessary or it can be installed as a separate piece of equipment at any intermediate point in the line. However, the separate piece of equipment is desirable on economical or technical grounds to increase the voltage at an intermediate point in a line rather than at the ends as with tap changing transformer. As shown in Fig. 14.9, the secondary of the booster transformer is connected in series with the line whose voltage is to be controlled.

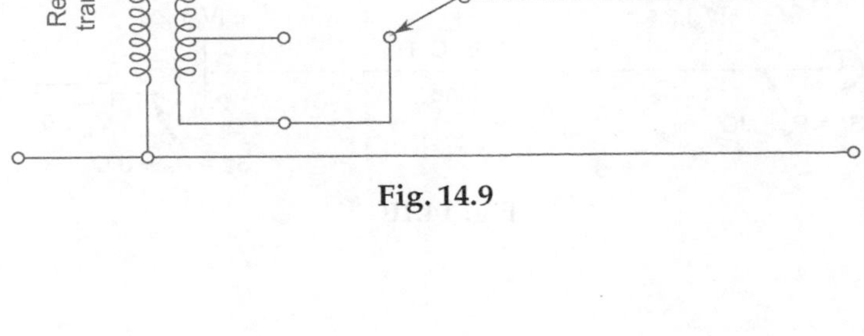

Fig. 14.9

The primary of booster transformer is energized by means of a regulating transformer and the secondary of which is provided with tappings along with tap-changing gear. The booster transformer is connected in such a way that its secondary injects a voltage in phase with the line voltage.

The main advantages of booster transformer is:

(*a*) When it is used along with a fixed ratio transformer, it can be taken out for inspection without affecting much the system.

(*b*) This transformer can be used at any intermediate point in the system.

The main disadvantages of booster transformer is:

(*a*) It is less efficient due to losses in the booster.

(*b*) It is more expensive than the on-load tap-changing transformer.

(*c*) More floor space is required.

14.3.8 Static VAR Sources (SVS)

Switchable shunt reactors and shunt capacitors have inherent slow speed. Circuit-breakers require at least 3 cycles (60 milliseconds) and protection requires 1 cycle (20 millisecond) resulting in total minimum time of 4 cycles. Secondly, circuit-breakers are not suitable for repeated switching during voltage variations. These limitations of switchable shunt compensation have been overcome by static VAR sources (SVS). In SVS system, thyristors are used as switching devices instead of circuit-breakers. Thyristors are fast and the compensation can be varied within a few tens of milliseconds. A stepless variable compensation is possible by thyristorised control of shunt capacitors and shunt reactors.

During heavy loads, the thyristors of capacitors control are made to conduct for a longer duration in each cycle. During low loads, the thyristors in reactor circuit are made to conduct for longer duration in each cycle.

The main advantages of SVS are:

(*a*) Better stepless, fast, accurate voltage control of substation buses over a wide range of loads by supplying-Q (capacitive) or + Q (Inductive) into the substation bus.

(*b*) Improving transient stability of power system by rapid voltage control.

(*c*) When a large load is switched off due to any reason like fault or frequency control, the receiving bus voltage rises rapidly. SVS provides rapid change in reactive power compensation and regulates the voltage.

(*d*) Damping of sub-synchronous resonance frequency oscillations in power systems.

(*e*) With fluctuating reactive loads like arc furnaces, SVS are used for rapid change in reactive power compensation in accordance with varying load. This reduces lamp flicker and voltage dips.

14.4 POWER FLOW THROUGH A TRANSMISSION LINE

Figure 14.10 shows a transmission line with sending end quantities with subscript S and receiving end quantities with subscript R.

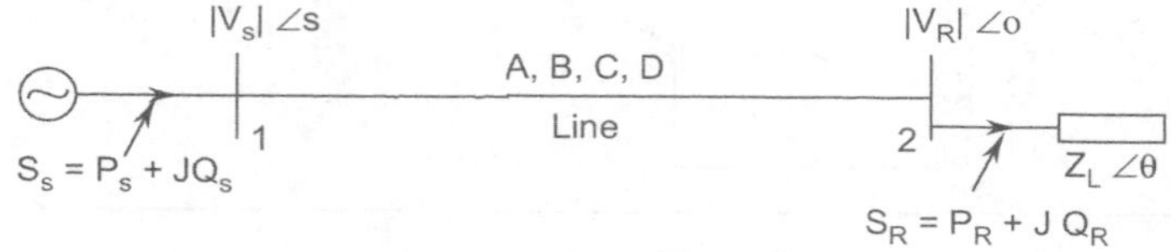

Fig. 14.10

In Fig. 14.10,

$$S_s = \text{Sending end power} = P_s + J\,Q_s$$
$$S_R = \text{Receiving end power} = P_R + J\,Q_R$$

The complex power delivered by the receiving end and received by the sending end of the transmission line can be expressed as:

$$S_R = P_R + J\,Q_R = V_R\,I_R^*$$
$$S_S = P_S + J\,Q_S = V_S\,I_S^*$$

As already discussed, we know :

$$\begin{bmatrix} V_S \\ I_S \end{bmatrix} = \begin{bmatrix} A & B \\ C & D \end{bmatrix} \begin{bmatrix} V_R \\ I_R \end{bmatrix}$$

and

$$\begin{bmatrix} V_R \\ I_R \end{bmatrix} = \begin{bmatrix} D & -B \\ -C & A \end{bmatrix} \begin{bmatrix} V_S \\ I_S \end{bmatrix}$$

From above equations, we obtain I_R in terms of V_R and V_S :

$$V_S = A\,V_R + B\,I_R$$

or

$$I_R = \frac{V_S - A\,V_R}{B}$$

or

$$I_R = \frac{1}{B} V_S - \frac{A}{B} V_R$$

Also

$$I_S = \frac{D}{B} V_S - \frac{1}{B} V_R$$

Let the transmission line constants A, B and D can be written as:

$$A = |A| \angle\alpha$$

$$B = |B| \angle\beta$$

$$D = |D| \angle\alpha \qquad \text{(Since A = D)}$$

Taking V_R as reference vector at $\angle 0°$, V_s is at angle δ.

$\therefore$

$$V_R = |V_R| \angle 0°$$

$$V_S = |V_S| \angle\delta$$

Thus,

$$I_R = \frac{|V_S|}{|B|} \angle\delta - \beta - \frac{|A| \cdot |V_R|}{|B|} \angle\alpha - \beta$$

The complex power $V_R\,I_R^*$ at the receiving end is:

$$S_R = P_R + J\,Q_R$$

$$= \frac{|V_S| \cdot |V_R|}{|B|} \angle\beta - \delta - \frac{|A| \cdot |V_R|^2}{|B|} \angle\beta - \alpha$$

Hence, real and reactive power at the receiving end are:

$$P_R = \frac{|V_S| \cdot |V_R|}{|B|} \cos(\beta - \delta) - \frac{|A| \cdot |V_R|^2}{|B|} \cos(\beta - \alpha)$$

and
$$Q_R = \frac{|V_S| \cdot |V_R|}{|B|} \sin(\beta - \delta) - \frac{|A| \cdot |V_R|^2}{|B|} \sin(\beta - \alpha)$$

The expression for complex power $S_R = P_R + J\,Q_R$ can be plotted on complex plane as shown in Fig. 14.11 with P on *x*-axis (watts) and Q on *y*-axis (VArs). The point which has P_R and Q_R as coordinates more along a circle called power circle diagram for receiving end:

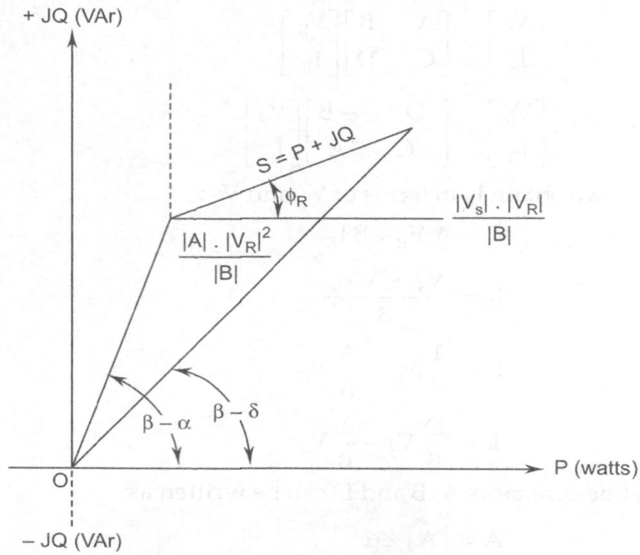

Fig. 14.11

Similarly, the real and reactive powers at sending-end are:

$$P_s = \left|\frac{D}{B}\right| |V_S|^2 \cos(\beta - \alpha) - \frac{|V_S||V_R|}{|B|} \cos(\beta + \delta)$$

and
$$Q_S = \left|\frac{D}{B}\right| |V_S|^2 \sin(\beta - \alpha) - \frac{|V_S||V_R|}{|B|} \sin(\beta - \delta)$$

There is a limit to the power that can be transmitted to the receiving end of the line for specified magnitudes of sending and receiving end voltages. Maximum power will be delivered when δ equals β. When $\delta = \beta$,

$$P_{R\,(max.)} = \frac{|V_S| \cdot |V_R|}{|B|} - \frac{|A| \cdot |V_R|^2}{|B|} \cos(\beta - \alpha)$$

and the corresponding Q_R is:

$$Q_R = -\frac{|A| \cdot |V_R|^2}{|B|} \sin(\beta - \alpha)$$

Thus, the load must draw this much leading VAr in order to receive the maximum real power.

Consider the case of a short line with a series impedance Z. Now:

$$A = D = 1 \angle 0$$

$$B = Z = |Z| \angle \theta$$

Substituting these in above equations,

For receiving-end power,

$$P_R = \frac{|V_S| \cdot |V_R|}{|Z|} \cos(\theta - \delta) - \frac{|V_R|^2}{|Z|} \cos\theta \quad \text{...(1)}$$

and $$Q_R = \frac{|V_S| \cdot |V_R|}{|Z|} \sin(\theta - \delta) - \frac{|V_R|^2}{|Z|} \sin\theta \quad \text{...(2)}$$

For sending-end power,

$$P_S = \frac{|V_S|^2}{|Z|} \cos\theta - \frac{|V_S||V_R|}{|Z|} \cos(\theta + \delta)$$

and $$Q_S = \frac{|V_R|^2}{|Z|} \sin\theta - \frac{|V_S||V_R|}{|Z|} \sin(\theta + \delta)$$

From equation (1), the maximum receiving end power is received when $\delta = \theta$, so that:

$$P_{R\,(max.)} = \frac{|V_S| \cdot |V_R|}{|Z|} - \frac{|V_R|^2}{|Z|} \cos\theta$$

As $$\cos\theta = R/|Z|$$

∴ $$P_{R\,(max.)} = \frac{|V_S| \cdot |V_R|}{|Z|} - \frac{|V_R|^2}{|Z|^2} \cdot R$$

Normally the resistance of a transmission line is small compared to its reactance, thus:

$$Z = R + J X_L \approx J X_L$$

and $$\theta = \tan^{-1} \frac{X_L}{R} \approx 90°$$

The receiving-end equations (1) and (2) can be approximated as:

$$P_R = \frac{|V_S| \cdot |V_R|}{X_L} \sin\delta \quad \text{...(3) } [\cos\theta = \cos 90° = 0]$$

and $$Q_R = \frac{|V_S| \cdot |V_R|}{X_L} \cos\delta - \frac{|V_R|^2}{X_L}$$

The above equation of Q_R can be further simplified by assuming $\cos\delta \approx 1$ since δ is normally small. Thus,

$$Q_R = \frac{|V_R|}{X_L} (|V_S| - |V_R|)$$

Let $|V_S| - |V_R| = \Delta V$ = Magnitude of voltage drop across the transmission line.

$$\therefore \quad Q_R = \frac{|V_R|}{X_L} \cdot |\Delta V| \quad \text{...(4)}$$

Hence, several important conclusion can be enumerated from equations (3) and (4) as:

1. For small R and small δ, the real power P_R is proportional to sin δ while the reactive power is proportional to the magnitude of the voltage drop across the line.
2. Maximum real power transfer occurs for δ = 90° and sin δ = 1, this is called steady state stability limit.
3. Maximum power depends on V^2. Therefore, maximum real power transferred for a given line (fixed X_L) can be increased by raising its voltage level.

Example 14.2. *A 275 KV transmission line has the following line constants:*

$$A = 0.85 \angle 5°,\ B = 200 \angle 75°$$

(a) Determine the power at unity power factor that can be received if the voltage profile at each end is to be maintained at 275 KV.

(b) What type and rating of compensation equipment would be required if the load is 150 MW at unity power factor with the same voltage profile as in part (a).

Solution.

(*a*) Given $|V_S| = |V_R| = 275$ KV

$$\alpha = 5°,\ \beta = 75°$$

Since the power is received at unity power factor.

$$\therefore \quad Q_R = 0$$

or
$$\frac{|V_S| \cdot |V_R|}{|B|} \sin(\beta - \delta) - \frac{|A|}{|B|} |V_R|^2 \sin(\beta - \alpha) = 0$$

or
$$\frac{275 \times 275}{200} \sin(75° - \delta) - \frac{0.85}{200} \times (275)^2 \sin(75° - 5°) = 0$$

or
$$\delta = 22°$$

$$P_R = \frac{|V_S| \cdot |V_R|}{|B|} \cos(\beta - \delta) - \frac{|A|}{|B|} |V_R|^2 \cos(\beta - \alpha)$$

or
$$P_R = \frac{275 \times 275}{200} \cos(75° - 22°) - \frac{0.85}{200} \times (275)^2 \cos(75° - 5°)$$

$$= 227.6 - 109.9 = 117.7 \text{ MW}$$

(*b*) Power demanded by load = 150 MW at UPF

$$\therefore \quad P_R = 150 \text{ MW}$$

$$150 = \frac{275 \times 275}{200} \cos(75° - \delta) - \frac{0.85}{200} \times (275)^2 \cos(75° - 5°)$$

or
$$\delta = 28.46°$$

Thus, $$Q_R = \frac{275 \times 275}{200} \sin (75° - 28.46°) - \frac{0.85}{200} \times (275)^2 \sin 70°$$

$$= 274.46 - 302 = - 27.56 \text{ MVAR}$$

In order to maintain 275 KV at the receiving end, $Q_R = - 27.56$ MVAR must be drawn along with the real power of $P_R = 150$ MW. The load being 150 MW at unity power factor, compensation equipment must be installed at the receiving end and the compensation equipment must feed positive VARs into the line.

14.5 POWER CIRCLE DIAGRAMS

A power circle diagram is drawn with real power P on X-axis and reactive power Q on Y-axis, on complex plain S. Following circle diagrams can be drawn:

(*a*) Receiving-end power circle diagram

(*b*) Sending-end power circle diagram.

14.5.1 Receiving-End Power Circle Diagram

For receiving-end complex power:

$$S_r = P_r + J\,Q_r$$

$$= \frac{-|A||V_R|^2}{|B|} \angle(\beta - \alpha) + \frac{|V_S| \cdot |V_R|}{|B|} \angle(\beta - \delta)$$

Thus, the power is sum of two phasors and the two components P and JQ are at quadrature with each-other. Out of two phasors, one is a constant phasor and other a phasor of fixed magnitude but variable angle. The loci for S_R would therefore be circle drawn from the tip of constant phasors as centres.

Hence, the centre of receiving-end circle is located at the tip of phasor

$$= \frac{-|A||V_R|^2}{|B|} \angle(\beta - \alpha)$$

In polar coordinates or in terms of rectangular coordinates.

Coordinates of the centre of the receiving-end circle are:

$$\text{X-axis (Horizontal)} = \frac{-|A||V_R|^2}{|B|} \cos (\beta - \alpha) \text{ Watts}$$

$$\text{Y-axis (Vertical)} = \frac{-|A||V_R|^2}{|B|} \sin (\beta - \alpha) \text{ VArs}$$

The radius of the receiving-end circle is:

$$= \frac{|V_S| \cdot |V_R|}{|B|} \text{ VA}$$

The receiving-end circle diagram is drawn in Fig. 14.12. The centre C_r is located by drawing OC_r equal to $\frac{|A||V_R|^2}{|B|}$ inclined at $\angle(\beta-\alpha)$ in the positive direction from the negative x-axis. From the centre C_r, the receiving end circle is drawn with the radius $|V_S|\ |V_R|/|B|$.

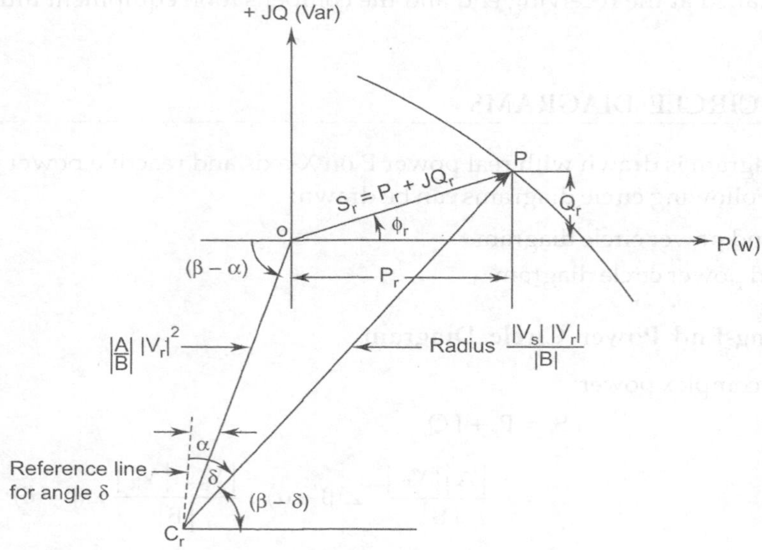

Fig. 14.12

The operating point P is located on the circle by means of the received real power P_R. The corresponding Q_R can be immediately read from the circle diagram. The torque angle δ can be read in accordance with the positive direction indicated from the reference line.

14.5.2 Sending-End Power Circle Diagram

For sending-end complex power:

$$S_S = P_S + J\,Q_S$$

$$= \frac{|D||V_S|^2}{|B|}\angle(\beta-\alpha) - \frac{|V_S||V_R|}{|B|}\angle(\beta+\delta)$$

Thus, the centre of the sending-end circle is located at the tip of phasor:

$$= \left|\frac{D}{B}\right||V_S|^2\angle(\beta-\alpha)$$

in the polar coordinates or in terms of rectangular coordinates.

Horizontal coordinate of the centre is:

$$= \left|\frac{D}{B}\right||V_S|^2\cos(\beta-\alpha)\ W$$

Vertical coordinate of the centre is:

$$= \left|\frac{D}{B}\right| |V_S|^2 \sin(\beta - \alpha) \text{ VAr}$$

The radius of the sending-end circle is:

$$= \frac{|V_S||V_R|}{|B|}$$

The sending-end circle diagram is shown in Fig. 14.13. The centre C_S is located by drawing OC_S equal to $\frac{|D||V_S|^2}{|B|}$ inclined at an angle $\angle(\beta - \alpha)$ to the positive x-axis. From the centre the sending-end circle is drawn with a radius $\frac{|V_s||V_r|}{|B|}$. The operating point M is located by measuring the torque angle δ in the direction indicated from the reference line.

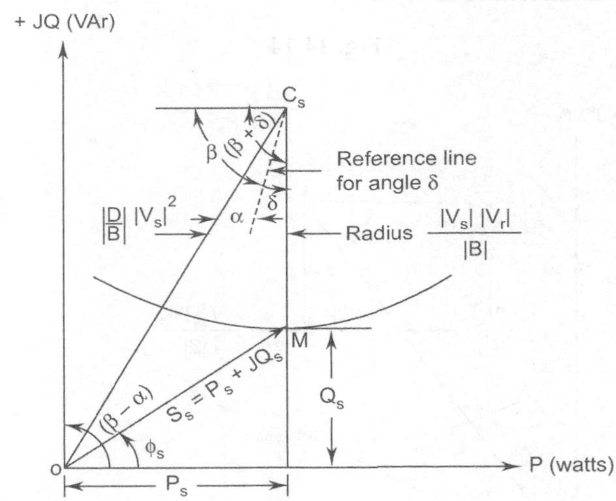

Fig. 14.13

For constant $|V_S|$, the centre C_S remains fixed and concentric circles result for varying $|V_R|$. However, if $|V_R|$ is fixed and $|V_S|$ varies, the centres of the circles move along the line OC_S and have radii in accordance to $|V_S|\ |V_r|/|B|$.

For the case of a short line with a series impedance $|Z|\ \angle\theta$, the simplified circle diagrams can be easily drawn by recognizing:

$$|A| = |D| = 1, \alpha = 0$$

$$|B| = |Z|, \beta = \theta$$

The corresponding receiving-end and sending-end circle diagrams are drawn in Fig. 14.14 and Fig. 14.15.

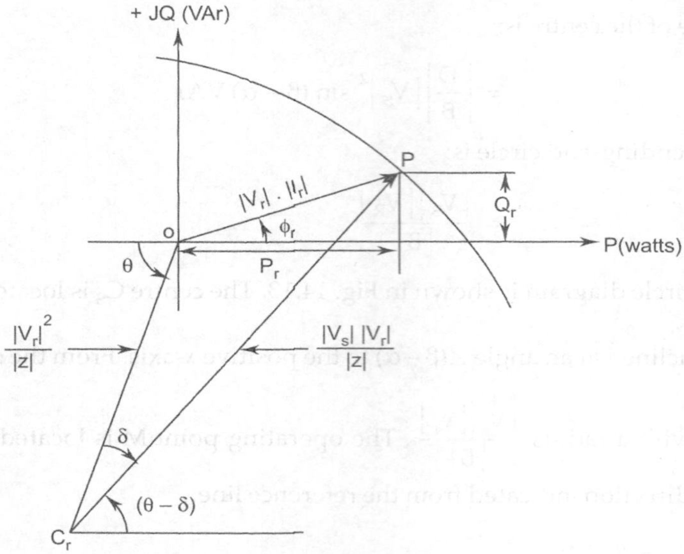

Fig. 14.14

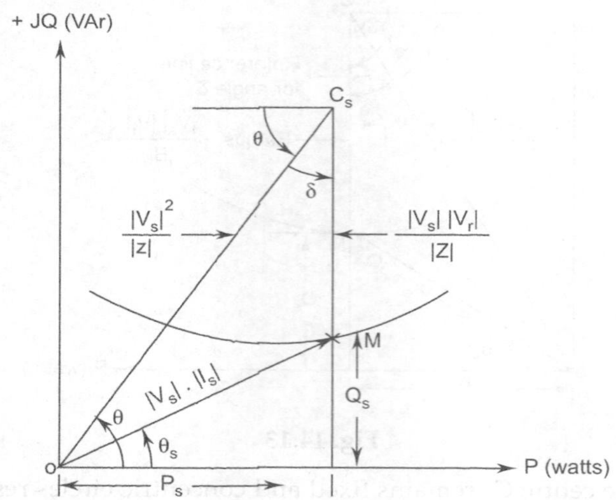

Fig. 14.15

Example 14.3. *A 50 Hz, three-phase, 275 KV, 400 km. transmission line has the following parameters:-*

Resistance = 0.035 Ω/km. per phase

Inductance = 1.1 mH/km. per phase

Capacitance = 0.012 μF/km. per phase

If the line is supplied at 275 KV, determine the MVA rating of a shunt reactor having negligible losses that would be required to maintain 275 KV at the receiving end when the line is delivering no load. Use nominal π-method.

Solution.

$$R = 0.035 \times 400 = 14\ \Omega$$

$$X = 2\pi f L = 314 \times 1.1 \times 10^{-3} \times 400 = 138\ \Omega$$

$$Z = R + J X = 14 + J\,138$$

$$= 138.7\ \angle 84.2°\ \Omega$$

$$Y = 314 \times 0.012 \times 10^{-6} \times 400\ \angle 90°$$

$$= 1.507 \times 10^{-3}\ \angle 90°\ \mho$$

$$A = 1 + \frac{1}{2}\ YZ$$

$$= 1 + \frac{1}{2} \times 1.507 \times 10^{-3} \times 138.7\ \angle 174.2°$$

$$= 0.896 + J\,0.0106$$

$$= 0.896\ \angle 0.7°$$

$$B = Z = 138.7\ \angle 84.2°$$

$$|V_S| = 275\ \text{kV},\ |V_r| = 275\ \text{kV}$$

$$\text{Radius of receiving-end circle} = \frac{|V_S||V_r|}{|B|}$$

$$= \frac{275 \times 275}{138.7} = 545.2\ \text{MVA}$$

Location of the centre of receiving-end circle,

$$\left|\frac{A}{B}\right||V_r|^2 = \frac{275 \times 275 \times 0.896}{138.7} = 488.5\ \text{MVA}$$

$$\angle(\beta - \alpha) = 84.2° - 0.7° = 83.5°$$

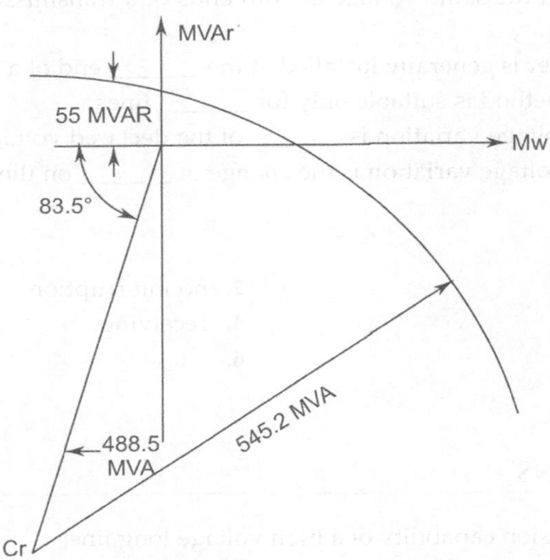

From the circle diagram, + 55 MVAr must be drawn from the receiving-end of the line in order to maintain a voltage of 275 KV. Thus rating of shunt reactor needed is 55 MVA.

Example 14.4. *A 3-phase overhead line has A = D = 0.9 ∠1° and B = 140 ∠84° ohms. It is operated with V_S = 240 KV and V_R = 220 KV. Find maximum power which can be received at the receiving end.*

Solution.

$$P_r = \frac{|V_S||V_r|}{|B|}\cos(\beta - \delta) - \frac{|A||V_r|^2}{|B|}\cos(\beta - \alpha)$$

For maximum P_r,
$$\frac{dP_r}{dS} = 0$$

$$\frac{dP_r}{dS} = -\frac{|V_S||V_r|}{|B|}\sin(\beta - \delta) = 0$$

or $\sin(\beta - \delta) = 0$

or $\beta = \delta$

Hence $\delta = 84°$

$$\therefore \quad P_r = \frac{(240)(220)}{140} - \frac{0.9\,(220)^2}{140}\cos(84° - 1°)$$
$$= 377.14 - 37.92$$
$$= 339.22 \text{ MW}$$

SELF TEST

Fill in the blanks by inserting appropriate words/figures:

1. The voltage control equipment is used at _______ in the power system.
2. In practice, tap-changing is performed on load so that there is _______ to supply.
3. It is _______ to maintain the same voltage at both ends of a transmission line by synchronous condenser method.
4. A synchronous condenser is generally installed at the _______ end of a transmission line.
5. The excitation control method is suitable only for _______ lines.
6. The statutory limit of voltage variation is _______ of the declared voltage at consumer's terminals.
7. The principal cause of voltage variation is the change of _______ on the system.

Answers to Self Test

1. more than one point
2. no interruption
3. not economical
4. receiving
5. short
6. ± 6%
7. load.

OBJECTIVE QUESTIONS

1. To increase the transmission capability of a high voltage long line:
   (*a*) the series reactance can be reduced
   (*b*) the shunt admittance can be reduced
   (*c*) the resistance can be decreased
   (*d*) the resistance can be increased.

2. Series capacitors are used to:
   (*a*) improve the voltage (*b*) reduce fault level
   (c) improve the power factor (d) compensate for line inductive reactance and improve the stability of the power system.
3. In *ac* distribution system the voltage can be controlled by using:
   (*a*) tap changing transformer (*b*) induction regulator
   (*c*) booster transformer (*d*) any of the above.
4. If a synchronous machine is overexcited, It takes lagging vars from the system when it is operated as a:
   (a) synchronous generator (b) synchronous motor
   (c) synchronous motor as well as generator (d) none of the above.
5. The capacity of a transmission line can be increased by reducing:
   (*a*) line inductance (*b*) sending end voltage
   (*c*) line capacitance (*d*) any of the above.
6. A synchronous generator is feeding an infinite bus. At the finite bus a shunt capacitor is added. The stability limit will:
   (*a*) increase (*b*) decrease
   (*c*) remain the same (*d*) any of above.
7. Series capacitor is used in a transmission line to:
   (*a*) reduce line loss (*b*) limit short circuit current
   (*c*) compensate voltage drop (*d*) improve load power factor.
8. Tap changing transformers are employed for:
   (*a*) stepping up the voltage
   (*b*) stepping down the voltage
   (*c*) supplying low voltage current to instruments
   (*d*) both stepping up and stepping down the voltage.
9. If a synchronous machine is underexcited it takes lagging vars from the system when it is operated as a:
   (*a*) synchronous generator (*b*) synchronous motor
   (*c*) synchronous motor as well as generator (*d*) none of the above.
10. The addition of a synchronous compensator in the system:
    (*a*) has no effect on stability (*b*) improves systems stability
    (*c*) decreases systems stability.
11. The series compensation:
    (*a*) increases stability limit (*b*) decreases stability limit
    (*c*) has no effect on stability limit.
12. For the same voltage boost, the reactive power capacity is more for a:
    (*a*) shunt capacitor (*b*) series capacitor
    (*c*) It is same for both series and shunt.
13. Series capacitors can be used in transmission lines to:
    (*a*) reduce the line losses
    (*b*) reduce the voltage drop
    (*c*) reduce the receiving end voltage under light load conditions
    (*d*) provide reactive power compensation.
14. If the excitation of the synchronous generator fails, it acts as a:
    (*a*) synchronous generator (*b*) synchronous motor
    (*c*) induction generator (*d*) induction motor.

15. Capacitors are used in power system to:
(*a*) improve voltage regulation
(*b*) improve supply power factor
(*c*) change the load characteristics
(*d*) all of the above.

16. The best location for use of a booster transformer in a transmission line is:
(*a*) at the sending end
(*b*) at the receiving end
(*c*) at the intermediate point
(*d*) anywhere in the line.

17. Series capacitors on transmission lines are of little use when the required reactive voltamperes are:
(*a*) large
(*b*) small
(*c*) fluctuating
(*d*) any of the above.

18. The effect of series capacitance compensation is:
(*a*) to decrease the effective length of the line
(*b*) to decrease the virtual surge impedance
(*c*) to increase virtual surge impedance loading
(*d*) all of the above.

19. In tap-changing transformers, the tappings are provided on:
(*a*) primary winding
(*b*) secondary winding
(*c*) tertiary winding
(*d*) high voltage winding.

20. Use of additional shunt capacitor can be made for increasing the capability of line as it:
(*a*) reduces surge impedance Z_0
(*b*) increases phase shift β
(*c*) increase in α
(*d*) all of above.

21. Shunt capacitors in a substation:
(*a*) consume active power
(*b*) deliver active power
(*c*) deliver lagging var
(*d*) consume lagging var.

22. For constant voltage transmission, the voltage drop along the line is maintained constant by installing.
(*a*) capacitors
(*b*) inductors
(*c*) resistors
(*d*) synchronous phase modifiers at the receiving end.

23. A synchronous generator connected to an infinite bus delivers power at a lag p.f. If its excitation is increased:
(*a*) the terminal voltage increases
(*b*) voltage angle δ increases
(*c*) current delivered increases
(*d*) (*b*) and (*c*).

24. A synchronous motor connected to an infinite bus takes power at a lag p.f. If its excitation is increased:
(*a*) the terminal voltage increases
(*b*) the load angle increases
(*c*) the p.f. of motor increases
(*d*) (*b*) and (*c*).

25. Power dispatch through a line can be increased by:
(*a*) installing shunt capacitors
(*b*) installing series capacitors
(*c*) installing series reactor
(*d*) installing shunt reactor.

Answers to Objective Questions

1. (*a*)	**2.** (*d*)	**3.** (*d*)	**4.** (*d*)
5. (*a*)	**6.** (*c*)	**7.** (*c*)	**8.** (*d*)
9. (*c*)	**10.** (*b*)	**11.** (*a*)	**12.** (*a*)
13. (*b*)	**14.** (*c*)	**15.** (*d*)	**16.** (*c*)
17. (*b*)	**18.** (*d*)	**19.** (*d*)	**20.** (*a*)
21. (*c*)	**22.** (*d*)	**23.** (*c*)	**24.** (*c*)
25. (*b*).			

EXERCISE

1. Discuss the method of voltage control in transmission system utilizing OLTC transformer.
2. What are the different methods used for voltage control of a power system?
3. How is voltage control achieved?
4. What is booster transformer? Where it is used?
5. Describe the off load tap changing transformer method of voltage control. What are the limitations of the method?
6. State whether the tap changer in a transformer is provided on hv side or lv side.
7. Explain with a neat sketch about Auto-transformer tap-changing.
8. Why is voltage control required in power systems?

15

Distribution System

Distribution System is that part of power system which distributes power to the consumers for utilization. AC 3-phase 4-wire is the standard distribution all over the world.

15.1 MAIN PARTS OF A DISTRIBUTION SYSTEM

The distribution system is the electrical system between the sub-station fed by the transmission system and the consumers meters. It generally consists of feeders, distributors and the service mains. Figure 15.1 shows the single line diagram of a typical low tension distribution system.

1. **Feeders:** A feeder is a conductor which connects the substation to the area where power is to be distributed. Feeders are not tapped in between so that current in it remains the same throughout. The main consideration in the design of a feeder is the current carrying capacity.

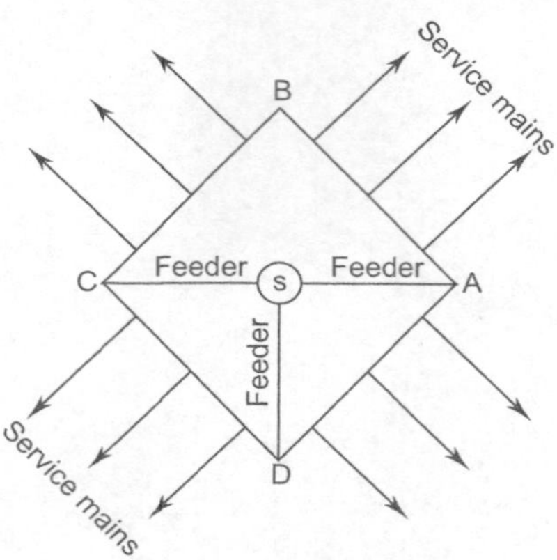

Fig. 15.1

2. **Distributor:** A distributor is a conductor from which tappings are taken for supply to the consumers. In Fig 15.1, AB, BC, CD and DA are the distributors. The current through a distributor is not constant because tappings are taken at various places along its length. The main consideration in the design of a distributor is the voltage drop across its length.

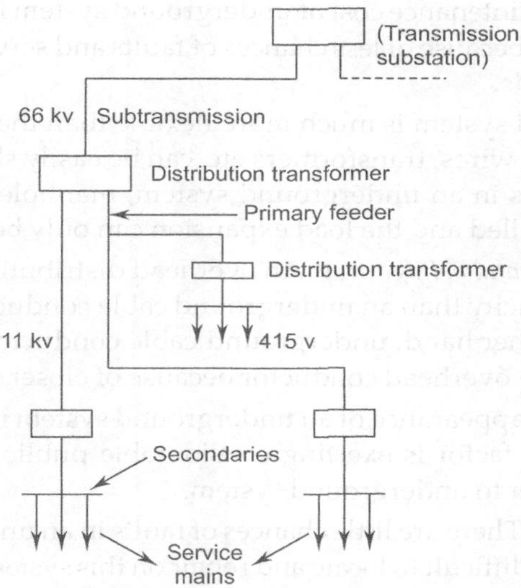

Fig. 15.2

3. **Service Mains:** A service main is a small cable which connects the distributor to the consumer terminals.

A typical distribution system showing its main parts is also shown in Fig. 15.2.

15.2 CLASSIFICATION OF DISTRIBUTION SYSTEMS

15.2.1 According to Type of Construction

It may be classified as:

(*a*) Overhead system

(*b*) Underground system.

Overhead lines are generally mounted on wooden, concrete or steel poles which are arranged to carry distribution transformers in addition to the conductors. The underground system uses conduits, cables and manholes under the surface streets and sidewalks.

The comparison between the both systems is stated as under:

1. *Faults:* The chances of faults in underground system are very rare as the cables are laid underground and are generally provided with better insulation.
2. *Public safety:* The underground system is more safe than overhead system because all distribution wiring is placed underground and there are little chances of any hazard.
3. *Initial cost:* The underground system is more expensive due to high cost of trenching, conduits, cables, manholes and other special equipment. The initial cost of an underground system may be 5 to 10 times than that of an overhead system.
4. *Useful life:* The useful life of underground system is much longer than that of an overhead system. An overhead system may have a useful life of 25 years whereas an underground system may have a useful life of more than 50 years.

5. *Maintenance cost:* The maintenance cost of underground system is very low as compared with that of overhead system because of less chances of faults and service interruptions from wind, ice and lightning hazards.
6. *Flexibility:* The overhead system is much more flexible than the underground system. In an overhead system, poles, wires, transformers etc. can be easily shifted to meet the changes in load conditions whereas in an underground system, manholes, duct lines etc. are permanently placed once installed and the load expansion can only be met by laying new lines.
7. *Current carrying capacity and voltage drop:* An overhead distribution conductor has a considerably higher current capacity than an underground cable conductor of the same material and cross-section. On the other hand, underground cable conductor has much lower inductive reactance than that of an overhead conductor because of closer spacing of conductors.
8. *Appearance:* The general appearance of an underground system is better as all the distribution lines are invisible. This factor is exerting considerable public pressure on electric supply companies to switch over to underground system.
9. *Fault location and repairs:* There are little chances of faults in an underground system. However if a fault does occur, it is difficult to locate and repair on this system. In an overhead system, the conductors are visible and easily accessible so that fault location and repairs can be easily made.
10. *Interference with communication circuits:* An overhead system causes electromagnetic interference with the telephone lines. The power line currents are superimposed on speech currents resulting in the potential of the communication channel being raised to an undesirable level. However, there is no such interference with the underground system.

15.2.2 Classification of Distribution System According to Nature of Current

According to nature of current, distribution may be classified as:

(*a*) DC distribution system

(*b*) AC distribution system.

AC distribution system is universally adopted for distribution of electric power as it is simpler and more economical than direct current method.

(*a*) **DC distribution system:** It may be:

(*i*) DC two-wire system

(*ii*) DC three-wire system.

(*i*) *DC two-wire system:* As the name implies, this system of distribution consists of two wires. One is the outgoing or positive wire and the other is the return or negative wire. The loads such as lamps, motors etc., are connected in parallel between the two wires as shown in Fig. 15.3

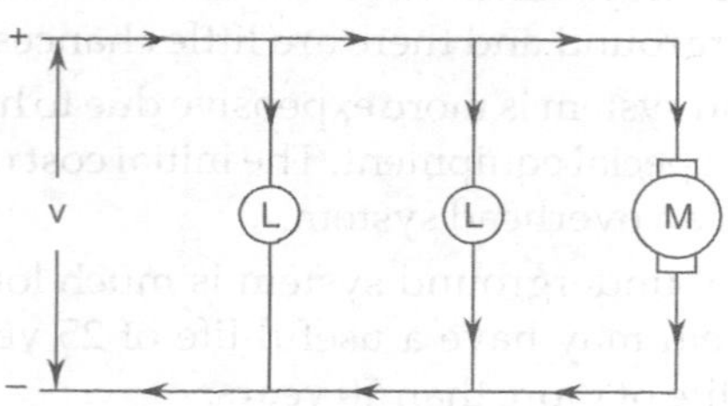

Fig. 15.3

A simple DC two-wire system is also shown in Fig. 15.4. In this, the neutral acts as the return conductor and carries the same current as the positive conductor.

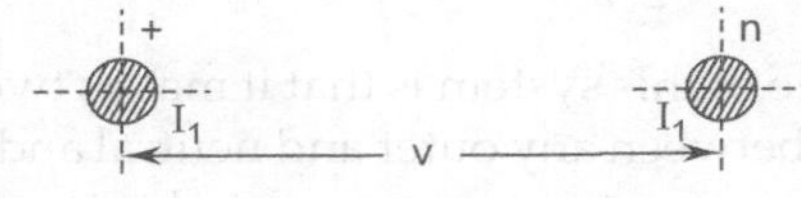

Fig 15.4

Line Current $I_1 = \frac{P}{V}$

If resistance of each conductor is R_1,

$$\text{Line losses} = 2\,I_1^2\,R_1$$

$$= \frac{2\,P^2 \cdot R_1}{V^2} \quad ...(1)$$

This system is never used for transmission purposes due to low efficiency but may be employed for distribution of DC power.

(*ii*) *DC three-wire system:* It consists of two outers and a middle or neutral wire which is earthed at the substation. The voltage between the outers is twice the voltage between either outer and neutral wire as shown in Fig. 15.5

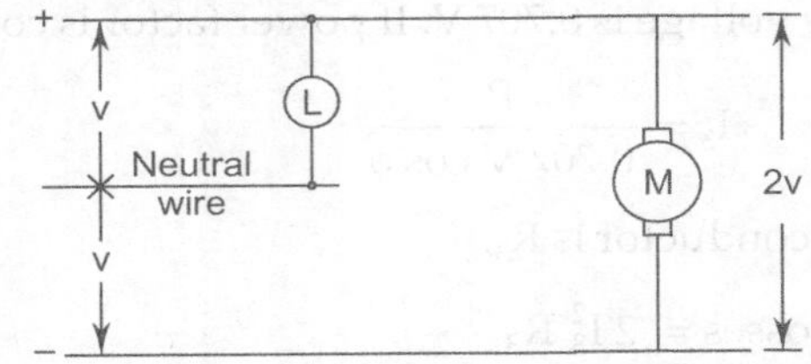

Fig. 15.5

A simplified DC three-wire system is also shown in Fig. 15.6. Since the voltages from the two outer conductors to the neutral are of opposite polarity, the currents flow in the outer conductor is in opposite direction.

$$I_2 = \frac{P}{2\,V}$$

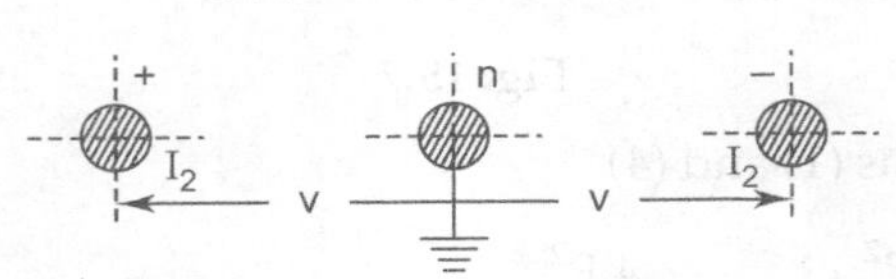

Fig. 15.6

If resistance of each conductor is R_2,

$$\text{Line losses} = 2\,I_2^2\,R_2$$

$$= 2\left(\frac{P}{2\,V}\right)^2 R_2$$

$$= \frac{P^2}{2\,V^2} . R_2 \qquad ...(2)$$

The principal advantage of this system is that it makes two voltages available at the consumer terminals viz. V between any outer and neutral and 2 V between the outers. Loads requiring high voltage (*eg.*, motors) are connected across the outers whereas lamps and heating circuits requiring less voltage are connected between either outer and the neutral.

Comparing equations (1) and (2)

$$2 \frac{P^2}{V^2} . R_1 = \frac{P^2}{2\,V^2} . R_2$$

or $$R_2 = 4\,R_1$$

The resistance is inversely proportional to the area of cross-section. Considering the area of cross-section of conductors in two-wire DC system A_1 and in three-wire DC. system by A_2.

$$A_2 = 0.25\,A_1 \qquad ...(3)$$

(*b*) **AC distribution system:** It may be,

(*i*) Single phase AC.

(*ii*) 3-phase, 4-wire AC.

(*i*) *Single phase AC:* Figure 15.7 shows this system. Since the peak voltage between the conductors is V, the rms voltage is 0.707 V. If power factor is cos ϕ.

$$I_3 = \frac{P}{0.707\,V \cos\phi}$$

If resistance of each conductor is R_3,

$$\text{Line Losses} = 2\,I_3^2\,R_3$$

$$= \frac{4\,P^2}{V^2 \cos^2\phi} . R_3 \qquad ...(4)$$

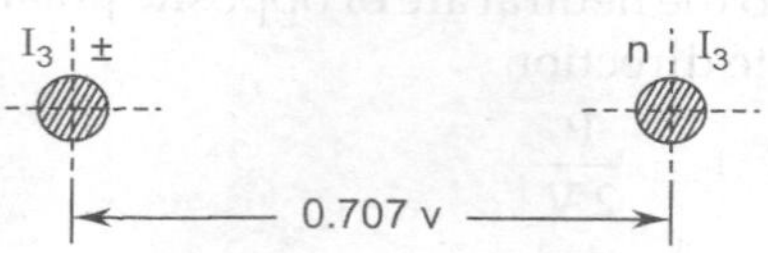

Fig. 15.7

Comparing equations (1) and (4)

$$\frac{2\,P^2}{V^2} R_1 = \frac{4\,P^2}{V^2 \cos^2\phi} . R_3$$

or $$R_3 = (0.5 \cos^2\phi)\,R_1$$

Consider area of cross-section of conductors in this system by A_3.

$$A_3 = \frac{2}{\cos^2\phi} A_1 \qquad ...(5)$$

(*ii*) *3-phase, 4-wire AC* : Figure 15.8 shows this system. The voltage between each phase conductor and neutral is 0.707 V. If the system is balanced, there would be no current in the neutral.

$$I_4 = \frac{P}{3\,(0.707\ V)\cos\phi}$$

If each conductor has a resistance R_4,

$$\text{Line Losses} = 3\,I_4^2\,R_4$$

$$= \frac{2\,P^2}{3\,V^2\cos^2\phi}\,.\,R_4 \qquad ...(6)$$

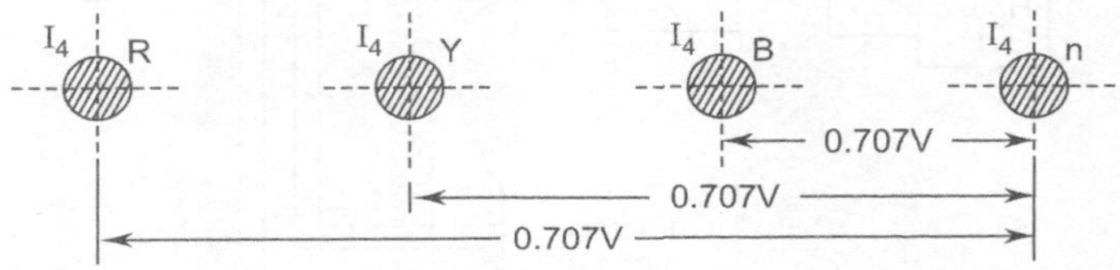

Fig. 15.8

Comparing equations (1) and (6)

$$\frac{2\,P^2}{V^2}\,R_1 = \frac{2\,P^2}{3\,V^2\cos^2\phi}\,R_4$$

or $$R_4 = (3\cos^2\phi)\,R_1$$

Consider the area of cross-section of each phase conductor by A_4,

$$A_4 = \frac{A_1}{3\cos^2\phi} \qquad ...(7)$$

If v_1, v_2, v_3 and v_4 denote the total conductor volume in above 4-cases, then

$$v_1 : v_2 : v_3 : v_4 = 1 : \frac{1}{4} : \frac{2}{\cos^2\phi} : \frac{1}{3\cos^2\phi}$$

A typical 3-phase, 4-wire distribution system is shown in Fig. 15.9. The primary distribution circuit delivers power to various sub-stations called distribution substations. This primary distribution is carried out by 3-phase, 3-wire system due to economic considerations. The substations are situated near the consumer's localities and contain step-down transformers. At each distribution substation, the voltage is stepped down to 400 V and power is delivered by 3-phase, 4-wire AC. system. The voltage between any two phases is 400 V and between any phase and neutral is 230 V. The single phase domestic loads are connected between any one phase and the neutral whereas 3-phase, 400 V motor loads are connected across 3-phase lines directly.

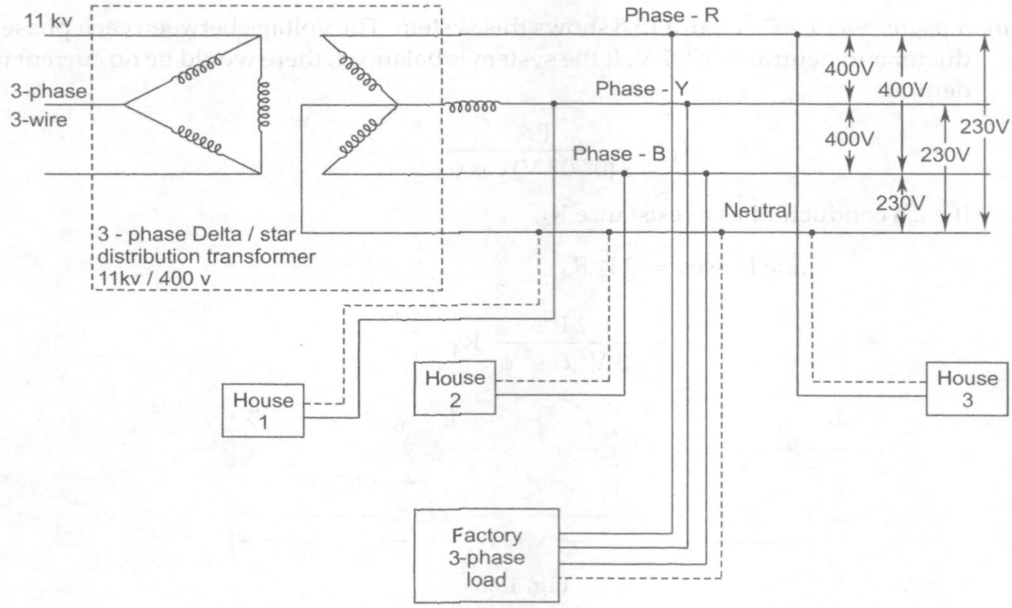

Fig. 15.9

15.2.3 Classification of Distribution System According to Scheme of Connection

It may be classified as:

(*a*) Radial system

(*b*) Ring main system

(*c*) Inter-connected system.

(*a*) *Radial system:* It is the simplest and most commonly used system. It is used extensively to supply small and medium residential, commercial, industrial and non-critical loads. In this system, separate feeders radiate from a single substation and feed the distributors at one end only. Figure 15.10 shows a single line diagram of a radial system for DC. distribution where a feeder OC supplies a distributor AB at point A. Obviously, the distributor is fed at one end only *i.e.*, point A in this case.

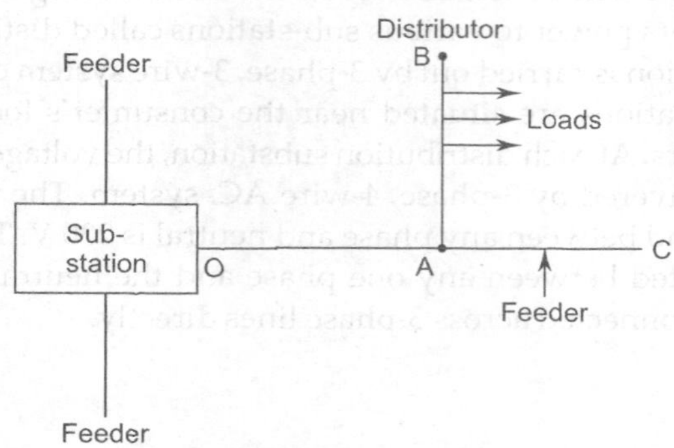

Fig. 15.10

Figure 15.11 shows a single line diagram of a radial system for AC. distribution.

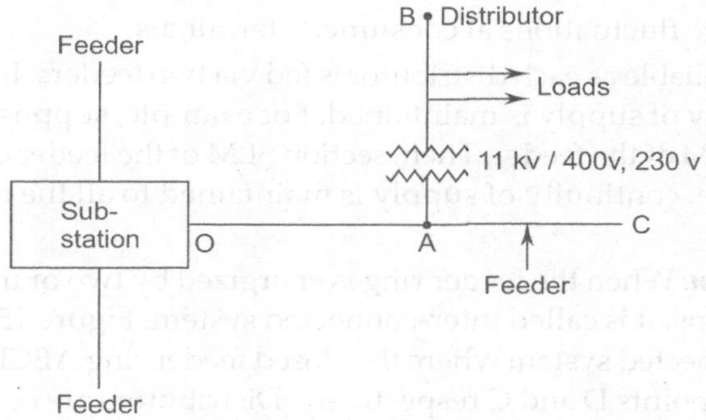

Fig. 15.11

The radial system is employed only when power is generated at low voltage and the sub-station is located at the centre of the load. The radial system is most economical and has the lowest initial cost.

This system has following drawbacks:

(*i*) The consumers are dependent on a single feeder. A fault in the feeder means disruption of supply.

(*ii*) The consumer at the distant end of the distributor would be subjected to serious voltage fluctuations when the load on the distributor changes.

(*b*) *Ring main system:* In this system, the primaries of distribution transformers form a loop. The loop circuit starts from the sub-station bus bars, makes a loop through the area to be served and return to the sub-station. Figure 15.12 shows the single line diagram of ring main system for AC. distribution where substation supplies to the closed feeder LMNOPQRS. The distributors are tapped from different points M, O and Q of the feeder through distribution transformers.

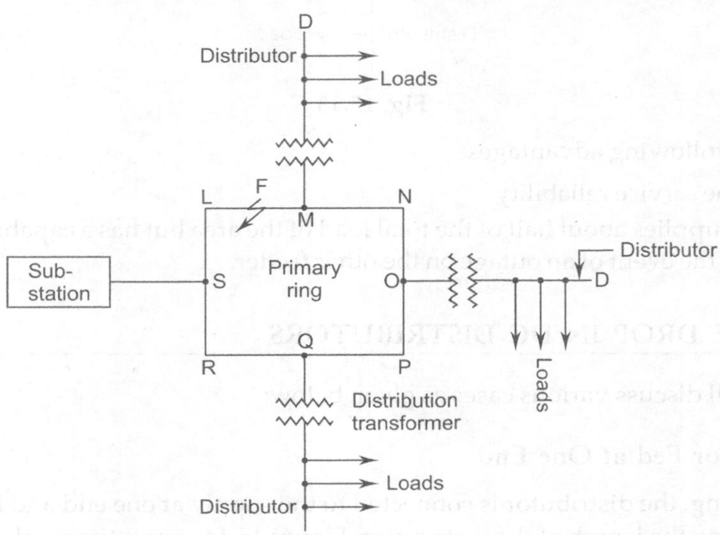

Fig. 15.12

This system has following advantages:

(*i*) There are less voltage fluctuations at consumer's terminals.

(*ii*) The system is very reliable as each distributor is fed via two feeders. In the event of fault on any feeder, the continuity of supply is maintained. For example, suppose that fault occurs at any point F of section SLM of the feeder. Then, section SLM of the feeder can be isolated for repairs and at the same time, continuity of supply is maintained to all the consumers via the feeder SRQPONM.

(*c*) *Inter-connected system:* When the feeder ring is energized by two or more than two generating stations or substations, it is called inter-connected system. Figure 15.13 shows the single line diagram of interconnected system where the closed feeder ring ABCD is supplied by two sub-stations S_1 and S_2 at points D and C respectively. Distributors are connected to points O, P, Q and R of the feeder ring through distribution transformers.

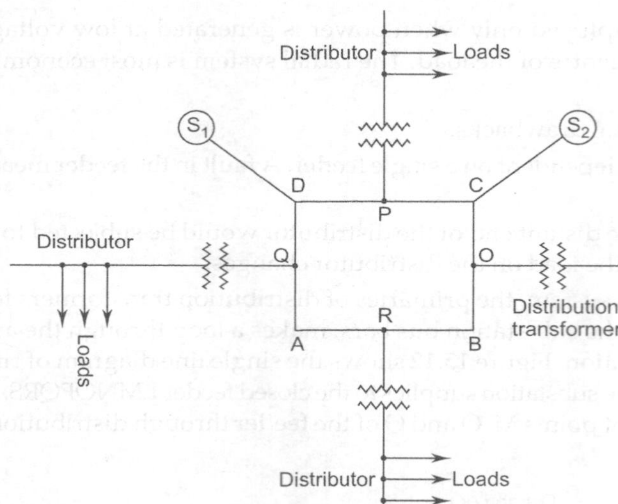

Fig. 15.13

This system has following advantages:

(*i*) It increases the service reliability.

(*ii*) Each feeder supplies about half of the total load of the area but has a capability to supply the entire load in the event of an outage on the other feeder.

15.3 VOLTAGE DROP IN DC DISTRIBUTORS

In this topic, we will discuss various cases as given below:

15.3.1 Distributor Fed at One End

In this type of feeding, the distributor is connected to the supply at one end and loads are taken at different points along the length of the distributor. Figure 15.14 shows the single line diagram of a DC distributor AB fed at one end A and I_1, I_2, I_3 are the load currents tapped at different points and r is the resistance per unit length of the distributor. (go and return):

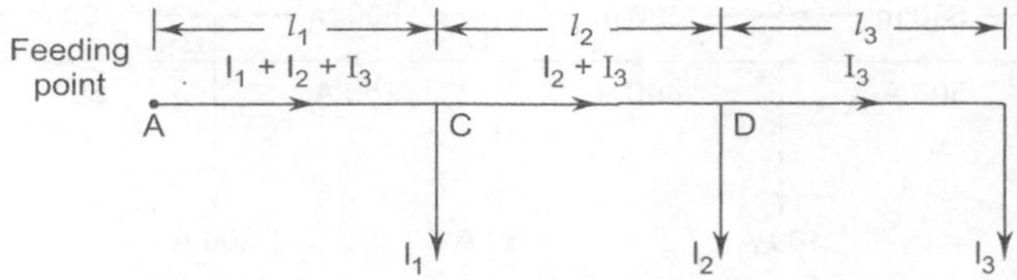

Fig. 15.14

Voltage drop in section AC $= (I_1 + I_2 + I_3)\, l_1\, r$

Voltage drop in section CD $= (I_2 + I_3)\, l_2\, r$

Voltage drop in section BD $= I_3\, l_3\, r$

$\therefore$ Total voltage drop in the distributor

$$= (I_1 + I_2 + I_3)\, l_1\, r + (I_2 + I_3)\, l_2\, r + I_3\, l_3\, r$$
$$= I_1\, l_1\, r + I_2\, (l_1 + l_2)\, r + I_3\, (l_1 + l_2 + l_3)\, r$$

Following points may be noted from above case:

(*a*) Minimum potential will occur at point B which is farthest from the feeding point A.

(*b*) The current in the various sections of the distributor away from feeding point goes on decreasing. Thus, current in section AC is more than the current CD and the current in the section CD is more than the current in section DB.

Example 15.1. *A 2-wire d.c. distributor cable AB is 2 km. long and supplies loads of 100 A, 150 A, 200 A and 50 A situated 500 m, 1000 m, 1600 m and 2000 m. from the feeding point A. Each conductor has a resistance of 0.01 Ω per 1000 m. Calculate potential difference at each load point if a potential difference of 300V is maintained at point A.*

Solution.

Resistance per 1000 m of distributor $= 2 \times 0.01 = 0.02\ \Omega$

Resistance of section AC, $R_{AC} = 0.02 \times \dfrac{500}{1000} = 0.01\ \Omega$

$$R_{CD} = 0.02 \times \frac{500}{1000} = 0.01\ \Omega$$

$$R_{DE} = 0.02 \times \frac{600}{1000} = 0.012\ \Omega$$

$$R_{EB} = 0.02 \times \frac{400}{1000} = 0.008\ \Omega.$$

From Fig. 15.15,

$$I_{EB} = 50\text{ A}$$
$$I_{DE} = 50 + 200 = 250\text{ A}$$
$$I_{CD} = 250 + 150 = 400\text{ A}$$
$$I_{AC} = 400 + 100 = 500\text{ A}$$

Potential difference at load point C, $V_c = V_A - \text{Voltage drop in AC}$

$$= V_A - I_{AC}\, R_{AC} = 300 - 500 \times 0.01 = 295\text{ V}$$

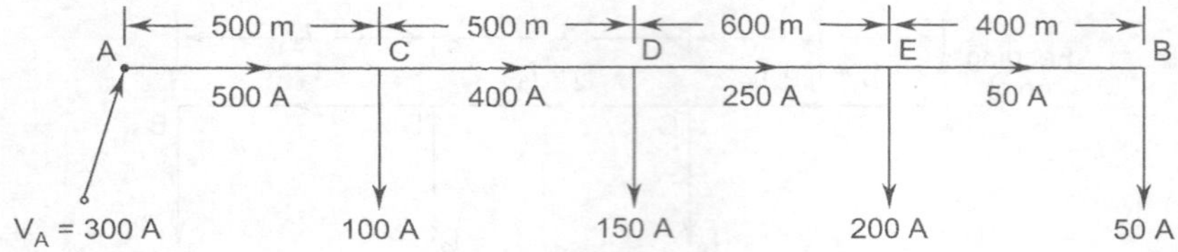

Fig. 15.15

Potential difference at load point D,

$$V_D = V_C - I_{CD} R_{CD}$$
$$= 295 - 400 \times 0.01 = 291 \text{ V}$$

Potential difference at load point E,

$$V_E = V_D - I_{DE} R_{DE}$$
$$= 291 - 250 \times 0.012 = 288 \text{ V}$$

Potential difference at load point B,

$$V_B = V_E - I_{EB} R_{EB}$$
$$= 288 - 50 \times 0.008 = 287.6 \text{ V}$$

Example 15.2. *A simple DC distributor as shown in Fig. 15.16 is loaded as under, $I_1 = 100$ A, $I_2 = 150$ A, $I_3 = 200$ A. The lengths are $l_1 = 150$ m, $l_2 = 100$ m, $l_3 = 100$ m. The resistance of both the conductors (go and return) is 0.1 Ω per 1000 m. Find voltages at points C, D and B if $V_A = 200$ V.*

Solution.

Resistance of section AC, $R_{AC} = 0.1 \times \dfrac{150}{1000} = 0.015\ \Omega$

$$R_{CD} = 0.1 \times \frac{100}{1000} = 0.01\ \Omega$$

$$R_{DB} = 0.1 \times \frac{100}{1000} = 0.01\ \Omega$$

From Fig. 15.16,

$$I_{DB} = 200 \text{ A}$$
$$I_{CD} = 200 + 150 = 350 \text{ A}$$
$$I_{AC} = 350 + 100 = 450 \text{ A}$$

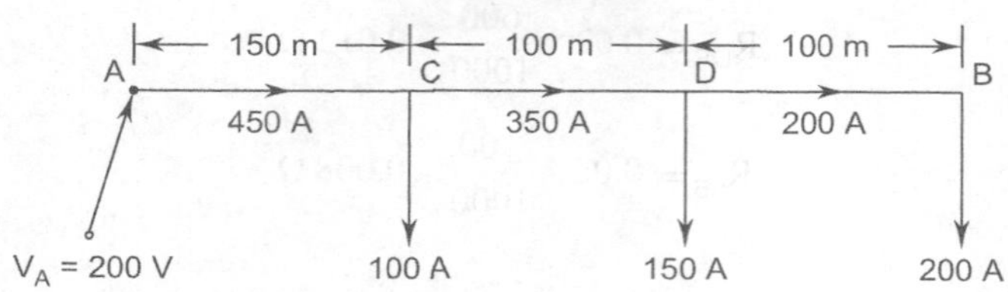

Fig. 15.16

Potential difference at load point C,

$$V_C = V_A - I_{AC} \cdot R_{AC}$$
$$= 200 - 450 \times 0.015 = 193.25 \text{ V}$$

Potential difference at load point D,

$$V_D = V_C - I_{CD} \cdot R_{CD}$$
$$= 193.25 - 350 \times 0.01 = 189.75 \text{ V}$$

Potential difference at load point B,

$$V_B = V_D - I_{DB} \cdot R_{DB}$$
$$= 189.75 - 200 \times 0.01 = 187.75 \text{ V}$$

Example 15.3. *A 2-wire DC distributor AB is 300 m long. It is fed at point A. The various loads and their positions are given below:*

At point	*Distance from A in meters*	*Concentrated load in amperes*
C	40	30
D	100	40
E	150	100
F	250	50

If the maximum permissible voltage drop is not to exceed 10 V, find the cross-sectional area at the distributor. Take $\rho = 1.98 \times 10^{-8}\ \Omega m$.

Solution.

The arrangement is shown in Fig. 15.17. Suppose that resistance of 100 m length of distributor is r ohms.

$\therefore$ Resistance of section AC – $R_{AC} = 0.4\, r\ \Omega$

$$R_{CD} = 0.6\, r\ \Omega$$
$$R_{DE} = 0.5\, r\ \Omega$$
$$R_{EF} = r\ \Omega$$

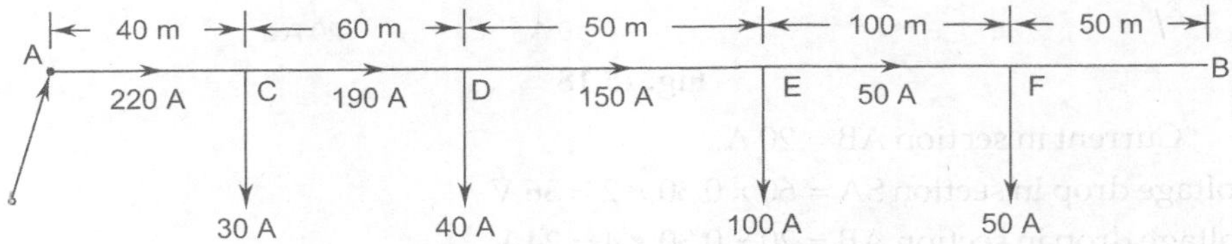

Fig. 15.17

As shown in Fig. 15.17, the currents in various sections are:

$$I_{AC} = 220 \text{ A}$$
$$I_{CD} = 190 \text{ A}$$
$$I_{DE} = 150 \text{ A}$$
$$I_{EF} = 50 \text{ A}$$

Total voltage drop over the distributor

$$= I_{AC} \cdot R_{AC} + I_{CD} \cdot R_{CD} + I_{DE} \cdot R_{DE} + I_{EF} \cdot R_{EF}$$
$$= 220 \times 0.4\, r + 190 \times 0.6\, r + 150 \times 0.5\, r + 50 \times r$$
$$= 327\, r$$

As maximum permissible drop in distributor is 10 V,

$\therefore$ $$10 = 327\, r$$

or $$r = \frac{10}{327} = 0.03058\ \Omega$$

As $$R = \frac{\delta l}{A}$$

or $$A = \text{Cross-sectional area of conductor}$$

$$= \frac{\delta l}{r/2}\left[\because R = \frac{r}{2}\right]$$

$$= \frac{1.98 \times 10^{-8} \times 100}{\frac{0.03058}{2}}$$

$$= 129.4 \times 10^{-6}\ \text{m}^2 = 1.294\ \text{cm}^2.$$

Exmple 15.4: *Two tram cars (A & B) are 2 km and 6 km away from a sub-station & return 40 A and 20 A respectively to the rails. The sub-station voltage is 600 V DC. The resistance of trolley wire is 0.25 Ω/km and that of track is 0.05 Ω/km. Calculate the voltage across each tram car.*

Solution.

Figure 15.18 shows the arrangement.

Resistance of trolley wire and track/km.

$$= 0.25 + 0.05 = 0.30\ \Omega$$

Current in section SA = 40 + 20 = 60 A

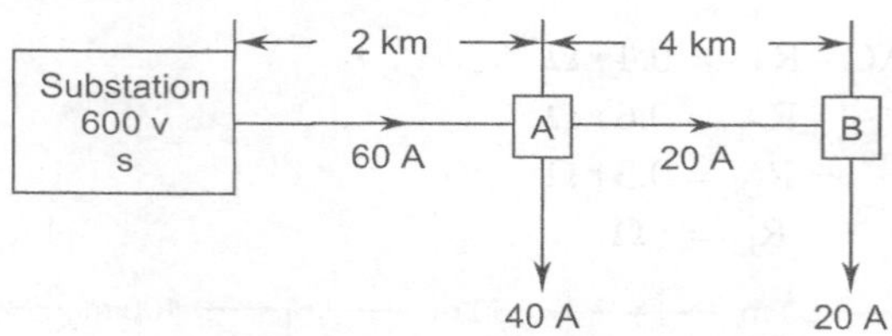

Fig. 15.18

Current in section AB = 20 A.

Voltage drop in section SA = $60 \times 0.30 \times 2 = 36$ V

Voltage drop in section AB = $20 \times 0.30 \times 4 = 24$ V

∴ Voltage across tram A = 600 – 36 = 564 V

Voltage across tram B = 564 – 24 = 540 V.

15.3.2 Uniformly Loaded Distributor Fed at One End

Figure 15.19 shows a uniformly loaded distributor fed at one end. A and loaded uniformly with i amperes per unit length. The total length of the distributor is l and r is the resistance per unit length of both wires of the distributor (go and return).

Consider a point C on the distributor at *a*

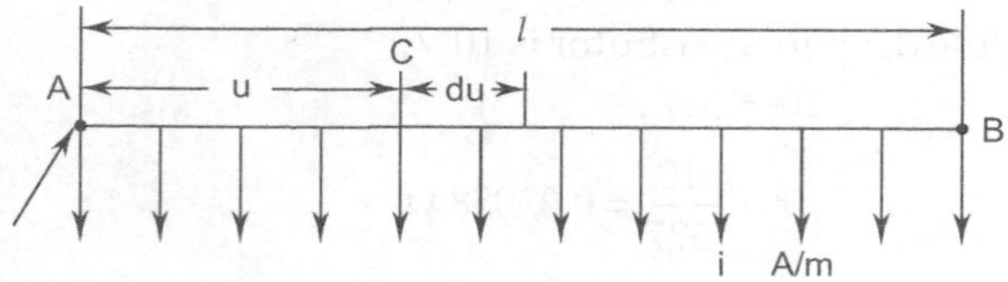

Fig. 15.19

Distance u metres from the feeding point A. Then, the current at point C is:

$$= il - iu \text{ Amps.}$$

$$= i(l-u) \text{ amp.}$$

Consider a small length du near point C. Its resistance is $r.\,du$ and voltage drop over length du is,

$$dv = i(l-u)\,r.\,du$$

$$= ir(l-u)\,du$$

Total voltage drop in distributor upto point C:

$$v = \int_0^u ir(l-u)\,du$$

$$= ir\left(lu - \frac{u^2}{2}\right)$$

Total voltage drop over the length l upto point B can be obtained by substituting $u = l$ in above equation.

∴ Voltage drop over the distributor AB:

$$= irl^2 - \frac{1}{2}irl^2$$

$$= \frac{irl^2}{2} = \frac{1}{2}(il)(rl)$$

$$= \frac{1}{2}\text{I.R}$$

where I = il, the total current supplied by the distributor and R = rl, the total resistance of the distributor.

Example 15.5. *A 2-wire DC distributor 200 m. long is uniformly loaded with 4 A/metre. Resistance of single wire is 0.3 Ω/km. If distributor is fed at one end, calculate.*

(*a*) *voltage drop upto a distance of 150 m from the feeding point.*

(*b*) *the maximum voltage drop.*

Solution.

Given, $i = 4$ A/m.

Resistance of distributor per metre run

$$r = 2 \times \frac{0.3}{1000} = 0.0006\ \Omega$$

$$l = 200 \text{ m.}$$

(*a*) Voltage drop upto distance u from feeding point is

$$= ir\left(lu - \frac{u^2}{2}\right)$$

$$= 4 \times 0.0006\left(200 \times 150 - \frac{150 \times 150}{2}\right) \qquad [\because u = 150 \text{ m.}]$$

$$= 45 \text{ V}$$

(*b*) Total current entering the distributor,

$$\text{I} = il = 4 \times 200 = 800 \text{ A}$$

Total resistance of the distributor:

$$R = rl = 0.0006 \times 200 = 0.12\ \Omega$$

∴ Total voltage drop over the distributor

$$= \frac{1}{2} IR = \frac{1}{2} \times 800 \times 0.12 = 48\ V$$

Example 15.6. *A uniform 2-wire DC distributor 500 m long is loaded with 0.4 A/m and is fed at one end. If the maximum permissible voltage drop is not to exceed 10 V, find the cross-sectional area of the distributor. Take* $\rho = 1.98 \times 10^{-6}\ \Omega$ *cm.*

Solution.

Current entering the distributor,

$$I = i \times l = 0.4 \times 500 = 200\ A$$

Let r be resistance per meter length of distributor (both wires)

∴ $$\text{Maximum voltage drop} = \frac{1}{2} IR$$

or $$10 = \frac{1}{2} IRl$$

or $$r = \frac{2 \times 10}{200 \times 500} = 0.2 \times 10^{-3}\ \Omega$$

∴ Area of cross-section of distributor conductor:

$$A = \frac{\rho l}{r/2} = \frac{1.98 \times 10^{-6} \times 100 \times 2}{0.2 \times 10^{-3}} \qquad [l = 1m. = 100\ cm.]$$

$$= 1.98\ cm^2.$$

15.3.3 Distributor Fed at Both Ends

In this type, the distributor is connected to the supply mains at both ends and loads are tapped off at different points along the length of the distributor. The voltage at the feeding point may or may not be equal.

Such a distributor is used when the length of the distributor is long and the load currents are high. If such a distributor is fed at only one end, the voltage drop along the distributor may be rather high.

Figure 15.20 shows a distributor AB fed at the ends A and B and loads of I_1, I_2, I_3 and I_4 tapped off at points C, D, E and F respectively. Here, the load voltage goes on decreasing as we move away from one feeding point say A, reaches minimum value and then again starts rising and reaches maximum value when we reach the other feeding point B. The minimum voltage occurs at some load point and is never fixed. The currents I_A and I_B can be calculated by the application of Kirchoff's Laws. The total voltage drop along the distributor must be equal to difference of voltages V_A and V_B.

Following points may be noted from above case:

(*a*) If a fault occurs on any feeding point of the distributor, the continuity of supply is maintained from the other feeding point.

(*b*) The area of cross-section required for a doubly fed distributor is much less than that of a singly fed distributor.

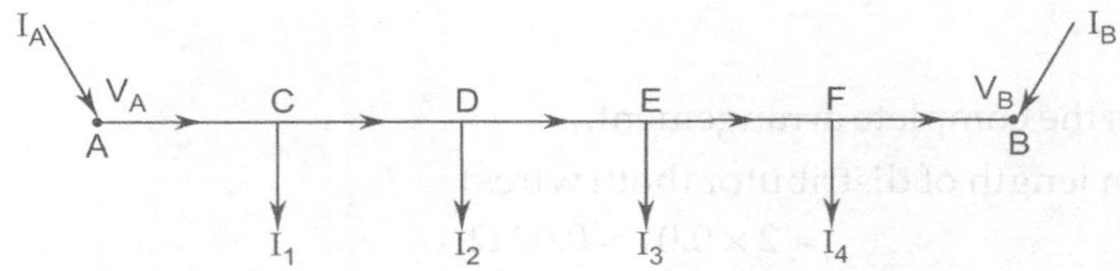

Fig. 20

Example 15.7. *Figure 15.21 shows a two-wire distributor fed at both ends A and B at 255 V and 250 V. The loads and resistances of distributor (both wires) are shown in Fig. 15.12. Find the current fed at A and B and voltages at points C and D.*

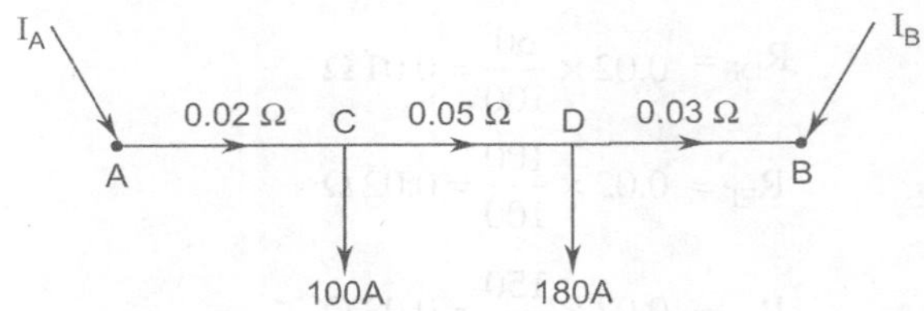

Solution.

The above Fig. 15.21 can be drawn as shown below indicating currents in various sections.

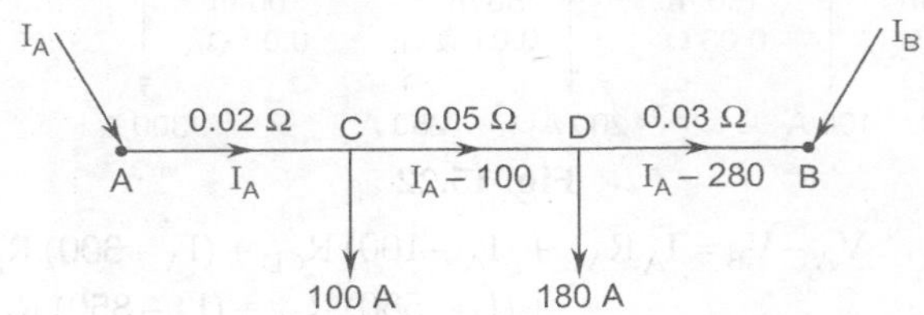

Fig. 15.21

$$I_A \times 0.02 + (I_A - 100) \times 0.05 + (I_A - 280) \times 0.03 = V_A - V_B$$

or $I_A \times 0.02 + (I_A - 100) \times 0.05 + (I_A - 280) \times 0.03 = 5$

or $$I_A = 184 \text{ A}$$

$$I_B = -(I_A - 280)$$
$$= -(184 - 280) = 96 \text{ A}$$

$$\text{Voltage at C} - V_C = V_A - I_A \times 0.02$$
$$= 255 - 184 \times 0.02 = 251.32 \text{ V}$$
$$V_D = 251.32 - (I_A - 100) \times 0.05$$
$$= 251.32 - (184 - 100) \times 0.05 = 247.12 \text{ V.}$$

Example 15.8. *A two-wire DC distributor AB, 600 m long is loaded as under*

Distance from A (m.):	150	300	350	450
Loads in Amperes:	100	200	250	300

The feeding point A is maintained at 440 V and that of B at 430 V. If each conductor has a resistance of 0.01 Ω per 100 m. Calculate:

(a) the currents supplied from A and B.

(b) the power dissipated in the distributor.

Solution.

Figure 15.22 shows the complete arrangement.

Resistance of 100 m length of distributor (both wires)

$$= 2 \times 0.01 = 0.02\ \Omega$$

$$\text{Resistance of section AC, } R_{AC} = 0.02 \times \frac{150}{100} = 0.03\ \Omega$$

$$R_{CD} = 0.02 \times \frac{150}{100} = 0.03\ \Omega$$

$$R_{DE} = 0.02 \times \frac{50}{100} = 0.01\ \Omega$$

$$R_{EF} = 0.02 \times \frac{100}{100} = 0.02\ \Omega$$

$$R_{FB} = 0.02 \times \frac{150}{100} = 0.03\ \Omega$$

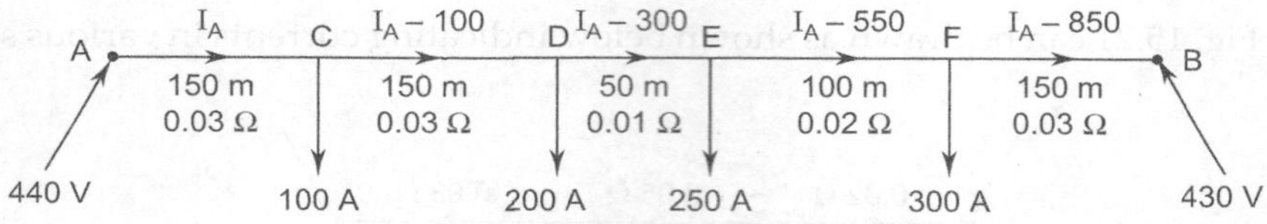

Fig. 15.22

From Fig. 15.22

$$V_A - V_B = I_A R_{AC} + (I_A - 100) R_{CD} + (I_A - 300) R_{DE} + (I_A - 550) R_{EF} + (I_A - 850) R_{FB}$$

$$440 - 430 = 0.03\, I_A + 0.03\, (I_A - 100) + 0.01\, (I_A - 300) + 0.02\, (I_A - 550) + 0.03\, (I_A - 850)$$

or

$$I_A = 437.5 \text{ A}$$

The actual distribution of current in various section after putting value of I_A is shown below:

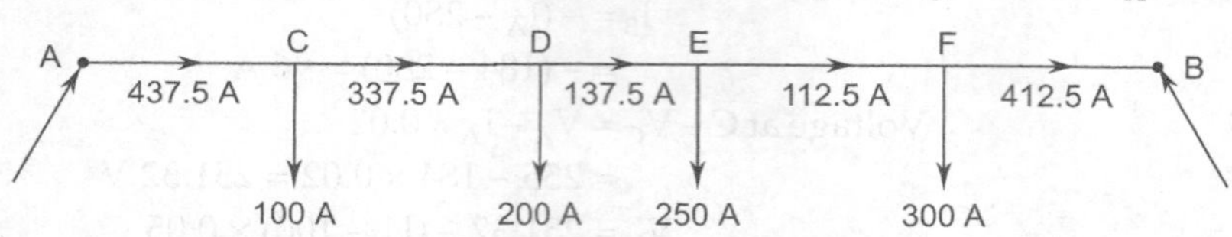

It is clear that current is coming to point E from both sides *i.e.,* D and F. Hence, E is the point of minimum potential.

(*b*) Power loss in the distributor:

$$= I_{AC}^2 R_{AC} + I_{CD}^2 R_{CD} + I_{DE}^2 R_{DE} + I_{EF}^2 R_{EF} + I_{FB}^2 R_{FB}$$

$$= (437.5)^2 \times 0.03 + (337.5)^2 \times 0.03 + (137.5)^2 \times 0.01 + (112.5)^2 \times 0.02 + (412.5)^2 \times 0.03$$

$$= 5742 + 3417 + 189 + 253 + 5104$$

$$= 14705 \text{ W}$$

$$= 14.705\ kW$$

Example 15.9. *A 2-wire DC distributor AB is fed from both ends. At feeding point A, the voltage is maintained at 235 V and at B 240 V. The total length of the feeder is 200 m and loads are tapped off as under.*

25 A at 50 m from A

50 A at 75 m from A

30 A at 100 m from A.

40 A at 150 m from A.

The resistance per km. of one conductor is 0.3 Ω. Calculate:

(a) Currents in various sections of the distributor.

(b) Minimum voltage and the point at which it occurs.

Solution.

Figure 15.23 shows the distributor with its tapped currents.

Resistance of 1000 m length of distributor (both wires)

$$= 2 \times 0.3 = 0.6\ \Omega$$

$$\text{Resistance of section AC, } R_{AC} = 0.6 \times \frac{50}{1000} = 0.03\ \Omega$$

$$R_{CD} = 0.6 \times \frac{25}{1000} = 0.015\ \Omega$$

$$R_{DE} = 0.6 \times \frac{25}{1000} = 0.015\ \Omega$$

$$R_{EF} = 0.6 \times \frac{50}{1000} = 0.03\ \Omega$$

$$R_{FB} = 0.6 \times \frac{50}{1000} = 0.03\ \Omega$$

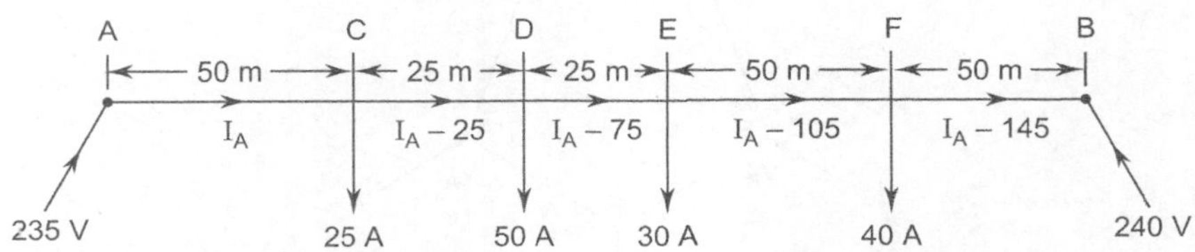

Fig. 15.23

From Fig. 15.23

$$V_B = V_A - [I_A R_{AC} + (I_A - 25) R_{CD} + (I_A - 75) R_{DE} + (I_A - 105) R_{EF} + (I_A - 145) R_{FB}]$$

$$240 = 235 - [0.03 I_A + 0.015 (I_A - 25) + 0.015 (I_A - 75) + 0.03 (I_A - 105) + 0.03 (I_A - 145)]$$

$$= 235 - [0.12 I_A - 9]$$

or

$$I_A = \frac{244 - 240}{0.12} = 33.34\ \text{A}$$

(*a*) Current in section AC, $I_{AC} = I_A = 33.34$ A

$I_{CD} = I_A - 25 = 33.34 - 25 = 8.34$ A

$I_{DE} = I_A - 75 = 33.34 - 75 = -41.66$ A from D to E = 41.66 A from E to D.

$I_{EF} = I_A - 105 = 33.34 - 105 = -71.66$ A from E to F = 71.66 from F to E.

$I_{FB} = I_A - 145 = 33.34 - 145 = -111.66$ A from F to B = 111.66 A from B to F.

The actual distribution of currents is shown in Fig. 15.24

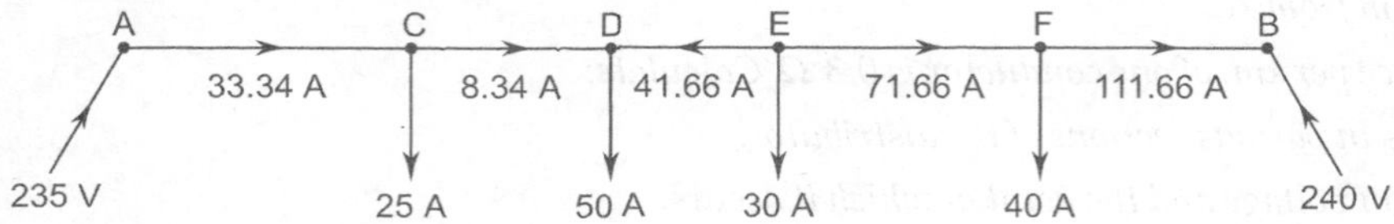

Fig. 15.24

Load point D is the point of minimum potential as the current is coming to load point D from both sides of the distributor.

$$\therefore \quad \text{Voltage at D, } V_D = V_A - [I_{AC} R_{AC} + I_{CD} \cdot R_{CD}]$$
$$= 235 - [33.34 \times 0.03 + 8.34 \times 0.015]$$
$$= 235 - 1.125 = 233.875 \text{ V}$$

15.3.4 Ring Main Distributor

A distributor arranged to form a closed loop and fed at one or more points is called a ring distributor. Such a system has better flexibility and reliability. In case a fault develops in any one section, it can be isolated from both sides so that only that section suffers discontinuity of supply. The continuity of supply is maintained for the remaining sections. Figure 15.25 shows a ring main distributor.

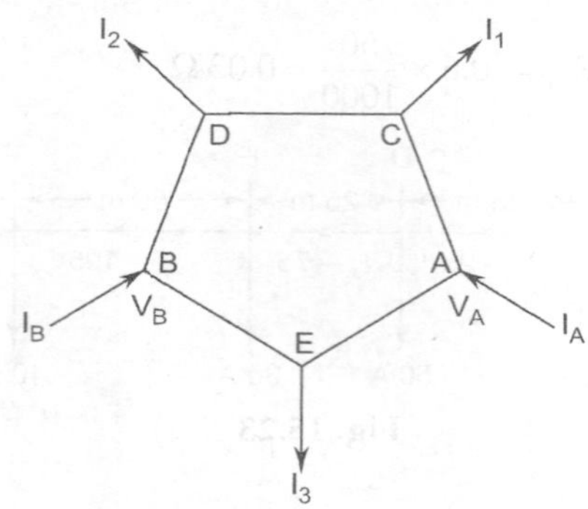

Fig. 15.25

Example 15.10 *A 2-wire DC distributor ABCDEA in the form of a ring main is fed at point A at 220 V and is loaded as under.*

10 A at B, 20 A at C, 30 A at D and 10 A at E.

The resistance of various sections (go and return) are:

AB = 0.1 Ω, BC = 0.05 Ω, CD = 0.01 Ω, DE = 0.025 Ω and EA = 0.075 Ω. Determine:

(*a*) *the point of minimum potential.*

(*b*) *current in each section of distributor.*

Solution.

Figure 15.26 shows the arrangement of ring main distributor,

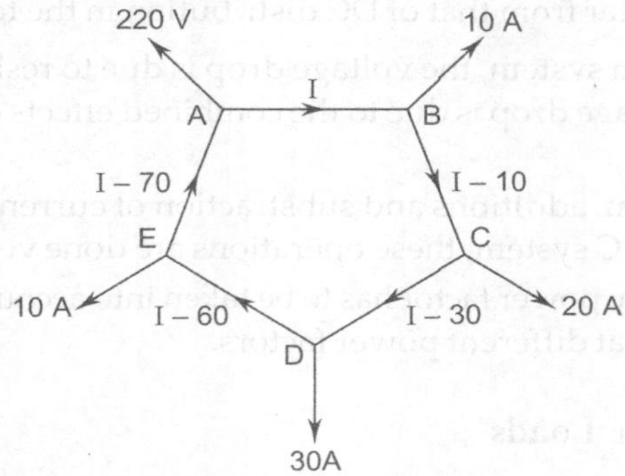

Fig. 15.26

(*a*) According to Kirchhoff's law,

$$= I_{AB} R_{AB} + I_{BC} R_{BC} + I_{CD} R_{CD} + I_{DE} R_{DE} + I_{EA} R_{EA} = 0$$

$$= 0.1\,I + 0.05\,(I - 10) + 0.01\,(I - 30) + 0.025\,(I - 60) + 0.075\,(I - 70) = 0$$

or $$0.26\,I = 7.55$$

or $$I = \frac{7.55}{0.26} = 29.04 \text{ A}$$

(*b*) Current in section AB, $I_{AB} = I = 29.04$ A from A to B.

$$I_{BC} = I - 10 = 29.04 - 10 = 19.04 \text{ A from B to C.}$$

$$I_{CD} = I - 30 = 29.04 - 30 = -0.96 \text{ A}$$

$$= 0.96 \text{ A from D to C.}$$

$$I_{DE} = I - 60 = 29.04 - 60 = -30.96 \text{ A}$$

$$= 30.96 \text{ A from E to D.}$$

$$I_{EA} = I - 70 = 29.04 - 70 = -40.96 \text{ A}$$

$$= 40.96 \text{ A from A to E.}$$

The actual distribution of current is shown in Fig. 15.27

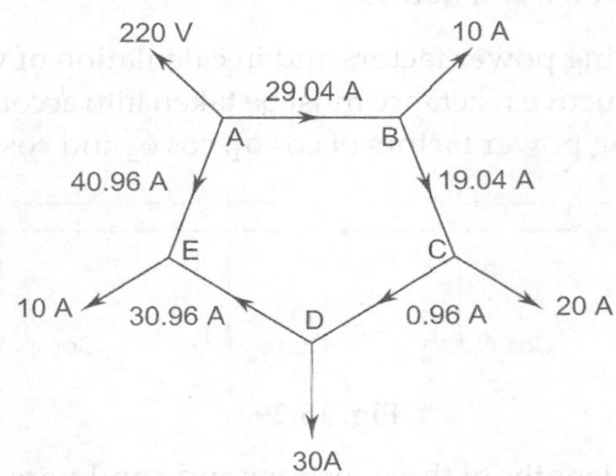

Fig. 15.27

From Fig. 15.27, it is clear that C is the point of minimum potential.

15.4 VOLTAGE DROP IN AC DISTRIBUTORS

AC distribution calculation differ from that of DC distribution in the following respects:

1. In case of DC distribution system, the voltage drop is due to resistance alone. However in an AC distribution, the voltage drop is due to the combined effects of resistance, inductance and capacitance.
2. In DC distribution system, additions and substraction of currents or voltages are done arithmetically but in case of AC system, these operations are done vectorially.
3. In AC distribution system, power factor has to be taken into account. Loads tapped off from the distributor are generally at different power factors.

15.4.1 Unity Power Factor Loads

Figure 15.28 shows a single phase distributor supplying three unity power factor loads. If l_1, l_2, l_3 are the lengths, r is the resistance and x is inductive reactance per unit length of the distributor, V_s is voltage at feeding point and V_r is the voltage at far end. Then,

$$V_s = V_r + (I_1 + I_2 + I_3)\, l_1 (r + Jx) + (I_2 + I_3)\, l_2 (r + Jx) + I_3 l_3 (r + Jx)$$
$$= V_r + I_1 l_1 (r + Jx) + I_2 (l_1 + l_2)(r + Jx) + I_3 (l_1 + l_2 + l_3)(r + Jx)$$
$$= V_r + I_1 l_1 r + I_2 (l_1 + l_2) r + I_3 (l_1 + l_2 + l_3) r + J [I_1 l_1 x + I_2 (l_1 + l_2) x + I_3 (l_1 + l_2 + l_3) x]$$

or
$$V_s \approx V_r + I_1 l_1 r + I_2 (l_1 + l_2) r + I_3 (l_1 + l_2 + l_3) r$$

$\therefore$
$$\text{Voltage Drop } (V_s - V_r) = I_1 l_1 r + I_2 (l_1 + l_2) r + I_3 (l_1 + l_2 + l_3) r$$

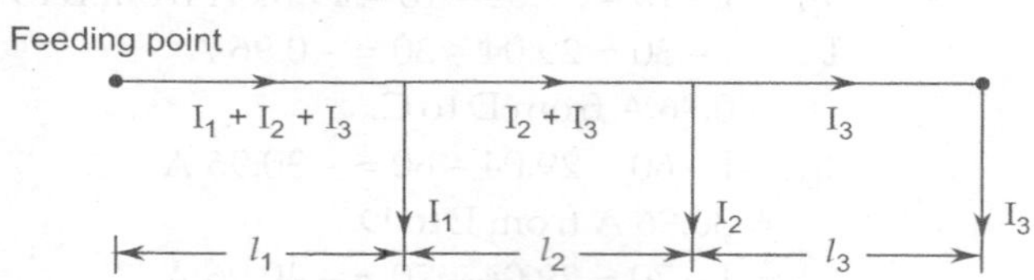

Fig. 15.28

15.4.2 Loads of Different Power Factors

The industrial loads have lagging power factors and in calculation of voltage drop in the distributors feeding such loads, the inductive reactance must be taken into account. Figure 15.29 shows three load currents I_1, I_2 and I_3 having power factors of $\cos \phi_1$, $\cos \phi_2$ and $\cos \phi_3$ tapped off a distributor.

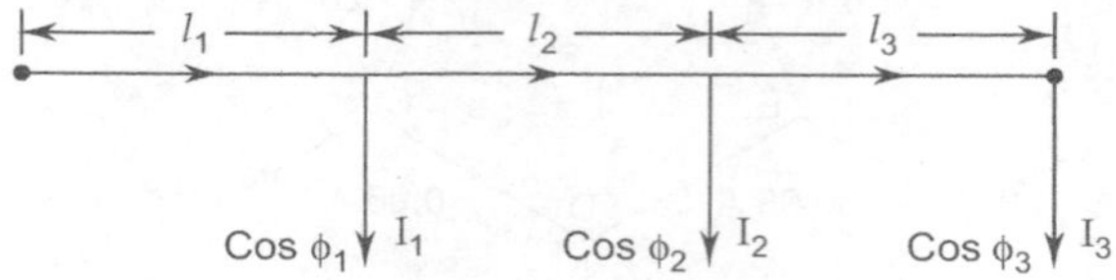

Fig. 15.29

Consider l_1, l_2, and l_3 are the lengths of three sections and r and x are the resistance and inductive reactance per unit length. Replacing r by $(r + Jx)$ and current by $I(\cos \phi - J \sin \phi)$ in equation. of unity power factor load, then we get voltage drop. Thus, the voltage drop upto the far end is given by:

$$\text{Voltage drop} = I_1 (\cos \phi_1 - J \sin \phi_1)\, l_1 (r + Jx) + I_2 (\cos \phi_2 - J \sin \phi_2) \times (l_1 + l_2)(r + Jx) + I_3 (\cos \phi_3 - J \sin \phi_3)(l_1 + l_2 + l_3)(r + Jx)$$
$$\approx I_1 l_1 (r \cos \phi_1 + x \sin \phi_1) + I_2 (l_1 + l_2)(r \cos \phi_2 + x \sin \phi_2) + I_3 (l_1 + l_2 + l_3)(r \cos \phi_3 + x \sin \phi_3)$$

Example 15.11. *A 400 m long distributor has following unity power factor loads tapped off R-phase 100 A load at 100 m. from feeding point, 120 A load at 250 m. from the feeding point and 80 A load at 400 m. from the feeding point. The resistance of the conductor is 0.50 ohm per km length. Neglecting the voltage drop in the neutral conductor, find the voltage across the load at the far end if the voltage at feeding point is 250 V.*

Solution. The arrangement is shown in Fig. 15.30.

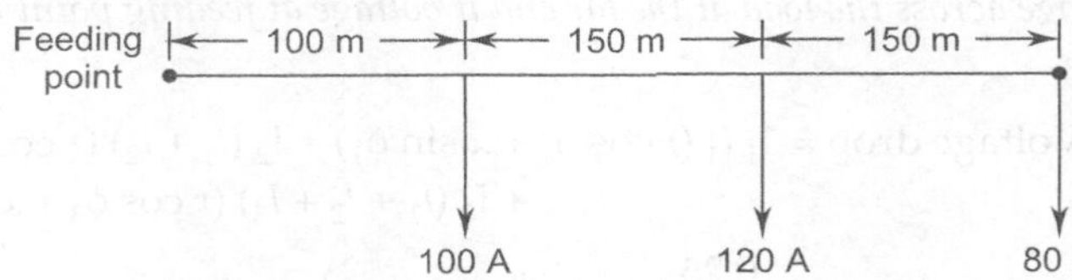

Fig. 15.30

The total voltage drop from the feeding point to the far end is equal to

$$= \frac{0.50}{1000}[100 \times 100 + 120 \times 250 + 80 \times 400]$$
$$= 36 \text{ V}$$

$\therefore$ Voltage at the far end = 250 – 36 = 214 V

Example 15.12. *A single phase AC distributor AB 300 m long is fed from end A and is loaded as under:*

(a) 100 A at 0.707 p.f. lagging 200 m. from point A

(b) 200 A at 0.8 p.f. lagging 300 m. from point A.

The total resistance and reactance of the distributor is 0.2 Ω and 0.1 Ω per km. Calculate the total voltage drop in the distributor.

Solution.

Figure 15.31 shows the arrangement of distributor.

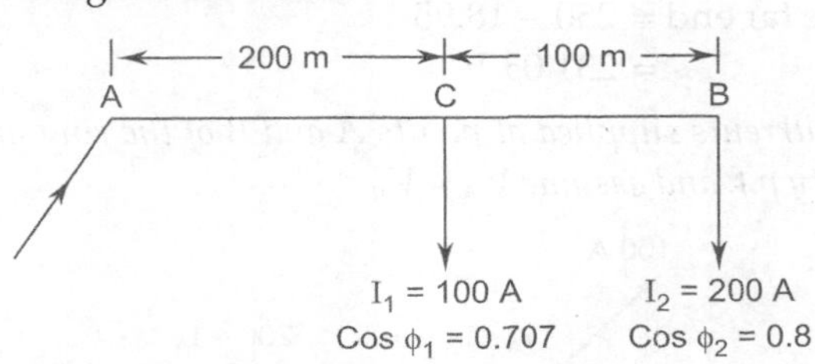

Fig. 15.31

$$\text{Voltage Drop is given by} = I_1 l_1 (r \cos \phi_1 + x \sin \phi_1) + I_2 (l_1 + l_2)(r \cos \phi_2 + x \sin \phi_1)$$

Given $r = \frac{0.2}{1000}$ ohm per m.

$$x = \frac{0.1}{1000} \text{ ohm per m.}$$

$\cos \phi_1 = 0.707,$ $\sin \phi_1 = 0.707$

$\cos \phi_2 = 0.8,$ $\sin \phi_2 = 0.6$

$$\therefore \quad \text{Voltage drop} = 100 \times 200\left[\frac{0.2}{1000} \times 0.707 + \frac{0.1}{1000} \times 0.707\right]$$
$$+ 200 \times 300\left[\frac{0.2}{1000} \times 0.8 + \frac{0.1}{1000} \times 0.6\right]$$
$$= 4.242 + 13.2$$
$$= 17.442 \text{ V.}$$

Example 15.13. *The loads on the R-phase of the distributor are as under:- 100 A, 0.707 lagging p.f. load at 100 m., 120 A unity p.f. load at 250 m. and 80 A, 0.8 lagging p.f. load at 400 m. from the feeding point. The resistance and inductive reactance per km. length are 0.25 and 0.125 ohm. Neglecting the voltage drop in neutral wire, find the voltage across the load at the far end if voltage at feeding point is 250 V.*

Solution.

$$\text{Voltage drop} = I_1 l_1 (r \cos \phi_1 + x \sin \phi_1) + I_2 (l_1 + l_2)(r \cos \phi_2 + x \sin \phi_2) + I_3 (l_1 + l_2 + l_3)(r \cos \phi_3 + x \sin \phi_3)$$

Given

$$r = \frac{0.25}{1000} \text{ ohm per m.}$$
$$x = \frac{0.125}{1000} \text{ ohm per m.}$$
$$\cos \phi_1 = 0.707, \ \sin \phi_1 = 0.707$$
$$\cos \phi_2 = 1.0, \ \sin \phi_2 = 0$$
$$\cos \phi_3 = 0.8, \ \sin \phi_3 = 0.6$$

$$\therefore \quad \text{Voltage drop} = 100 \times 100\left(\frac{0.25}{1000} \times 0.707 + \frac{0.125}{1000} \times 0.707\right)$$
$$+ 120 \times 250\left(\frac{0.25}{1000} \times 1.0 + \frac{0.125}{1000} \times 0\right)$$
$$+ 80 \times 400\left(\frac{0.25}{1000} \times 0.8 + \frac{0.125}{1000} \times 0.6\right)$$
$$= 18.96 \text{ V}$$
$$\text{Voltage at the far end} = 250 - 18.95$$
$$= 231.05 \text{ V}$$

Example 15.14. *Find the currents supplied at points A and B of the ring main distributor as shown in Fig. 15.32. All loads are at unity p.f. and assume $V_A = V_B$.*

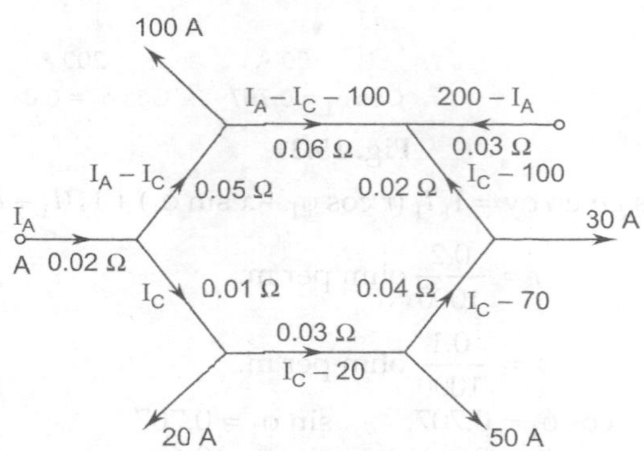

Fig. 15.32

Solution.

As $V_A = V_B$

Therefore, voltage drop from point A to point B in either direction is zero.

$\therefore\ 0.02\, I_A + 0.05\, (I_A - I_C) + 0.06\, (I_A - I_C - 100) - 0.03\, (200 - I_A) = 0$

or $\quad 0.16\, I_A - 0.11\, I_C = 12$

or $\quad 16\, I_A - 11\, I_C = 1200$

$0.02\, I_A + 0.01\, I_C + 0.03\, (I_C - 20) + 0.04\, (I_C - 70) + 0.02\, (I_C - 100) - 0.03\, (200 - I_A) = 0$

or $\quad 5\, I_A + 10\, I_C = 1140$

Solving above both eqns:

$$I_A = 114\ \text{A}$$

$$I_B = 200 - I_A = 200 - 114 = 86\ \text{A}$$

Example 15.15. *A single phase ring distributor ABC is fed at point A. The impedances of sections AB, BC, AC are 2 + J 1, 2 + J 3, 1 + J 2 ohms. Load at B is 40 A at 0.8 p.f. lagging while at C is 60 A at 0.6 p.f. lagging both w.r.t. voltage at A. Find currents in the three sections wrt voltage at A.*

Solution.

Figure 15.33 shows the arrangement. Assume current $(x - Jy)$ in section AB. The currents in other two sections are shown in Fig. 15.33

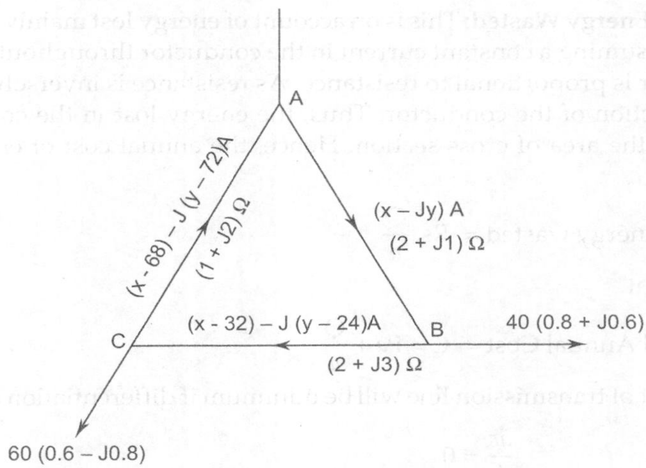

Fig. 15.33

Equating the total voltage drop around the ring to zero. Thus,

$(x - Jy)\,(2 + J\,1) + [(x - 32) - J\,(y - 24)]\,(2 + J\,3) + [(x - 68) - J\,(y - 72)]\,(1 + J\,2) = 0$

Equating the real and imaginary parts to zero,

$$5x + 6y = 348$$

$$6x - 5y = 112$$

Solving above equations

$$x = 39.54\ \text{A}$$

$$y = 25.05\ \text{A}$$

Current in section AB = 39.54 – J 25.05 A from A to B.

Current in section AC = 28.46 – J 46.95 A from A to C.

Current in section BC = 7.54 – J 1.05 A from B to C.

15.5 KELVIN'S LAW

The cost of conductor material is generally a very considerable part of the total cost of a transmission line. Therefore, the determination of proper size of conductor for the line is of vital importance. The most economical area of the conductor is that for which the total annual cost of transmission line is minimum. This is known as Kelvin's law after Lord Kelvin who first stated it in 1881. The total annual cost of a transmission line can be divided broadly into two parts:

(*a*) Annual charges on capital outlay.

(*b*) Annual cost of energy wasted.

(*a*) **Annual Charges on Capital Outlay:** This is on account of interest and depreciation on the capital cost of purchase and installation of transmission line. For a given length of the line, the weight and therefore the cost of the conductor is proportional to the area of cross-section of the conductor. Thus, the annual cost due to interest and depreciation is also proportional to the cross-section. It can be written as-

Annual charge = P a

where P is a constant and

a is the area of cross-section.

(*b*) **Annual Cost of Energy Wasted:** This is on account of energy lost mainly in the conductor due to I^2 R losses. Assuming a constant current in the conductor throughout the year, the energy lost in conductor is proportional to resistance. As resistance is inversely proportional to the area of cross-section of the conductor. Thus, the energy lost in the conductor is inversely proportional to the area of cross-section. Hence, the annual cost of energy wasted can be expressed as:

$$\text{Annual cost of energy wasted} = \text{Rs. } \frac{Q}{a}$$

where Q is a constant.

$$\text{Total Annual Cost} = C = Pa + \frac{Q}{a}$$

The total annual cost of transmission line will be minimum if differentiation of C *wrt* a is zero *i.e.*,

$$\frac{dc}{da} = 0$$

$$\frac{d}{da}\left(Pa + \frac{Q}{a}\right) = 0$$

or
$$P - \frac{Q}{a^2} = 0$$

or
$$a = \left(\frac{Q}{P}\right)^{1/2}$$

For this value of cross-section both the components of the total cost becomes equal, each being equal to $\sqrt{PQ}$. Hence, the Kelvin's law can also be stated in an another way *i.e.*, the most economical area of conductor is that for which the annual value of interest and depreciation of the conductor is equal to the annual cost of the energy wasted in the conductor.

Graphical illustration of Kelvin's law is shown in Fig. 15.34. In this, the straight line (1) shows the relationship between the annual charge and area of cross-section a of the conductor. The rectangular hyperbola (2) gives the relation between annual cost of energy wasted and cross-sectional area a. The total cost is minimum at the point where the graphs of Pa and $\frac{Q}{a}$ intersect with each-other.

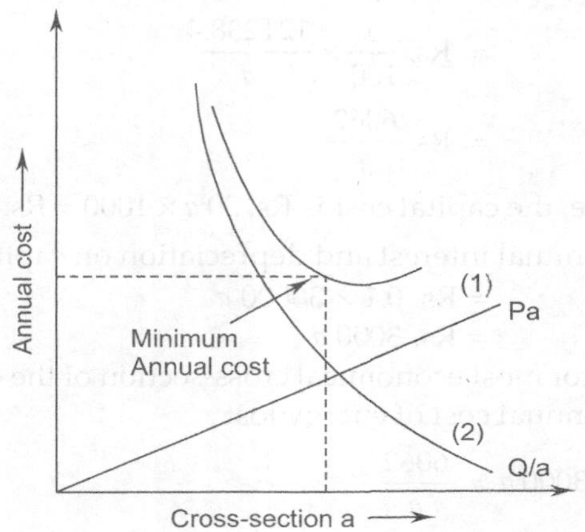

Fig. 15.34

Limitations of Kelvin's Law: Although theoretically Kelvin's law holds good, there is often considerable difficulty in applying it to a proposed scheme of power transmission. The limitations of this law are:

(*a*) This law does not take into account several physical factors like safe current density, mechanical strength, corona loss etc.

(*b*) The cost of conductor is not only the cost which varies with the cross-section. The cost of insulation also varies with the conductor cross-section at least to some extent. It is difficult to express the cost of insulator in terms of conductor cross-section.

(*c*) Besides I^2 R losses in the conductor, dielectric loss in metallic sheath also takes place all the time. It is very difficult to take all the losses into account while calculating the economical cross-section.

(*d*) Interest and Depreciation on the capital outlay cannot be determined accurately.

Example 15.16. *A 2-conductor cable 1 km long is required to supply a constant current of 200 A throughout the year. The cost of cable is Rs. 30 a per meter where a is the area of cross-section of the conductor in cm^2. The cost of energy is 5 P per kwh and interest and depreciation charges amount to 10%. Calculate the most economical conductor size. Assume resistivity of conductor material to be 1.73 u Ωcm.*

Solution.

$$\text{Resistance of one conductor, } R = \frac{\rho l}{a}$$

$$= \frac{1.73 \times 10^{-6} \times 10^{5}}{a} = \frac{0.173}{a}\ \Omega$$

$$\text{Energy lost per annum} = \frac{2I^2\,Rt}{1000}\ \text{kwh}$$

$$= \frac{2 \times (200)^2 \times 0.173 \times (365 \times 24)}{1000 \times a} kwh$$

$$= \frac{121238.4}{a} kwh$$

$$\text{Annual cost of energy lost} = \text{Cost per } kwh \times \text{Annual energy loss}$$

$$= \text{Rs. } \frac{5}{100} \times \frac{121238.4}{a}$$

$$= \text{Rs. } \frac{6062}{a}$$

For 1 km. length of the cable, the capital cost is Rs. 30 $a \times 1000$ = Rs. 30000 a

Variable annual charge = Annual interest and depreciation on capital cost of cable

$$= \text{Rs. } 0.1 \times 30000\, a$$

$$= \text{Rs. } 3000\, a$$

According to Kelvin's law, for most economical cross-section of the conductor,
Variable annual charge = Annual cost of energy lost

or
$$3000\, a = \frac{6062}{a}$$

or
$$a = \sqrt{\frac{6062}{3000}} = 1.42 \text{ cm}^2.$$

Example 15.17. *The cost of a 3-phase overhead transmission line is Rs. (30000 a + 3000) per km. where a is the area of cross-section of each conductor in cm^2. The line is supplying a load of 5 MW at 33 kV and 0.8 p.f. lagging assumed to be constant throughout the year. Energy costs 4 P per kwh and interest and depreciation charges amount to 10% per annum. Find the most economical size of the conductor. Assume the resistivity of the conductor material to be 2 u Ω cm.*

Solution.

Resistance of one conductor,

$$R = \frac{\rho l}{a} = \frac{0.2 \times 10^{-6} \times 10^5}{a} = \frac{0.2}{a} \Omega$$

Line Current,
$$I = \frac{P}{\sqrt{3}\, V \cos\phi} = \frac{5 \times 10^6}{\sqrt{3} \times 33 \times 10^3 \times 0.8}$$

$$= 109.35 \text{ A}$$

$$\text{Energy lost per annum} = \frac{3 I^2 R t}{1000} \text{ kWh}$$

$$= \frac{3 \times (109.35)^2 \times 0.2 \times (365 \times 24)}{1000 \times a} = \frac{62848}{a} \text{ kWh}$$

$$\text{Annual cost of energy lost} = \text{Rs. } \frac{4}{100} \times \frac{62848}{a} = \text{Rs. } \frac{2514}{a}$$

$$\text{Variable annual charge} = \text{Rs. } 0.1 \times 30000\, a$$

$$= \text{Rs. } 3000\, a$$

According to Kelvin's Law

Variable annual charge = Annual cost of energy lost

$$3000\,a = \text{Rs. } \frac{2514}{a}$$

or $$a = \sqrt{\frac{2514}{3000}} = 0.91 \text{ cm}^2$$

Example 15.18. *Determine the most economical cross-section for a 3-phase transmission line 1 km long to supply at a constant voltage of 110 KV the following daily load cycle.*

6 hrs., 20 MW at p.f. 0.8 lagging,

12 hrs., 5 MW at p.f. 0.8 lagging,

6 hrs., 6 MW at p.f. 0.8 lagging.

The line is used for 365 days yearly. The cost per km. of line including erection is Rs. (7000 + 5000 a) where a is area of cross-section of conductor in cm². The annual rate of interest and depreciation is 10% and the energy costs 6 P per kwh. The resistance per km. of each conductor is 0.176/a.

Solution.

The load currents at various loads are:

(*a*) $$I_1 = \frac{20 \times 10^6}{\sqrt{3} \times 110 \times 10^3 \times 0.8} = 131.2 \text{ A}$$ [At 20 MW]

(*b*) $$I_2 = \frac{5 \times 10^6}{\sqrt{3} \times 110 \times 10^3 \times 0.8} = 32.8 \text{ A}$$ [At 5 MW]

(*c*) $$I_3 = \frac{6 \times 10^6}{\sqrt{3} \times 110 \times 10^3 \times 0.8} = 39.36 \text{ A}$$ [At 6 MW]

Given $$R = \frac{0.176}{a}\ \Omega$$

$$\text{Energy lost per annum} = 3 \times \frac{0.176}{a} \times 365 \times \frac{1}{1000} [(131.2)^2 \times 6 + (32.8)^2 \times 12 + (39.36)^2 \times 6]$$

$$= \frac{24184.9}{a}\ kwh$$

$$\text{Annual cost of energy} = \text{Rs. } \frac{6}{100} \times \frac{24184.9}{a}$$

$$= \text{Rs. } \frac{1451}{a}$$

Variable annual charge = Rs. 0.1 × 5000 a = Rs. 500 a

According to Kelvin's Law

Variable annual charge = Annual cost of energy

or $$500\,a = \frac{1451}{a}$$

or $$a = \sqrt{\frac{1451}{500}} = 1.70 \text{ cm}^2.$$

SELF TEST

Fill in the blanks by inserting appropriate words/figures:

1. The overhead system is __________ flexible than underground system.
2. The chances of faults in underground system are __________ as compared to overhead system.
3. A distributor arranged to form a closed circuit and fed at one or more than one point is called the __________.
4. The point of minimum potential of a uniformly loaded distributor fed at both ends with equal voltages will occur at __________.
5. Underground system is __________ expensive in comparison to overhead system.
6. If in a 3-wire DC system, the current in the neutral wire is zero, then voltage between any outer and neutral is __________.
7. The most common system for secondary distribution is __________ 3-phase, __________wire system.
8. Maintenance cost of underground system is quite __________ in comparison with that of overhead system.
9. The main consideration in the design of a feeder is the __________.
10. The voltage drop in underground system is very __________ as compared to that in overhead system.
11. The underground system has __________ initial cost than the overhead system.
12. In a 3-wire d.c. system, the load on + ve side is 400 A and on negative side it is 300 A. Then current in neutral wire is__________.
13. The distribution transformer links the primary and __________ distribution systems.
14. In a single fed distributor, if fault occurs on any section, the supply to all consumers has to be __________.
15. A ring main system of distribution is __________ reliable than the radial systems.
16. The statutory limit for voltage variations at the consumer's terminals is __________% of rated value.

Answers to Self Test

1. more
2. less
3. ring distributor
4. mid-point
5. more
6. the same
7. 400/230 V, 4
8. low
9. current carrying capacity
10. low
11. more
12. 100 A
13. secondary
14. shut off
15. more
16. ± 6.

OBJECTIVE QUESTIONS

1. In a distribution system, which of the following items shares the major cost?
   (*a*) Insulators
   (*b*) Earthing system
   (*c*) Conductors
   (*d*) Distribution transformer.
2. Which of the following distribution systems is preferred for good efficiency and high economy?
   (*a*) single phase, 2-wire system
   (*b*) 2-phase, 3-wire system
   (*c*) 3-phase, 3-wire system
   (*d*) 3-phase, 4-wire system.

3. Transmission and distribution of electric power by underground system is superior to overhead system in respect of:
   (*a*) appearance and public safety
   (*b*) frequency of faults, power failure and accidents
   (*c*) maintenance cost
   (*d*) all of above.
4. Which of the following is usually not the generating voltage?
   (*a*) 13.2 KV (*b*) 12.5 KV
   (*c*) 11 KV (*d*) 6.6 KV.
5. Distributors are designed from the point of view of:
   (*a*) operating voltage (*b*) operating frequency
   (*c*) voltage drop in it (*d*) its current carrying capacity.
6. For a given amount of power to be transmitted over a certain distance with fixed power loss, the volume of copper required is:
   (*a*) inversely proportional to the square of voltage and pf of the load
   (*b*) directly proportional to the square of voltage and pf of the load
   (*c*) inversely proportional to voltage
   (*d*) directly proportional to voltage.
7. The voltage and current in a circuit are given by:
   $v = 10 \sin (\omega t - \pi/6)$
   $i = 10 \sin (\omega t + \pi/6)$
   The power consumed is given as
   (*a*) 25 watts (*b*) 50 watts
   (*c*) 86.6 watts (*d*) 100 watts.
8. Which of the following dc distribution systems is the lowest and cheapest in first cost?
   (*a*) Ring main (*b*) Radial
   (*c*) interconnected (*d*) none of these.
9. Which of the following dc distribution systems is most reliable?
   (*a*) Tree (*b*) Ring main
   (*c*) Radial (*d*) All are equally reliable.
10. Which of the following distribution systems is the most economical?
   (*a*) DC system (*b*) Single phase ac system
   (*c*) Three phase, 3-wire ac systems. (*d*) Three phase, 4-wire ac systems.
11. A 3-phase, 4-wire system is commonly used for:
   (*a*) Primary transmission (*b*) Secondary transmission
   (*c*) Primary distribution (*d*) Secondary distribution.
12. In a circuit the voltage and current are given by $v = (10 + j\,5)$ and $i = (6 + j\,4)$. The power is:
   (*a*) 20 watts (*b*) 70 watts
   (*c*) 80 watts (*d*) 60 watts.
13. In the above problem (Ques-12), the circuit is:
   (*a*) resistive (*b*) capacitive
   (*c*) inductive (*d*) it could be any of these.

14. 750 kV fall under the category of:

(*a*) high voltage (*b*) extra high voltage
(*c*) ultra high voltage (*d*) medium high voltage.

15. With same maximum voltage between conductors, the ratio of copper volumes needed in 3-phase, 3-wire system and 1-phase, 2-wire system is:

(*a*) 3/4 (*b*) 3/5
(*c*) 4/3 (*d*) 5/3.

16. Overhead system can be designed for operation upto:

(*a*) 400 KV (*b*) 220 KV
(*c*) 132 KV (*d*) 66 KV.

17. In a DC 2-wire feeder, voltage drop per wire is 2.5%, transmission efficiency of the feeder will be:

(*a*) 93.75% (*b*) 95%
(*c*) 97.5% (*d*) 98.75%.

18. The main drawback(s) of underground system over overhead system is/are:

(*a*) exposure to lightning
(*b*) exposure to atmospheric hazards such as smoke, ice, wind etc
(*c*) heavy initial cost
(*d*) inductive interference between power and communication circuits.

19. The conductor connecting consumer's terminals to the distributor is called the:

(*a*) feeder (*b*) service main
(*c*) distributor (*d*) none of these.

20. As per IE rules the maximum allowable variation between declared and actual voltage at consumer's premises should be:

(*a*) ± 6% (*b*) ± 8%
(*c*) ± 5% (*d*) ± 4%.

21. A uniformly loaded dc distributor is fed at both ends with equal voltages. In comparison to a similar distributor fed at one end only, the drop at the mid-point will be:

(*a*) 1/2 (*b*) 1/3
(*c*) 1/4 (*d*) 1/6.

22. A uniformly loaded dc distributor is fed at both ends with equal voltages. In comparison to a similar distributor fed at one end only, the maximum voltage drop will be:

(*a*) 1/2 (*b*) 1/3
(*c*) 1/4 (*d*) 1/6.

23. The loads on 3-phase, 4-wire distributors are usually:

(*a*) balanced (*b*) unbalanced
(*c*) either of above (*d*) none of above.

24. In a transmission system the feeder supplies power to:

(*a*) transformer substations (step-up) (*b*) service mains
(*c*) distributors (*d*) all of above.

25. Feeder is designed mainly from the point of view of:

(*a*) voltage drop in it (*b*) operating voltage
(*c*) operating frequency (*d*) its current carrying capacity.

26. The distributors in residential areas are:
(*a*) single phase, 2-wire (*b*) three phase, 3-wire
(*c*) three phase, 4-wire (*d*) two phase, 4-wire.

27. In a transmission system, the weight of copper used is proportional to:
(*a*) E (*b*) E^2
(*c*) 1/E (*d*) $1/E^2$.

28. For the same voltage drop, increasing the voltage of a distributor *n*-times:
(*a*) reduces the *x*-section of the conductor by *n*-times
(*b*) increases the *x*-section of the conductor by *n*-times
(*c*) reduces the *x*-section of the conductor by n^2-times
(*d*) increases the *x*-section of the conductor by n^2-times.

29. The cross-sectional area of the neutral wire in a 3-wire dc distribution system is usually the:
(*a*) same as that of outer (*b*) one-half of that of outer
(*c*) one-fourth of that of outer (*d*) double of that of outer.

30. Transmitting power remaining the same, if the supply voltage of a dc 2-wire distributor is doubled saving in copper will be:
(*a*) 25% (*b*) 50%
(*c*) 75% (*d*) 100%.

31. The underground system cannot be operated above:
(*a*) 11 KV (*b*) 33 KV
(*c*) 66 KV (*d*) 220 KV.

32. The volume of copper required for an ac transmission line is inversely proportional to:
(*a*) current (*b*) pf
(*c*) voltage (*d*) both (b) & (c).

33. If variable part of annual cost on account of interest and depreciation on the capital outlay is equal to the annual cost of electrical energy wasted in the conductors, the total annual cost will be minimum and the corresponding size of conductor will be the most economical. This statement is known as:
(*a*) Ohm's Law (*b*) Kelvin's Law
(*c*) Lenz's Law (*d*) Faraday's Law.

Answers to Objective Questions

1. (*d*)	2. (*d*)	3. (*d*)	4. (*b*)	5. (*c*)
6. (*a*)	7. (*a*)	8. (*b*)	9. (*b*)	10. (*a*)
11. (*d*)	12. (*c*)	13. (*b*)	14. (*c*)	15. (*a*)
16. (*a*)	17. (*b*)	18. (*c*)	19. (*b*)	20. (*c*)
21. (*b*)	22. (*c*)	23. (*b*)	24. (*c*)	25. (*d*)
26. (*c*)	27. (*d*)	28. (*a*)	29. (*b*)	30. (*b*)
31. (*c*)	32. (*d*)	33. (*b*).		

EXERCISE

1. Write a short note on various systems of DC distribution.
2. What are the advantages of a doubly fed distributor over singly fed distributor?

3. What are the different types of feeders and distributors? Bring out their relative advantages and disadvantages.
4. What are the advantages of 3-wire distribution over 2-wire distribution?
5. Discuss why open loop distributors are preferred to radial distributors.
6. Explain the difference between radial distribution system and ring-main distribution system.
7. Derive an expression for the voltage drop for a uniformly loaded distributor fed at one end.
8. Why AC transmission circuits are 3-phase 3-wire but AC distribution circuits are 3-phase 4-wire?
9. Compare the 2-wire and 3-wire systems of DC distribution in respect of weights of copper conductors, all other conditions being the same.
10. A 2-wire DC distributor cable AB is 2 km long and supplies loads of 100 A, 150 A, 200 A and 50 A situated 500 m, 1000 m, 1600 m and 2000 m from the feeding point A. Each conductor has a resistance of 0.01 W per 1000 m calculate voltage at each load point if a voltage of 300 V is maintained at point A. [**Ans:** 295 V, 291 V, 288 V and 284 V]
11. **The DC distribution cable is 1000 m long and is loaded a shown in figure. The resistance of each conductor is 0.05 W per 1000 m calculate the voltage at each load point.**

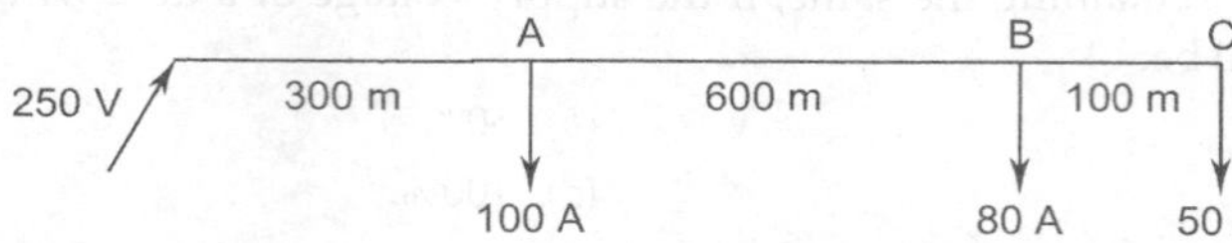

[**Ans:** $V_A = 243.1$ V, $V_B = 235.3$ V, $V_C = 234.8$ V]

12. Following figure shows a sing main supplied at point X and loaded at points A and B. The resistances of the various sections (both go and return) are as shown. Find (*a*) the current in AB and voltages at load points A and B (*b*) the voltage drop between X and A when load at B is removed.

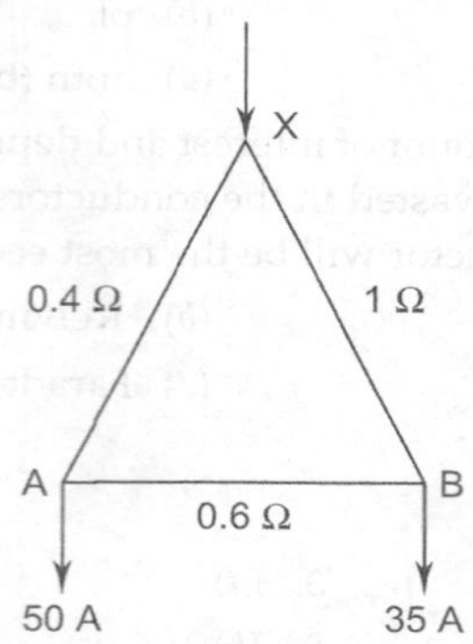

[**Ans:** (*a*) 7.50 amps from B to A; $V_{AX} = -23$ V; $V_{BX} = -27.50$ V, (*b*) $V_{AX} = -16$ V]

13. A 2-wire DC distributor AB is 300 m long. It is fed at point A. The various loads and their positions are given below:

At Point	*Distance from A (in Meters)*	*Concentrated Loads (in Amperes)*
C	40	30
D	100	40
E	150	100
F	250	50

If the maximum permissible voltage drop is not to exceed 10 V, find the x-sectional area of the distributor. Take $\rho = 1.78 \times 10^{-8}\ \Omega\text{-m}$. **[Ans:** $1.163\ cm^2$]

14. A two-wire distributor is loaded as shown in the figure, the two ends A and F being fed at 248 and 250 volts respectively. If the resistance of the distributor per meter is 400 micro-ohms for both lines calculate the minimum voltage and show the position where it occurs. Distances are shown in meters and the loads in amperes.

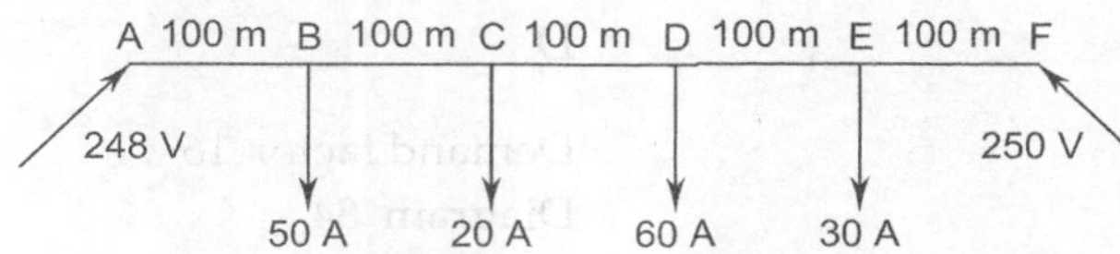

INDEX

H

I

L

M

N

O

P

R